Editor biographies

Dakeshwar Kumar Verma holds a Ph.D. degree and is Assistant Professor of Chemistry at Govt. Digvijay Autonomous Postgraduate College in Rajnandgaon, Chhattisgarh, India. Driven by a profound passion for scientific exploration, his research focuses primarily on preparing and designing organic compounds for diverse applications. With an impressive track record, Dr. Verma has authored/co-authored 70 research papers/review articles and 38 book chapters. He has one design patent and two published patents under IPR, India. He serves as a reviewer for over 50 high-impact journals. A testament to his academic influence, Dr. Verma has been the lead editor for 22 published books and 7 upcoming books slated to be published by renowned international publishers. With a cumulative citation count of over 3300, an h-index of 32, and an i-10 index of 52, Dr. Verma's impact is evident through the recognition and relevance of his work in the scientific community. In addition, one research scholar from the National Institute of Technology Raipur completed her Ph.D. under his co-supervision, and two research scholars from Govt. Digvijay Autonomous Postgraduate College in Rajnandgaon, Chhattisgarh, submitted their Ph.D theses recently. More than 30 M.Sc. chemistry students completed their projects under his guidance. He got 50 lakhs for the project from DST, Delhi.

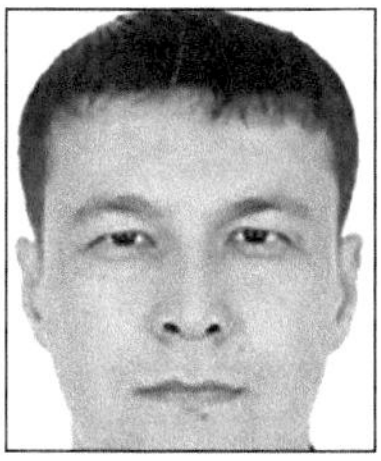

Berdimurodov Elyor is Associate Professor at the National University of Uzbekistan and Akfa University. He was enrolled in the Ph.D. program at Tianjin University and the National University of Uzbekistan. He was a researcher at Tianjin University, Karshi State University, and Changchun Applied Chemistry Institute (Chinese Academy of Science). Elyor was a participant in several international and domestic study and research programs (Chinese Government Scholarships, Chinese Great Belt program, National Fundamental, and practical grants). He has published around 50 articles in Scopus and Web of Science journals. The h-index is 30 in Scopus, 25 in Web of Science, and 32 (i-10 index of 50) in Google Scholar. He has contributed over 20 book chapters and participated in over 15 international conferences in China, India, and other developed countries. Dr. Berdimurodov has served as the editor for five books, including scientific monographs and handbooks in his research areas. His research areas are corrosion science, electrochemistry, quantum chemistry, material science, green chemistry, and nanochemistry.

Fahmida Khan is Professor at the Department of Chemistry, National Institute of Technology Raipur, Raipur, Chhattisgarh 492010, INDIA. Her main research interests include organic chemistry/analytical chemistry, nuclear chemistry, environmental chemistry, nanotechnology and luminescence/drug designing, and DNA binding. She has published around 60 articles in Scopus and Web of Science journals. The h-index is 19, and the i10 index is 35, with total citations of 1216 in Google Scholar. She has contributed over 6 book chapters and participated in over 30 international conferences in China, Russia, UK, India, and other developed countries. She has successfully guided more than 12 research scholars.

Contributors

A. Bekhta
Laboratory of Organic Chemistry
Catalysis and Environment, Faculty of Sciences
Ibn Tofail University, PO Box 133, 14000
Kenitra, Morocco

Elyor Berdimurodov
Faculty of Chemistry
National University of Uzbekistan
Tashkent, 100034, Uzbekistan

Hüseyin Fatih Çetinkaya
Department of Environmental Engineering
Faculty of Engineering, Sivas Cumhuriyet University
Sivas, Turkey

Serap Çetinkaya
Department of Molecular Biologyand Genetics
Science Faculty, Sivas Cumhuriyet University
Sivas, Turkey

Halil İbrahim Çetintaş
Advanced Technology Research and Application Center (CUTAM)
Sivas Cumhuriyet University
Sivas, Turkey

Veenu Chandraker
Department of Science and Humanities
Government Polytechnic Mahasamund
Chhattisgarh-493445, India

Lakha Chopda
Government Engineering College
Bhuj -370 001 (Gujarat), India

Omar Dagdag
Department of Mechanical Engineering
Gachon University
Seongnam 13120, Republic of Korea.

Pragnesh Dave
Department of Chemistry
Sardar Patel University
Vallabh Vidynagar-388 120 (Gujarat)

Ali Dehghani
Department of Chemical Engineering
Faculty of Engineering, Golestan University
Aliabad Katoul, Iran

David J. Edwards
Department of the Built Environment
Birmingham City University, City Centre Campus, Millennium Point
Birmingham B4 7XG, United Kingdom
Faculty of Engineering and the Built Environment,
University of Johannesburg
South Africa

Dhriti Sundar Ghosh
Department of Physics
Indian Institute of Technology
Dharwad, India

Rajesh Haldhar
School of Chemical Engineering
Yeungnam University
Gyeongsan 38541, Republic of Korea

Fatma Azra Haspolat
Department of Molecular Biology and Genetics
Science Faculty, Sivas Cumhuriyet University
Sivas, 58140, Turkey

R. Hsissou
Laboratory of Organic Chemistry, Bioorganic and Environment
Chemistry Department, Faculty of Sciences
Chouaib Doukkali University
BP 20, 24000, El Jadida, Morocco

Vijayasri K.
Department of Chemistry
Bhilai Mahila Mahavidyalaya
Bhilai

Hülya Kübra Kiliç
Department of Molecular Biology and Genetics
Science Faculty, Sivas Cumhuriyet University
Sivas, 58140, Turkey

Hansang Kim
Department of Mechanical Engineering
Gachon University
Seongnam 13120, Republic of Korea

B Gopal Krishna
Photonics Research Laboratory
School of Studies in Electronics & Photonics
Pt. Ravishankar Shukla University
Raipur, India

Sandeep Kumar
Department of Chemistry
Swami Atmanand Govt. English Medium Model College
Durg (CG), India

B. Lakhrissi
Laboratory of Organic Chemistry
Catalysis and Environment, Faculty of Sciences
Ibn Tofail University
PO Box 133, 14000, Kenitra, Morocco

Amir Hossein Jafari Mofidabadi
Department of Chemical Engineering, Faculty of Engineering
Golestan University
Aliabad Katoul, Iran

Amir Hossein Mostafatabar
Department of Chemical Engineering, Faculty of Engineering
Golestan University
Aliabad Katoul, Iran

M. Oubaaqa
Laboratory of Advanced Materials and Process Engineering
Faculty of Sciences, Ibn Tofaïl University
POBox 133, 14000, Kenitra, Morocco

Shokoh Parham
Universiti Teknologi Malaysi
Ibnu Sina Institute for Scientific and Industrial Research (ISI-SIR)
Center for Sustainable Nanomaterials, Johor Bahru Campus
81310 Skudai, Johor, Malaysia

M Jagannadha Rao
Department of Geology
Andhra University
Visakhapatnam, Andhra Pradesh, India

M. Rbaa
The Higher Institute of Nursing Professions and Health Techniques of Casablanca, Morocco
Laboratory of Materials, Nanotechnology, and Environment
Faculty of Sciences, Mohammed V University in Rabat
P.O. Box. 1014 Agdal-Rabat, Morocco

Rakesh Kumar Sahu
Department of Chemistry
BCS Government PG College, Dhamtari, Chhattisgarh-493773, India

Swati Sahu
Institute of Renewable Energy Technology & Management
School of Studies in Electronics & Photonics
Pt. Ravishankar Shukla University
Raipur, India

Hadi Sarvari
Islamic Azad University
Department of Civil Engineering, Isfahan (Khorasgan) Branch
Isfahan, Iran
The Hong Kong Polytechnic University
Department of Building and Real Estate, Hung Hom, Kowloon
Hong Kong, China

Sanjay Tiwari
Photonics Research Laboratory
School of Studies in Electronics & Photonics
Pt. Ravishankar Shukla University
Raipur, India

Sanjay Tiwari
Institute of Renewable Energy Technology & Management
School of Studies in Electronics & Photonics
Pt. Ravishankar Shukla University
Raipur, India

Burak Tüzün
Plant and animal production department
Technical Sciences Vocational School of Sivas, Sivas Cumhuriyet University
Sivas, 58140, Turkey

Gamze Tüzün
Department of Chemistry, Faculty of Science
Sivas Cumhuriyet University
Sivas, 58140, Turkey

Nishi Verma
Department of Chemistry
Bhilai Mahila Mahavidyalaya
Bhilai, Chhattisgarh, India

Abdelkader Zarrouk
Laboratory of Materials, Nanotechnology, and Environment
Faculty of Sciences, Mohammed V University in Rabat, P.O. Box. 1014 Agdal
Rabat, Morocco

1 Corrosion

Fundamental aspects, economic impact, and its mitigation

Fatma Azra Haspolat, Hülya Kübra Kiliç, Gamze Tüzün, and Burak Tüzün

1.1 WHAT IS CORROSION?

Corrosion is the condition in which metals are damaged as a result of oxidation or abrasion by other chemical effects. However, it is known that environmental conditions other than metals have similar effects on other materials. The main cause of corrosion is that metals are too unstable in their pure form [1], as given in Figure 1.1. Materials that are converted to metallic form by energizing corrode with a tendency to return to their stable state in nature, and this causes irreversible damage. Corrosion

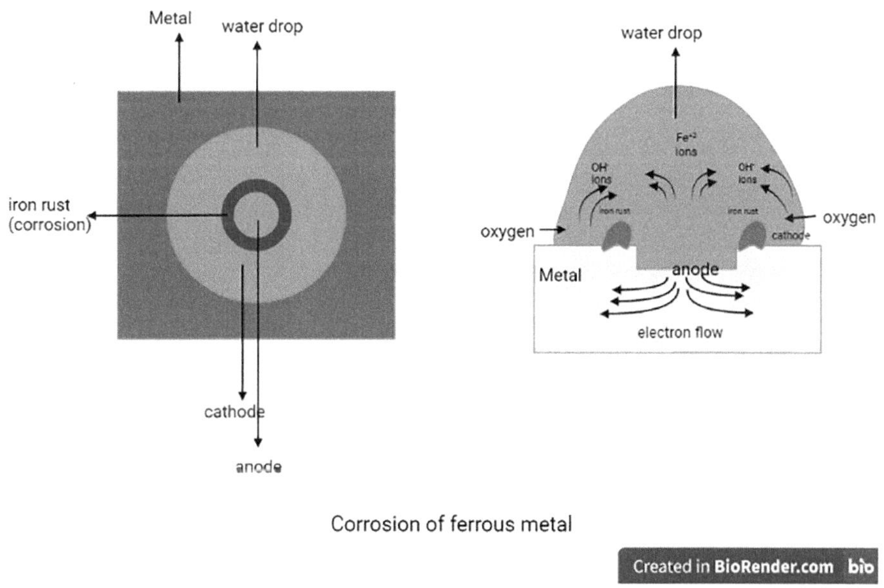

FIGURE 1.1 Corrosion of ferrous metal.

DOI: 10.1201/9781003456025-1

under the insulation, on the other hand, means that as a result of not choosing a suitable insulation material or an effective water vapor barrier, the insulation material gets wet and causes deterioration in the installation.

1.1.1 General corrosion reactions

In the corrosion process, anodic (electron-donating-oxidation) reactions and cathodic (electron-reduction-reduction) reactions occur together. Four basic factors are required for corrosion to start: the anode (corrosive area), cathode (protected area), electrolyte (conducting medium), and an electrolyte path connecting the anode and cathode [2].

The anodic reaction refers to the oxidation of the metal to its own ions. It is usually shown as:

$$X \rightarrow X^{n+} + ne^{-} \tag{1}$$

$$such as,\ Al \rightarrow Al^{3+} + 3e^{-} \tag{2}$$

The cathodic reaction is called the electrode reaction, in which a negative charge is transferred from the metal atoms to the electrolyte. Cathodic reactions are usually reduction reactions. Some metals are more active, while others are less active. The order from active metals to less active metals is as follows: Magnesium (Mg), Aluminum (Al), Zinc (Zn), Iron (Fe), Lead (Pb), Tin (Sn), Copper (Cu), Silver (Ag), Platinum (Pt), and Gold (Au). When two metals are brought together, the less active metal acts as the (-) electrode, while the other becomes the (+) electrode. While the metal in the (-) electrode is corroded, the metal in the other electrode is protected [3].

The places where corrosion occurs are generally the areas where chemical substances accumulate, temperature differences exist, and humidity is high. Chlorides and industrial wastes in the electrolyte can lead to the formation of an anodic field. In addition, residues on the metal surface before the coating or insulation is applied can also lead to the initiation of corrosion [4].

Among the factors affecting corrosion, many factors such as the effect of the environment, the effect of temperature, material selection, system design, the oxygen concentration of the environment where the system is located, and the electrical specific resistance of the ground can be counted [5]. These factors can accelerate or slow down the corrosion process.

1.1.2 Visual defects caused by corrosion

Corrosion can cause direct product loss (such as product loss in a punctured tank or transmission line pipes), but also contaminate the product and render it unusable. The presence of corrosion can be detected by the detection of the following visual defects [6]:

- Shiny metal surfaces become dull.
- Rust forms on iron. Rust means that the iron is corroded. It is a brown-colored compound formed as a result of oxidation of ferrous metal.

- Zinc is covered with a white and dull layer. When zinc is used as a protective coating against corrosion, it corrodes over time as a white and matte layer.
- A green layer forms on copper. Copper can react with compounds in the atmosphere to form a green patina layer.
- Silver tarnishes. When silver comes into contact with sulfur compounds, it starts to darken and turns black over time, which means corrosion.
- Platinum and gold remain shiny. Precious metals such as platinum and gold are highly resistant to corrosion and generally retain their luster.

These visual defects are indications that corrosion is present and that the materials are corroding. Detecting corrosion is important for taking preventive measures and minimizing damage [7].

1.1.3 Measures to be taken against corrosion

Prevention or limitation of corrosion can be achieved with conscious control [8], and this control starts at the design stage, starting with the use of the material. Variables related to design, material, environment, and interface must be considered to prevent corrosion.

By examining these variables, the factors that can cause corrosion should be determined, and many economic measures should be taken to eliminate them and thus prevent corrosion [9]. Surface coating processes are one of the most important methods of protecting the material. However, the coating made on a poorly cleaned surface cannot be long-lasting, so attention should be paid to surface preparation. The rate of corrosion can be slowed down by changes in the environment. Generally, the possibility of reducing the corrosive effect of the atmosphere in the open environment is very limited. In this case, it may be necessary to make the material used more durable or to change the metal-media interface. The best measure that can be taken in indoor environments is to reduce relative humidity and to use restraints [10]. Brakes reduce or prevent the metal from reacting with its environment. In aqueous environments, anodic inhibitors make the anodic reaction difficult by acting on the metal surface, while cathodic inhibitors make the cathodic reaction difficult. There are also dual brakes with both anodic and cathodic braking properties. Condensation prevention measures should be taken to prevent corrosion [11]. This can be achieved with sufficient and necessary thermal insulation. In areas where there is a possibility of condensation in the construction, precautions should be taken at the design stage, and corrosion should be prevented with the right insulation solutions.

Consequences such as direct product losses and product contamination may be encountered due to corrosion. In addition, corrosion products formed under the insulation may accumulate on the surface, worsening the heat transfer coefficient and causing a decrease in efficiency. In order to prevent corrosion under insulation, insulation materials with anti-corrosion properties should be selected, the correct insulation thickness should be calculated, the joints should be sealed, and when necessary, a high-performance vapor barrier should be used [12].

1.2 CORROSION MECHANISM

Corrosion refers to the losses that occur in materials due to the chemical and electrochemical effects of the environment. Corrosion can be of two types [13]: direct corrosion is loss caused by chemical melts, and electrochemical corrosion occurs continuously under the influence of environmental conditions.

Electrochemical corrosion takes place without the need for a special melt. Although corrosion is generally known as the rusting of iron, it is actually an event that affects all metals. Since ceramics and polymers are compounds with strong atomic bonds and do not contain free electrons, they are not affected by electrochemical corrosion, but acids can directly corrode ceramics and solvent polymers [14]. In metals, a metal oxide film layer is formed by the effect of moist air and oxygen. This layer is usually solid. When the density of the oxide layer is higher than the base metal, the contact of the base metal with air decreases and the corrosion rate slows down as the thickness of the oxide layer increases. However, in some metals (e.g., Fe and Mg), the oxide layer formed is more porous than the base metal, and corrosion continues at a linear rate [15].

Electrochemical reactions include the oxidation reaction (1) in the anode region and the reduction reaction (2) that occurs in the cathode region.

Electrochemical reactions include the oxidation reaction (1) in the anode region and the reduction reaction (2) that occurs in the cathode region.

(1) Anode Region

$$M \rightarrow Mn^{+} + ne^{-} \text{ (oxidation reaction)}$$

In this reaction, metal M donates electrons to form metal ions (Mn^{+}) and electrons (ne^{-}). This phenomenon is called oxidation in the anode region, because oxidation occurs when the metal becomes an ion.

(2) Cathode Region

$$Mn^{+} + ne^{-} \rightarrow M \text{ (reduction reaction)}$$

In this reaction, the metal ion (Mn^{+}) gains electrons and becomes metal again. This phenomenon is called reduction in the cathode region, because reduction occurs when the metal ion gains electrons [16].

These two reactions work together to form a system called an electrochemical battery or a galvanic battery. While the metal corrodes in the oxidation reaction in the anode region, the metal ions turn back into metal in the reduction reaction in the cathode region [17]. In this process, electrons flow from the anode to the cathode, resulting in an electric current.

As seen in the earlier reactions, the metal ion coming from the anode accumulates on the cathode surface as a result of the cathode reaction, and the metal in the anode state is corroded. Thus, an electrochemical or galvanic battery is formed, as given in Figure 1.2 [18].

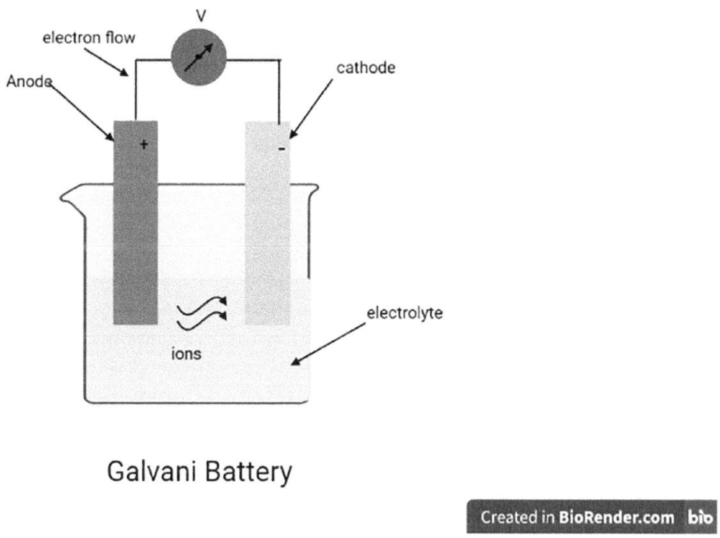

FIGURE 1.2 Electrode potentials of metal ions.

1.3 TYPES OF CORROSION

Corrosion can be classified and named in different ways. These classifications can be made according to the nature of the galvanic batteries, the appearance of corrosion, or the type of environment in which they are found. In this study, the classification made according to the appearance of corrosion is briefly emphasized [19]. In addition, the corrosion of reinforcing steel in reinforced concrete structures is also discussed, as given in Figure 1.3.

Corrosion types can be examined under eight main headings according to the classification made according to their appearance. These are homogeneous corrosion, pitting corrosion, crevice or crevice corrosion, undercoat corrosion, galvanic corrosion, selective corrosion, scrape corrosion, and intergranular corrosion [20]:

Homogeneous Corrosion: The metal surface is affected by the corrosive environment at the same level, and the section thickness of the metal decreases over time [21]. For example, atmospheric corrosion and corrosion of steel or zinc in sulfuric acid are examples of this category. In homogeneous corrosion, the life of the material can be predicted.

Pitting Corrosion: It is the most common example of regional corrosion types. This type of corrosion creates pits of various depths and shapes on the surface. Pitting corrosion usually occurs on metal surfaces in a solution or in moist soil [22]. In metals with a protective film layer, pitting corrosion develops where the protective layer is weakest. Usually, chloride ions cause pitting corrosion.

Gap or Slit Corrosion: This occurs in gaps or crevices at the junction or mounting points of metal surfaces. This type of corrosion can be associated with dirt, dust, and mud deposits. The formation mechanism is similar to pitting corrosion [23]. Crevice corrosion usually occurs on metal surfaces covered with oxide film. These different types of corrosion can lead to deterioration and damage to structural materials. Therefore, the effect of corrosion can be reduced or prevented by taking appropriate measures at the design stage.

Under Coating Corrosion: It is a special type of corrosion that occurs just under organic or metal coatings on metals such as steel, zinc, aluminum, and magnesium. This type of corrosion is affected by the relative humidity in the atmosphere and usually occurs in the range of 65–80%. When the relative humidity exceeds 90%, corrosion may appear as blisters. In undercoating corrosion, water leaks under the coating via osmosis due to the density of Fe^{+2} ions formed in the anode region under the coating [24]. Then it is separated from the corroded loose part, and the corrosion continues. The mechanism of this type of corrosion is related to the occurrence of anodic reactions in an acidic region with low oxygen density and the flow of oxygen toward the cathode region, where reduction occurs.

Galvanic Corrosion: It is a type of corrosion that develops depending on environmental conditions, electrolyte type, and concentration [25]. This type of corrosion occurs when dissimilar metal parts come together. The metal in the anode state, which has lower corrosion resistance, corrodes by losing electrons. For example, galvanic corrosion can occur when nails, screws, or rivets made of different metals are joined to the base metal. In order to prevent this situation, it is preferred that these parts are made of the same material as the base metal.

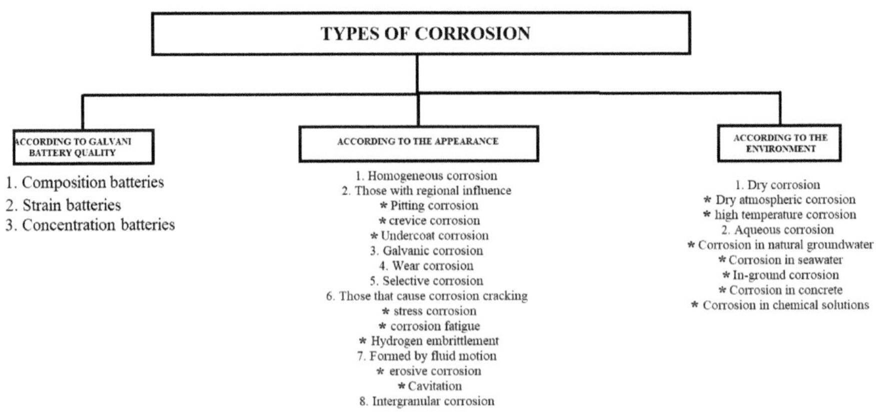

FIGURE 1.3 Types of corrosion.

Selective Corrosion: It occurs as a result of faster corrosion of the more active metals in an alloy than the other. For example, zinc in brass is more active than copper, and therefore selective corrosion occurs in brass. In this case, the brass turns into a red copper color, and the strength of the alloy decreases [26].

Scratch Corrosion: This occurs between surfaces moving on each other. In this type of corrosion, corrosion is seen in the form of pits and lines [27]. Often, machine parts are subject to this type of corrosion.

Intergranular Corrosion: Corrosion occurs at the grain boundaries of metal materials. This type of corrosion causes a loss of strength and ductility in the material. In particular, alloys such as aluminum, copper, and 18–8 stainless steel can be affected by intergranular corrosion. Progressive corrosion at the grain boundaries weakens the grain boundaries and may eventually lead to their disintegration [28]. These different types of corrosion can cause deterioration and damage to materials. Therefore, taking appropriate protection measures and being careful in material selection play an important role in reducing the effects of corrosion.

Three types of corrosion that cause corrosion cracking are stress corrosion, hydrogen embrittlement, and corrosion fatigue [20, 26].

Stress Corrosion: It is a type of corrosion that occurs with the combination of tensile stresses and chemical effects in the metal material. It starts with stresses caused by cracks and irregularities in the material surface. It may progress rapidly over time and cause sudden ruptures at values exceeding the carrying capacity of the material. Therefore, it is an important type of corrosion. Stress corrosion can last from a few minutes to several years, depending on the environment and stress level of the metal [29].

Hydrogen Brittleness: It occurs during plastic deformation of alloys in contact with hydrogen gas and depends on the loading rate [30]. Hydrogen embrittlement begins inside the material, while stress corrosion begins at the material surface.

Corrosion Fatigue: It refers to the fractures that occur on the metal material at low load levels due to the combined effect of repetitive stresses and corrosion. It grows faster in corrosive environments [31]; it starts and progresses at weak points on the metal surface. For example, it can occur in pits or areas where intergranular corrosion occurs. Generally, cracks form around the main crack, but fracture occurs in the main crack.

Types of corrosion caused by fluid movements cause mechanical damage to the metal surface. These are erosion corrosion and cavitation corrosion [32].

Erosive Corrosion: It causes corrosion by destroying the protective film layer on the metal surface with the turbulent movements of the fluid. In this type of corrosion, there may be homogeneous damage, as well as often pitted areas on the surface. Soft metals such as copper and lead are more susceptible to erosion corrosion [33]. Systems exposed to moving fluids, such as piping systems, can be affected by this corrosion.

Cavitation Corrosion: It is a subtype of erosive corrosion [34]. This type of corrosion creates cavities and pits as a result of the explosion of gas bubbles formed by the movement of the liquid on the metal surface. It is usually seen in places where there is rapid fluid flow, such as ship propellers.

These various types of corrosion can cause materials to crack and become damaged. Therefore, it is important to take appropriate precautions and be careful about material selection.

1.4 METHODS OF PROTECTION FROM CORROSION

The electrochemical corrosion cell consists of four main elements [35]. The anode, where electron loss occurs as a result of the oxidation reaction; the cathode, where electron gain is achieved as a result of the reduction reaction; the ionic current path formed by the anode and the solution in which the cathode is located; and the electron current path connecting these elements with a conductor. If any of these four elements is missing, corrosion does not occur or stops. For example, corrosion can be prevented by removing or inactivating the anode metal. Corrosion can also be prevented by blocking or eliminating the ionic current path. For example, in the absence of moisture in the metal environment, no ionic current path is formed and corrosion is prevented [36]. In galvanic corrosion, in the case of the combination of two different metals, corrosion is prevented because the separating layers to be placed at the junction of the metals are electrically insulated. In addition, preventing the presence of the cathode is one of the most effective corrosion control methods, as given in Figure 1.4.

The most widely used method of corrosion protection in industry is cathodic protection. Cathodic protection is accomplished by applying current to the cathode by an external source to prevent the flow of electrons from the anode to the cathode. Two methods can be used for cathodic protection [37]. In the first method, a more active metal is immersed in the same electrolyte to act as an anode. This creates a large-scale galvanic battery, protecting the cathode. The second method is to apply external direct current to the system by using an inert anode. This protection system is a widely used method in industry. Especially

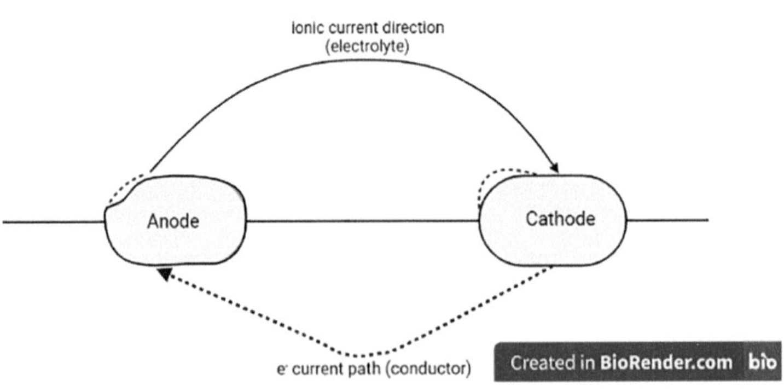

FIGURE 1.4 Formation scheme of the electrochemical corrosion cell.

underground pipelines, water, soil, and metal structures containing electrolytes, such as concrete, prefer the cathodic protection method. Cathodic protection is based on the principle of stopping the anode reaction by providing the electrons required for the cathode reaction from an external source. If there is not enough oxygen at the cathode, corrosion will not occur because the cathode reaction cannot occur. Electrons given from the external source allow the anode reactions to proceed at the auxiliary anode in the cathodic protection circuit, while completely stopping the anode reactions on the protected metal surface. In this way, the protected metal surface becomes completely cathodic [38].

External current-sourced cathodic protection is based on the principle of stopping the anode reactions with the electrons given to the protected metal surface and increasing the rate of the cathode reaction. The anode reactions take place at the auxiliary anode, while the protected metal surface is now fully cathodic.

Cathodic protection with galvanic anode is achieved by connecting a more active metal (galvanic anode) to the protected metal. In this case, the electrons required for the cathode reaction are provided by the spontaneous oxidation reaction of the galvanic anode. In this way, the anode reactions on the protected metal surface are completely stopped [39]. Galvanic anode cathodic protection is also based on this basic principle.

The purpose of cathodic protection is to protect the metal from any kind of corrosion by being more negatively charged than the soil or sea below a fixed potential threshold for any metal in a certain environment, as given in Figure 1.5. When this protection is provided, a continuous current flows per unit area from each point of

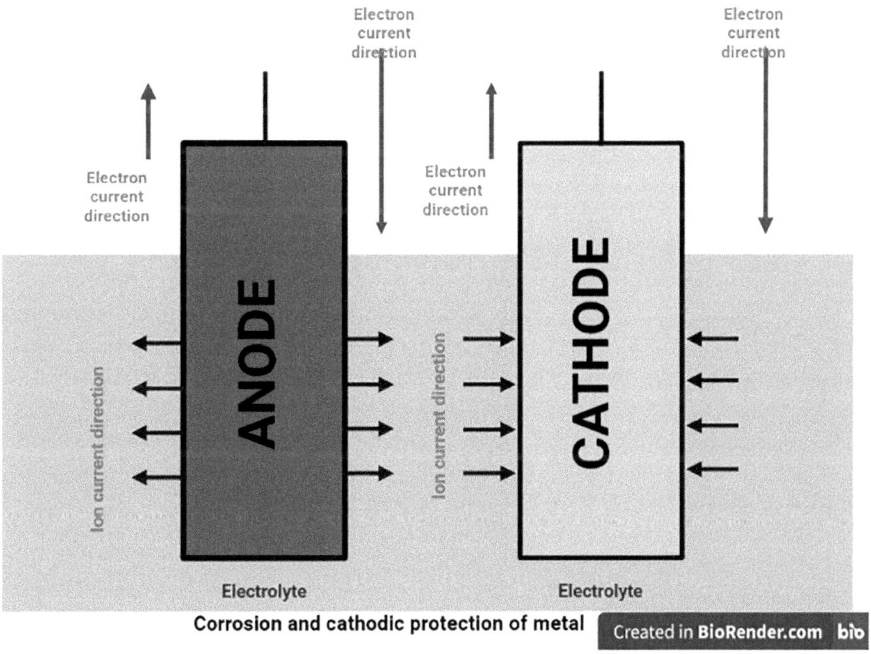

FIGURE 1.5 Metal corrosion and catalytic protection.

the metal structure [40]. The current density depends on the resistance of the coating and the potential difference in that area. Measurements made with reference to the copper/copper sulfate ($Cu/CuSO_4$) electrode are important in terms of meeting the criteria that determine the reliability of the cathodic protection.

Among the various methods used in corrosion protection, from chemical methods, inhibitors play an important role. Inhibitors are substances or mixtures used in low concentrations to prevent or minimize corrosion in aggressive environments. Inhibitors can passivate the metal by interacting with the metal surface or provide protection by reacting with aqueous corrosive compounds in the environment. Inhibitors can be added to the solution at low rates or applied by adding them to the paint. Inhibitors can be of organic or inorganic origin. Inorganic inhibitors include chromates, nitrites, silicates, carbonates, phosphates, and arsenates, while organic inhibitors show a wider variety [41]. Examples of organic inhibitors include amines, heterocyclic nitrogen compounds, sulfur compounds, and natural compounds. According to the protection mechanism of inhibitors, they are classified as anodic inhibitors, cathodic inhibitors, and mixed-type inhibitors. While anodic inhibitors provide protection by covering the anodic areas on the metal surface, cathodic inhibitors prevent corrosion by slowing down the reduction reaction rate of the metal in the electrochemical cell. Mixed-type inhibitors, on the other hand, form a protective layer on the metal surface in both the anodic and cathodic regions [42]. The use of corrosion inhibitors in reinforced concrete systems is also common. In these systems, anodic inhibitors, cathodic inhibitors, or protective coatings can be used to prevent reinforcement corrosion. The quality of the concrete, the water/cement ratio, the quality of the aggregates, and the pozzolan additives are also effective factors in the formation of protective concrete against corrosion. Today, studies are carried out on plant-derived materials as green corrosion inhibitors. Measures to be taken to prevent reinforcement corrosion include methods such as reducing the permeability of concrete, coating the reinforcement, using inhibitors, applying protective coating, and applying cathodic protection. The advantages and disadvantages of these measures should also be considered [43].

Distributed control infrastructure (DCI) [44] plays an important role in the effect of corrosion inhibitors. Used especially for reinforced concrete structures, the DCI corrosion inhibitor acts as a calcium nitrite-based solution to control the corrosion of steel. Calcium nitrite is a widely known inhibitor for systems requiring corrosion protection. Studies on structures that have served for a long time with field and laboratory tests prove the effectiveness of this inhibitor in concrete [45]. Corrosion inhibitors slow or stop corrosion by changing the surface chemistry of the steel. The DCI inhibitor does not prevent chlorides from penetrating concrete but protects the steel despite the presence of high concentrations of chloride. DCI is a simple application method that enters fresh concrete as a calcium nitrite solution. This method provides full corrosion protection for built-in steel and other materials by properly dispersing. Calcium nitrite, unlike other inhibitors, does not have any negative effect on the physical properties of concrete.

The DCI corrosion inhibitor is classified as an ASTM C494 type C additive [46]. Calcium nitrite helps to naturally protect the already existing steel in reinforced concrete structures. Steel forms an oxide layer consisting of iron (Fe), oxygen (O), and hydroxide (OH) ions in an alkaline environment. This oxide layer first starts with ferrous ions. In an alkaline medium, ferrous (Fe^{++}) ions are converted to ferric (Fe^{+++}) ions. Ferrous oxide on reinforced concrete is less resistant to chloride ions and forms the site where corrosion begins. Chlorides start corrosion by attacking the iron atoms. Calcium nitrite remains in the concrete throughout the life of the structure and does not react with other elements [47]. Nitrite and chloride ions form complexes with ferrous ions, preventing chloride attack. Nitrite elevates the ferrous oxide, enabling it to turn into protective ferric oxide. In addition, nitrite anions are chemically absorbed on the iron surface, forming a solid ferric oxide protective layer. In this way, ferrous oxide defects are surrounded by nitrite ions and are less likely to be exposed to chloride attack.

DCI protects steel against chloride-induced corrosion up to a certain chloride concentration and significantly increases the service life of structures [48]. However, over time, the amount of chloride in the steel reaches a level sufficient to initiate corrosion. Corrosion cells grow and corrosion products accumulate. The DCI inhibitor continues to inhibit corrosion through the buildup surrounding the steel. In this way, the corrosion rate is lower compared to unprotected steels.

As a result, DCI corrosion inhibitor provides effective protection for reinforced concrete structures containing calcium nitrite. Calcium nitrite raises the ferrous oxide layer in concrete to prevent corrosion and forms a protective ferric oxide layer [49]. This inhibitor extends the life of structures and increases their resistance to corrosion.

A general way to prevent corrosion is to choose metals and alloys suitable for the characteristics of the environment used [50]. While making these choices, some rules must be followed.

- In non-oxidizing reducing conditions, for example, in airless acids and aqueous solutions, nickel-copper alloys are generally preferred.
- In oxide conditions, it would be appropriate to use alloys containing chromium.
- In very severe oxidizing conditions, titanium and titanium alloys may be preferred.

While alloying elements impart corrosion-resistant properties, they also increase the cost of the material [51]. Alloying is an economically expensive method because in order to form an alloy, the alloying element must be evenly distributed over the entire structure of the material. This leads to the high use of corrosion-resistant elements and therefore to an increase in cost. Stainless steels are one of the best examples of this situation.

Chromium added to stainless steels is used to prevent rusting. However, stainless steels are not completely stainless, and they are not the most resistant to corrosion [52]. While stainless steels are commonly used in moderately oxidizing corrosion environments (e.g., nitric acid), they are less resistant to chloride-containing solutions and stress corrosion than structural steels. Therefore, care should be taken not to use stainless steel where they are not suitable.

When choosing an alloy, it is aimed to choose the material that is most resistant to corrosion at the most cost-effective price. The materials that can be used in different corrosive environments are summarized next [53]:

- Airless acids and aqueous solutions: Nickel-copper alloys can be preferred.
- Oxide conditions: Chromium-containing alloys can be used.
- Very severe oxidizing conditions: Titanium and titanium alloys can be used.

By paying attention to this information, the effects of corrosion can be minimized by choosing the right material.

Considering the situations that may cause corrosion during the design phase of the systems [54] and taking appropriate measures helps prevent corrosion. Examples of appropriate system design are listed next [55]:

- High fluid velocity in pipelines may cause erosion corrosion at junction areas. Therefore, the fluid velocity should be reduced when approaching the junction areas.
- When designing automobile gas tanks, they should be designed in such a way that deposits do not form inside. This prevents the accumulation of water or other corrosive substances inside.
- When fasteners such as rivets and bolts come into contact with metal surfaces, they should be isolated from their surroundings or materials that will not form galvanic couples should be preferred. In this way, galvanic corrosion is prevented.
- Electrodes that will not form galvanic couples should be used as welding electrodes for the materials to be welded. This helps prevent galvanic corrosion.

In order to prevent local corrosion, the rules to be followed during the design and production stages are as follows [56]:

- Direct contact of dissimilar metals should be avoided. Galvanic corrosion can occur between dissimilar metals, so suitable insulation or interlayers must be used.
- Crevices should be avoided because they can be foci of corrosion and cause localized corrosion.
- It is important to prevent random currents from electrical machines. Such currents can increase the rate of corrosion, so appropriate insulation and protection measures must be taken.
- Welding should be preferred rather than rivets in connections. Rivets can cause crevice corrosion, so if rivets are required, rivets with cathodic properties should be preferred.
- Parts of a system that can undergo rapid corrosion should be easily replaceable. This means that parts exposed to corrosion are more accessible and reconfigurable.

- Since the thickness of the material will decrease over time due to corrosion, the minimum acceptable thickness of the material should be determined at the design stage. Usually, twice the thickness required for the desired life is taken.

Proper implementation of these measures makes systems more resistant to corrosion and helps prevent corrosion-related problems.

Protective coating [57] is a method that aims to provide protection by forming an insulating layer between the material surface in contact with the corrosive environment and substances. Corrosion is a process that leads to the deterioration of metals by reacting with various factors. Covering metal surfaces with protective coatings to prevent or slow down this process creates an insulating barrier between the environment and the structure, making it a more economical solution.

The quality of protective coatings is closely related to their electrical resistivity properties [58]. Coatings with high resistance values are considered more suitable in terms of corrosion. For example, coatings with a resistance greater than 10,000 Ohm/mm^2 reduce the risk of the material being affected by corrosion, resulting in longer life. Protective coatings can be used on different materials and are usually applied as paint, varnish, epoxy, polymer, or metallic coatings. These coatings protect metal surfaces from environmental factors, chemicals, water, salty environments, and other corrosion effects. In addition, coatings provide protection against damage caused by abrasion, impact, UV rays, and other external factors. The selection and application of protective coatings depend on factors such as the intended use of the material, environmental conditions, corrosion potential, and budget. These coatings have a wide range of uses in industrial and structural applications by extending the life of materials, reducing costs, and saving energy.

Paints are coatings that are generally made of organic materials and prevent water and oxygen from reaching the metal surface by isolating the metal surface from the environment. In the composition of these coatings, various pigments are used to prevent corrosion. It is often preferred in order to protect it from corrosion due to its ease of application and low cost, which is the reason for the widespread use of paints [59]. For this purpose, different types of paints have been developed, each differing in chemical composition and physical properties. However, the common purpose of all of them is to prevent the contact of the metal with its environment by forming a solid and impermeable shell [60]. Paints fulfill the task of preventing corrosion in three ways:

First, all paints form an impermeable layer between the metal surface and the environment, preventing the metal from reacting with its surroundings. This layer prevents water, oxygen, and other corrosion effects from reaching the metal surface.

Second, some paints contain pigments that act as inhibitors. These pigments help passivate the metal surface and form a protective layer. In this way, it is ensured that the metal becomes more resistant to corrosion.

Third, some paints contain zinc dust as a filler. Zinc protects the metal cathodically by creating a galvanic effect. In this way, the metal coated with zinc does not act as an anodic zone during the corrosion process, and a more effective protection is provided.

In particular, two-component polymer paints such as epoxy, polyurethane, and zinc dust can last up to 15–20 years even in the most intense corrosion environments. These paints form a solid coating, preventing the metal from interacting with its environment and providing long-term protection. Therefore, they are among the preferred paints in industrial and structural applications [61].

Metallic coatings are coatings applied to a metal surface to protect it. These coatings are either applied on a more durable noble metal or preferentially dissolve themselves to provide protection by preventing the metal they cover from dissolving. Examples of noble metal platings of the first type are gold or silver plating on copper, and tin, lead, or chromium plating on steel. Examples of the second type of active coating metals are zinc, cadmium, or aluminum coating on steel.

Defects or discontinuities (scratches, pores, cuts, or nicked edges) in noble metal coatings cause the underlying metal to be anodic relative to the noble metal. In these regions, a severe corrosion event is observed with the effect of a large cathode and a small anode [62]. Therefore, it is extremely important to prevent defects in this type of coating, as coating defects can have serious consequences.

In active coating metals, small discontinuities do not cause corrosion of the metal underneath. Because the active metal itself is consumed, it provides a kind of cathodic protection and provides protection of the bottom metal. Interesting methods such as coating the active metal surfaces with a chromate or phosphate layer can also be used [60, 61]. In this case, the dissolution of the active metal is greatly reduced.

Metallic coatings not only protect metals from corrosion but also provide decorative appeal and functional properties. Multiple layers of coatings can also be applied and provide both corrosion protection and decorative effects. The composition, structure, and impurities of metallic coatings affect their corrosion resistance [57, 60]. Many factors play a role in the selection of metal coatings, such as environmental conditions, the expected life of the coated metal part, decorative effect, metal type, shape and size of the part, subsequent manufacturing methods, and mechanical factors. In order to achieve optimum corrosion resistance, factors such as the environment in which the metallic coatings will be used, the type of substrate, the type of coating, the method of application, and whether there are other coatings available should be considered.

Inorganic coatings are coatings used to make the surface of steel smooth and durable. One such coating is porcelain coating, which is usually applied to steel surfaces. Porcelain coating is a thin layer of glass applied to the steel surface by boiling. This coating provides good adhesion with steel and has a structure compatible with the expansion coefficient of the steel [60].

Glass-coated steel containers are used in areas such as the chemical industry. This is because these coatings are easy to clean and have high corrosion resistance. The glass layer protects the surface of the steel from the environment and forms an effective barrier against oxidation and corrosion [59, 61]. In addition, the glass coating provides resistance to the effects of chemicals and supports the longevity of the steel.

Inorganic coatings are often preferred in applications that must withstand harsh conditions such as high temperatures, corrosive chemicals, or heavy operations. These coatings play an important role in protecting steel surfaces and improving their properties. In addition, the color stability and aesthetic appeal of inorganic coatings are also advantageous, so they can also be used for decorative purposes [60].

In general, inorganic coatings increase the functional properties of the steel while protecting its surface. Inorganic coatings, such as porcelain coatings, increase the durability, corrosion resistance, and service life of steels, making them widely used in various industrial applications.

Three-layer polyethylene (PE) coatings are coatings with long-term high resistance [63]. The structure of these coatings includes three different layers:

The first coat is used to provide corrosion protection. This layer is usually applied as an epoxy primer or powder coating. Epoxy primer provides excellent adhesion to the steel surface, forming a protective layer. Powder paint, on the other hand, is sprayed on the steel surface and adheres electrostatically and provides corrosion resistance [63]. These layers help prevent corrosion by forming a barrier between the steel surface and environmental influences.

The second layer is called the intermediate layer and consists of thermoplastic material. This layer is used to provide flexibility and durability. Thermoplastic material resists various environmental influences and protects the steel surface from mechanical damage. Moreover, this layer reinforces the structure of the three-layer PE coating and acts as a transition layer [63].

The third layer consists of PE material to provide mechanical protection. Polyethylene offers excellent impact resistance and abrasion resistance [63]. This layer protects the steel surface against external factors and provides resistance against mechanical damage. In addition, the PE layer forms the outer surface of the three-layer PE coatings used in steel pipelines, completing the appearance and protective properties.

Three-layer PE coatings are widely preferred in applications such as pipelines, tanks, and steel structures used in various industries [64]. These coatings provide effective protection against corrosion, mechanical damage, and damage caused by external factors. In addition, thanks to its three-layer structure, it offers advantages such as durability, flexibility, and longevity.

Biological coating is a method used to protect metal surfaces such as ship hulls from corrosion in seawater. This type of coating has an advantage over other coating methods because it does not have any toxic effects on marine life [65].

Corrosion in seawater occurs when metals react with minerals in the water to produce hydrogen sulfide. Widely used technologies today require the use of inhibitors and biocides to inhibit the formation of hydrogen sulfide [66]. However, the method of protecting metal surfaces with these chemicals is expensive and ineffective. In addition, these chemicals can adversely affect marine life.

Biological coating, on the other hand, offers a natural and environmentally friendly approach. In this method, organisms in seawater are allowed to settle on the surface of the metals and form a biofilm. This biofilm prevents metals from coming into direct contact with seawater and provides protection against corrosion. At the same time, this biofilm does not harm marine life and is environmentally friendly. One of the advantages of biocoating is its longevity. It is more durable than chemical-based coatings and provides longer protection [65]. In addition, maintenance and replacement costs are lower. It plays an important role both in protecting metal surfaces such as ship hulls and in maintaining the health of marine life.

Anodic protection is based on the principle of forming a protective passive film on the surfaces of metals by forcing them to dissolve at higher rates to prevent dissolution and thus slowing down corrosion. Anodic protection is accomplished by keeping the metal permanently passive [67]. The passive state is the state where the corrosion current is lowest. However, the lowest corrosion current may not always be low enough, and for some solutions this method may not provide a suitable solution. Potential and current control should be done very well in anodic protection applications. If a malfunction occurs in the system, the protected metal can corrode in a short time. For this reason, anodic protection is generally used to reduce the corrosion rate in environments where severe corrosion events are experienced.

Anodic protection is applicable only for passivating metals and alloys (such as Fe, Ni, Cr, Ti, and similar metals and alloys) [67]. A very small current density may be sufficient to keep the metal in the passive region, but the anodic protection system has complex equipment and high hardware costs to provide this current density.

For this reason, the anodic protection method is used to reduce the corrosion rate, especially in environments with strong corrosion events. However, factors such as potential and current control that must be applied need to be carefully managed.

1.5 THE EFFECT OF CORROSION ON THE ECONOMY

Environmental pollution has become a serious problem, especially with the development of industry and the rise in living standards. This problem both threatens human health and causes economic losses in living and non-living systems. One of the most important factors among these economic losses is metal corrosion. Environmental pollution plays a role in accelerating corrosion. Especially for structures open to the atmosphere, the atmosphere becomes more corrosive with the effect of pollutants, as well as being corrosive [68]. Pollutants such as sulfurous gases, nitrogen oxides, and ammonia in the atmosphere of the city and industry dissolve in the moisture layer condensed on the metal surface and form corrosive electrolyte solutions, thus increasing the corrosion rate of building materials and shortening their lifetime. Therefore, the corrosion effects of pollutants such as sulfur compounds (SCb) and hydrogen sulfide (HiS) on iron were investigated in laboratory conditions by analytical methods. The amount of iron in the corrosion products formed on the surface of the samples exposed to corrosion was determined by methods such as Atomic Absorption Spectroscopy (AAS). While the corrosive effect of the atmosphere is sufficient for ferrous materials, it has been observed that iron corrosion accelerates in the presence of pollutants such as SO_2 and H_2S.

Relative humidity is also of great importance in atmospheric corrosion. It has been stated that the atmosphere is corrosive at 60–70% relative humidity under normal conditions. However, in environments where hydrogen sulfide gas is present, the corrosive effect starts from 30% relative humidity [69].

Considering these data, the effect of corrosion on the economy is of great importance. Corrosion shortens the life of building materials, increasing rebuilding costs, increasing maintenance and repair requirements, and leading to productivity losses. Therefore, the prevention or reduction of corrosion is of great economic importance.

The relationship between environmental pollution and corrosion cannot be ignored. In order to prevent or reduce the damage of corrosion to the economy, environmental factors should be controlled [70]. These measures extend the service life of building materials, reduce costs, and minimize economic losses.

Here are some of the economic effects of corrosion [71]:

- **Reconstruction Costs**: Corrosion shortens the life of building materials, increasing rebuilding costs. Corroded buildings, bridges, pipelines, and other infrastructure systems must be repaired or replaced. This leads to huge costs for the government and the private sector.
- **Maintenance and Repair Expenses**: Corrosion causes extra labor and material costs that require regular maintenance and repair. Protective measures to combat corrosion mean regular maintenance and repair of deteriorated materials. This requires the use of economic resources.
- **Loss of Productivity**: Malfunctions or equipment breakdowns as a result of corrosion can cause production processes to stop. Production losses lead to a decrease in the income of enterprises. In addition, there may be an increase in production costs, as equipment used in production processes must be regularly maintained and renewed with protective measures against corrosion.
- **Loss of Energy Efficiency**: Corrosion also causes negative effects on energy generation and distribution systems. Corroded pipelines or equipment can cause energy losses and reduced efficiency. This leads to an increase in energy costs and a decrease in economic efficiency.
- **Environmental Impacts**: Leakages or waste products resulting from corrosion may cause environmental pollution. Harmful substances leaking from corroded storage tanks or pipelines can contaminate water sources, soils, and ecosystems. Environmental cleaning and restoration costs also impose a burden on the economy.
- **Loss of Employment**: The negative effects of corrosion on the economy can affect the financial condition of businesses and negatively affect employment. Large-scale corrosion problems can reduce the financial strength of enterprises and endanger the job security of employees.
- **Economic Losses**: All the aforementioned effects of corrosion lead to economic losses. Corrosion can affect a significant percentage of the national income of countries and cause waste of resources.

- **Safety Threat**: Corrosion can create equipment and structures that weaken structural integrity. This can endanger work safety and cause serious accidents.
- **Decreased Product Quality**: Corrosion can lead to damage to equipment and tanks used in production processes. This may result in contamination of products or loss of quality.
- **Cost Increase**: Corrosion shortens the life of building materials, increasing rebuilding costs. Corrosion on metal surfaces leads to material deterioration and replacement, which requires extra expenses.

When all these effects are taken into consideration, it is clear that corrosion imposes a significant burden on the economy. Combating corrosion, taking protective measures, carrying out regular maintenance and repair works, using corrosion-resistant materials [72], and applying conscious economic policies play an important role in reducing the effects of this problem.

1.6 CONCLUSIONS

Corrosion and its effects on the economy are discussed. Corrosion is the chemical or electrochemical degradation of metals naturally entering the environment. These reactions usually occur through the interaction of the metal with the reader and water, oxygen, or other chemical compounds. Corrosion groups generally consist of two basic processes: anode activation and cathode competition. At the anode outlet, the emergence of metal and release metal ions (positively charged ions). During this ascent, electrons are released. In the cathode effect, electron acceptors (later respiration or ions) in the electrical charge enter the duration of the electrons, and their reduction takes place. Corrosion events can occur in different ways depending on various factors, such as temperature, humidity, pH level, salinity, deterioration, and corrosion tendency. The main reason for us to separate the corrosion according to its types was to have the chance to examine the formation system in a more specific and detailed way. For example, there are different types of interference, such as general interference, localized interference (pitting corrosion, surface corrosion), and stress corrosion. Each type comes under different components and has different effects. The effect of corrosion on the economy is very important. Corrosion reduces the durability of protective measures, shortens the life of equipment and infrastructure lines, and requires repairs and replacements. As a result, economic losses occur. Corrosion protection methods play an important role in reducing these harmful effects. There are various protection methods used against corrosion. These include coating, anodic protection, cathodic protection, and inhibition. This is used to prevent corrosion of metal surfaces or to reduce attacks. In our study, we identified the most effective method of cathodic protection among these protection methods. Cathodic protection is a method of shielding the surface metal with the use of a more electronegative metal or a sacrificial grade. In this method, the protected desired metal acts as the cathode and is coupled with a more electrochemically active metal or anode. In this way, the metal to be secured is pressed on the anode, and fracture is prevented. The application to reduce the management of corrosion to the economy is of great importance in the industrial and infrastructure sectors.

REFERENCES

[1] Alamri, A. H. (2020). Localized corrosion and mitigation approach of steel materials used in oil and gas pipelines—An overview. *Engineering Failure Analysis*, 116, 104735.

[2] Rodrigues, R., Gaboreau, S., Gance, J., Ignatiadis, I., & Betelu, S. (2021). Reinforced concrete structures: A review of corrosion mechanisms and advances in electrical methods for corrosion monitoring. *Construction and Building Materials*, 269, 121240.

[3] Cheng, X., Xia, J., Wu, R. J., Jin, W. L., & Pan, C. G. (2022). Optimisation of sacrificial anode cathodic protection system in chloride-contaminated reinforced concrete structure. *Journal of Building Engineering*, 45, 103515.

[4] Korde, J. M., Sreekumar, A. V., & Kandasubramanian, B. (2020). Corrosion inhibition of 316L-type stainless steel under marine environments using epoxy/waste plastic soot coatings. *SN Applied Sciences*, 2, 1–13.

[5] Jin, S. Y., Wu, R. Z., Wang, J. X., Wang, G. X., Boris, K. R. I. T., & Betsofen, S. (2021). Corrosion behavior of Mg-Li alloys: A review. *Transactions of Nonferrous Metals Society of China*, 31(11), 3228–3254.

[6] Khayatazad, M., De Pue, L., & De Waele, W. (2020). Detection of corrosion on steel structures using automated image processing. *Developments in the Built Environment*, 3, 100022.

[7] Deif, S., & Daneshmand, M. (2019). Multiresonantchipless RFID array system for coating defect detection and corrosion prediction. *IEEE Transactions on Industrial Electronics*, 67(10), 8868–8877.

[8] Fayomi, O. S. I., Akande, I. G., & Odigie, S. (2019, December). Economic impact of corrosion in oil sectors and prevention: An overview. In *Journal of Physics: Conference Series* (Vol. 1378, No. 2, p. 022037). IOP Publishing.

[9] Bell, S., Steinberg, T., & Will, G. (2019). Corrosion mechanisms in molten salt thermal energy storage for concentrating solar power. *Renewable and Sustainable Energy Reviews*, 114, 109328.

[10] Storch, L., Hamatschek, C., Hesse, D., Feist, F., Bachmann, T., Eichler, P., & Grigoratos, T. (2023). Comprehensive analysis of current primary measures to mitigate brake wear particle emissions from light-duty vehicles. *Atmosphere*, 14(4), 712.

[11] Bandiera, M., Mauri, A., Bestetti, M., di Milano, P., Bonfanti, A., Mancini, A., & Bertasi, F. (2020, June). Corrosion phenomena in braking systems. In *NACE CORROSION* (pp. NACE-2020). NACE.

[12] Dopffel, N., Jansen, S., & Gerritse, J. (2021). Microbial side effects of underground hydrogen storage—Knowledge gaps, risks and opportunities for successful implementation. *International Journal of Hydrogen Energy*, 46(12), 8594–8606.

[13] Sun, F., Peng, X., Bai, X., Chen, Z., Xie, R., He, B., & Han, P. (2022). EIS analysis of the electrochemical characteristics of the metal—water interface under the effect of temperature. *RSC Advances*, 12(27), 16979–16990.

[14] Harsimran, S., Santosh, K., & Rakesh, K. (2021). Overview of corrosion and its control: A critical review. *Proceedings of Engineering Sciences*, 3(1), 13–24.

[15] Fattah-alhosseini, A., Babaei, K., & Molaei, M. (2020). Plasma electrolytic oxidation (PEO) treatment of zinc and its alloys: A review. *Surfaces and Interfaces*, 18, 100441.

[16] Deng, F., Olvera-Vargas, H., Zhou, M., Qiu, S., Sirés, I., & Brillas, E. (2023). Critical review on the mechanisms of Fe2+ regeneration in the electro-Fenton process: Fundamentals and boosting strategies. *Chemical Reviews*, 123(8), 4635–4662.
[17] Garg, N., Ramadesigan, V., & Tatiparti, S. S. V. (2023). Principles of electrochemistry. In *Handbook of Sodium-Ion Batteries: Materials and Characterization* (Vol. 33). Jenny Stanford Publishing.
[18] Asserghine, A., Ashrafi, A. M., Mukherjee, A., Petrlak, F., Heger, Z., Svec, P., . . . & Adam, V. (2021). In situ investigation of the cytotoxic and interfacial characteristics of titanium when galvanically coupled with magnesium using scanning electrochemical microscopy. *ACS Applied Materials & Interfaces*, 13(36), 43587–43596.
[19] Robuschi, S., Sumearll, J., Fernandez, I., & Lundgren, K. (2021). Bond of naturally corroded, plain reinforcing bars in concrete. *Structure and Infrastructure Engineering*, 17(6), 792–808.
[20] Saji, V. S., & Umoren, S. A. (Eds.). (2020). *Corrosion Inhibitors in the Oil and Gas Industry*. John Wiley & Sons.
[21] Silva, C. L., Soares, R. B., Pereira, P. H. R., Figueiredo, R. B., Lins, V. F., & Langdon, T. G. (2019). The effect of high-pressure torsion on microstructure, hardness and corrosion behavior for pure magnesium and different magnesium alloys. *Advanced Engineering Materials*, 21(3), 1801081.
[22] Zhao, Z., Zhang, H., Xian, L., & Liu, H. (2020). Tensile strength of Q345 steel with random pitting corrosion based on numerical analysis. *Thin-Walled Structures*, 148, 106579.
[23] Mamin, A. N., Kodysh, E. N., Bobrov, V. V., Reutsu, A. V., Volkova, O. V., & Chaganov, A. B. (2019, December). Durability of suspension facade systems with air gap. In *Journal of Physics: Conference Series* (Vol. 1425, No. 1, p. 012054). IOP Publishing.
[24] May, Z., Alam, M. K., & Nayan, N. A. (2022). Recent advances in nondestructive method and assessment of corrosion undercoating in carbon-steel pipelines. *Sensors*, 22(17), 6654.
[25] Kolesnikov, A., Kolek, M., Dohmann, J. F., Horsthemke, F., Börner, M., Bieker, P., . . . & Stan, M. C. (2020). Galvanic corrosion of lithium-powder-based electrodes. *Advanced Energy Materials*, 10(15), 2000017.
[26] Kong, D., Dong, C., Ni, X., & Li, X. (2019). Corrosion of metallic materials fabricated by selective laser melting. *NPJ Materials Degradation*, 3(1), 24.
[27] Hu, Y., Wu, S., Withers, P. J., Cao, H., Chen, P., Zhang, Y., . . . & Hutař, P. (2021). Corrosion fatigue lifetime assessment of high-speed railway axle EA4T steel with artificial scratch. *Engineering Fracture Mechanics*, 245, 107588.
[28] Pan, Y., Zhang, D., Liu, H., Zhuang, L., & Zhang, J. (2021). Precipitation hardening and intergranular corrosion behavior of novel Al-Mg-Zn (-Cu) alloys. *Journal of Alloys and Compounds*, 853, 157199.
[29] Mohtadi-Bonab, M. A. (2019). Effects of different parameters on initiation and propagation of stress corrosion cracks in pipeline steels: A review. *Metals*, 9(5), 590.
[30] Zvirko, O. I., Kryzhanivskyi, E. I., Nykyforchyn, H. M., & Krechkovska, H. V. (2021). Methods for the evaluation of corrosion-hydrogen degradation of steels of oil-and-gas pipelines. *Materials Science*, 56, 585–592.
[31] Olugbade, T. O., Ojo, O. T., Omiyale, B. O., Olutomilola, E. O., & Olorunfemi, B. J. (2021). A review on the corrosion fatigue strength of surface-modified stainless steels. *Journal of the Brazilian Society of Mechanical Sciences and Engineering*, 43(9), 421.
[32] Swain, B., Mohapatra, S. S., & Behera, A. (2021). Plasma spray coating: A weapon to fight with erosion and corrosion phenomena. In *Thermal Spray Coatings* (pp. 317–330). CRC Press.

[33] Zhao, Q., Lu, Z., Wu, Y., & Zhao, W. (2022). Designing strong interfacial adhesion between carbon fiber and epoxy resin via dopamine towards excellent protection ability under high hydrostatic pressure and severe erosion corrosion condition. *Composites Science and Technology*, 217, 109090.
[34] Mayer, A. R., Bertuol, K., Siqueira, I. B., Chicoski, A., Váz, R. F., de Sousa, M. J., & Pukasiewicz, A. G. (2020). Evaluation of cavitation/corrosion synergy of the Cr3C2–25NiCr coating deposited by HVOF process. *Ultrasonics Sonochemistry*, 69, 105271.
[35] Hausmann, J. N., Beltrán-Suito, R., Mebs, S., Hlukhyy, V., Fässler, T. F., Dau, H., . . . & Menezes, P. W. (2021). Evolving highly active oxidic iron (III) phase from corrosion of intermetallic iron silicide to master efficient electrocatalytic water oxidation and selective oxygenation of 5-hydroxymethylfurfural. *Advanced Materials*, 33(27), 2008823.
[36] Mahidashti, Z., Rezaei, M., & Asfia, M. P. (2020). Internal under-deposit corrosion of X60 pipeline steel upon installation in a chloride-containing soil environment. *Colloids and Surfaces A: Physicochemical and Engineering Aspects*, 602, 125120.
[37] Pourhashem, S., Saba, F., Duan, J., Rashidi, A., Guan, F., Nezhad, E. G., & Hou, B. (2020). Polymer/Inorganic nanocomposite coatings with superior corrosion protection performance: A review. *Journal of Industrial and Engineering Chemistry*, 88, 29–57.
[38] Angst, U. M. (2019). A critical review of the science and engineering of cathodic protection of steel in soil and concrete. *Corrosion*, 75(12), 1420–1433.
[39] Wang, H., Tan, R., Yang, Z., Feng, Y., Duan, X., & Ma, J. (2021). Stabilization perspective on metal anodes for aqueous batteries. *Advanced Energy Materials*, 11(2), 2000962.
[40] Jothi, V. R., Karuppasamy, K., Maiyalagan, T., Rajan, H., Jung, C. Y., & Yi, S. C. (2020). Corrosion and alloy engineering in rational design of high current density electrodes for efficient water splitting. *Advanced Energy Materials*, 10(24), 1904020.
[41] Pradhan, B. (2022). A study on effectiveness of inorganic and organic corrosion inhibitors on rebar corrosion in concrete: A review. *Materials Today: Proceedings*, 65, 1360–1366.
[42] Topçu, İ. B., & Uzunömeroğlu, A. (2020). Properties of corrosion inhibitors on reinforced concrete. *Journal of Structural Engineering and Applied Mechanics*, 3(2), 93–109.
[43] Goyal, A., Ganjian, E., Pouya, H. S., & Tyrer, M. (2021). Inhibitor efficiency of migratory corrosion inhibitors to reduce corrosion in reinforced concrete exposed to high chloride environment. *Construction and Building Materials*, 303, 124461.
[44] Rahman, M. L., Malakooti, A., Ceylan, H., Kim, S., & Taylor, P. C. (2022). A review of electrically conductive concrete heated pavement system technology: From the laboratory to the full-scale implementation. *Construction and Building Materials*, 329, 127139.
[45] Daniyal, M., & Akhtar, S. (2020). Corrosion assessment and control techniques for reinforced concrete structures: A review. *Journal of Building Pathology and Rehabilitation*, 5, 1–20.
[46] Valdez-Salas, B., Vazquez-Delgado, R., Salvador-Carlos, J., Beltran-Partida, E., Salinas-Martinez, R., Cheng, N., & Curiel-Alvarez, M. (2021). Azadirachta indica leaf extract as green corrosion inhibitor for reinforced concrete structures: Corrosion effectiveness against commercial corrosion inhibitors and concrete integrity. *Materials*, 14(12), 3326.
[47] Xu, P., Zhou, J., Li, G., Wang, P., Wang, P., Li, F., . . . & Chi, H. (2021). Corrosion inhibition efficiency of compound nitrite with D-sodium gluconate on carbon steel in simulated concrete pore solution. *Construction and Building Materials*, 288, 123101.

[48] Agboola, O., Kupolati, K. W., Fayomi, O. S. I., Ayeni, A. O., Ayodeji, A., Akinmolayemi, J. J., . . . & Oluwasegun, K. M. (2022). A Review on corrosion in concrete structure: Inhibiting admixtures and their compatibility in concrete. *Journal of Bio-and Tribo-Corrosion*, 8(1), 25.
[49] Pradipta, I., Kong, D., & Tan, J. B. L. (2019). Natural organic antioxidants from green tea form a protective layer to inhibit corrosion of steel reinforcing bars embedded in mortar. *Construction and Building Materials*, 221, 351–362.
[50] Abdel-Karim, A. M., & El-Shamy, A. M. (2022). A review on green corrosion inhibitors for protection of archeological metal artifacts. *Journal of Bio-and Tribo-Corrosion*, 8(2), 35.
[51] Su, C., Wang, J., Hu, H., Wen, Y., Liu, S., & Ma, K. (2021). Enhanced strength and corrosion resistant of Mg-Gd-Y-Al alloys by LPSO phases with different Al content. *Journal of Alloys and Compounds*, 885, 160557.
[52] Francis, R., & Byrne, G. (2021). Duplex stainless steels—alloys for the 21st century. *Metals*, 11(5), 836.
[53] Mei, D., Lamaka, S. V., Lu, X., & Zheludkevich, M. L. (2020). Selecting medium for corrosion testing of bioabsorbable magnesium and other metals—a critical review. *Corrosion Science*, 171, 108722.
[54] Tsai, Y. H., Wang, J., Chien, W. T., Wei, C. Y., Wang, X., & Hsieh, S. H. (2019). A BIM-based approach for predicting corrosion under insulation. *Automation in Construction*, 107, 102923.
[55] Kong, D., Wu, Z., Fu, Q., Zhang, H., Liu, N., Zhang, J., . . . & Zhang, K. (2019, May). A novel fanless energy efficient edge computing system architecture and engineering practice for Baidu PCDN application. In *2019 18th IEEE Intersociety Conference on Thermal and Thermomechanical Phenomena in Electronic Systems (ITherm)* (pp. 1–7). IEEE.
[56] Luo, F. W., Ran, R., & Wang, L. (2020). Study on corrosion law of large crude oil storage tank floor and risk-based inspection and maintenance technology. *Corrosion Science and Technology*, 19(2), 66–74.
[57] Kulyk, B., Freitas, M. A., Santos, N. F., Mohseni, F., Carvalho, A. F., Yasakau, K., . . . & Costa, F. M. (2022). A critical review on the production and application of graphene and graphene-based materials in anti-corrosion coatings. *Critical Reviews in Solid State and Materials Sciences*, 47(3), 309–355.
[58] Burduhos-Nergis, D. P., Vizureanu, P., Sandu, A. V., & Bejinariu, C. (2020). Evaluation of the corrosion resistance of phosphate coatings deposited on the surface of the carbon steel used for carabiners manufacturing. *Applied Sciences*, 10(8), 2753.
[59] Almuslem, A. S., Shaikh, S. F., & Hussain, M. M. (2019). Flexible and stretchable electronics for harsh-environmental applications. *Advanced Materials Technologies*, 4(9), 1900145.
[60] Kordas, G. (2019). Nanotechnology to improve the biofouling and corrosion performance of marine paints: From lab experiments to real tests in sea. *International Journal of Physics Research and Applications*, 2(1), 033–037.
[61] Udhayakumar, M., Radhika, N., & Arun, K. L. (2022). A comprehensive review on nanocrystalline coatings: Properties, challenges and applications. *Journal of Bio-and Tribo-Corrosion*, 8(3), 83.
[62] Eliaz, N. (2019). Corrosion of metallic biomaterials: A review. *Materials*, 12(3), 407.
[63] Zargarnezhad, H., Asselin, E., Wong, D., & Lam, C. C. (2021). A critical review of the time-dependent performance of polymeric pipeline coatings: Focus on hydration of epoxy-based coatings. *Polymers*, 13(9), 1517.

[64] Samimi, A., & Zarinabadi, S. (2011). An analysis of polyethylene coating corrosion in oil and gas pipelines. *Journal of American Science, USA*, 7(2011), 1032–1036.
[65] Ferrari, M., Benedetti, A., Santini, E., Ravera, F., Liggieri, L., Guzman, E., & Cirisano, F. (2015). Biofouling control by superhydrophobic surfaces in shallow euphotic seawater. *Colloids and Surfaces A: Physicochemical and Engineering Aspects*, 480, 369–375.
[66] Khatun, S., Hirani, H., & Roy, P. (2021). Seawater electrocatalysis: Activity and selectivity. *Journal of Materials Chemistry A*, 9(1), 74–86.
[67] Gittens, R. A., Olivares-Navarrete, R., Tannenbaum, R., Boyan, B. D., & Schwartz, Z. (2011). Electrical implications of corrosion for osseointegration of titanium implants. *Journal of Dental Research*, 90(12), 1389–1397.
[68] Saha, D., Pandya, A., Singh, J. K., Paswan, S., & Singh, D. D. N. (2016). Role of environmental particulate matters on corrosion of copper. *Atmospheric Pollution Research*, 7(6), 1037–1042.
[69] Li, C., Ma, Y., Li, Y., & Wang, F. (2010). EIS monitoring study of atmospheric corrosion under variable relative humidity. *Corrosion Science*, 52(11), 3677–3686.
[70] Kokilaramani, S., Al-Ansari, M. M., Rajasekar, A., Al-Khattaf, F. S., Hussain, A., & Govarthanan, M. (2021). Microbial influenced corrosion of processing industry by re-circulating waste water and its control measures-A review. *Chemosphere*, 265, 129075.
[71] Akpanyung, K. V., & Loto, R. T. (2019, December). Pitting corrosion evaluation: A review. In *Journal of Physics: Conference Series* (Vol. 1378, No. 2, p. 022088). IOP Publishing.
[72] Dong, B., Zeng, D., Yu, Z., Cai, L., Yu, H., Shi, S., . . . & Yi, Y. (2021). Major corrosion influence factors analysis in the production well of CO2 flooding and the optimization of relative anti-corrosion measures. *Journal of Petroleum Science and Engineering*, 200, 108052.

2 Carbon allotropes

Important properties, synthesis methods, characterization, and applications

Pragnesh Dave and Lakha Chopda

2.1 INTRODUCTION

Many elements are able to exist in different structural forms known as allotropes. Broadly, it is defined as a single chemical element that exists in polymorphic form with more than one similar or dissimilar physical form. It refers to the different forms of the same elements with the same or different physical states (solid, liquid, or gas). In other words, it is the structural modification or arrangement of an element in which atoms of the element are bonded together in different manners [1]. The Swedish scientist Baron JönsJakob Berzelius is the proponent of the concept of allotropy. The term originates from the Greek word "άλλοτροπία" (allotropia), meaning "variability" or "changeableness." Following the acceptance of Avogadro's hypothesis in 1860, it became evident that elements could exist as polyatomic molecules, with two allotropes of oxygen identified: O_2 and O_3 [2]. The allotropes are different structural forms of the same element that show distinct physical properties and chemical behavior. The changes in structure and forms of allotropes are governed by underlying conditions (forces of light, temperature, and pressure) [3–5]. The allotropes of carbon are well recognized as carbon is an important and amazing element that is capable of generating many compounds from it [6]. It can form various allotropes (structurally different forms of the same element) known as allotropes of carbon. Diamonds and graphite are the natural examples of carbon allotropes, and subsequently many more carbon allotropes have been found. Carbon allotropes are the pure form of carbon. Carbon is among the few elements that have varied allotropes owing to variable oxidation state or coordination number, which makes carbon a versatile element for the formation of various allotropes and other important catenation properties (forming of bonds with neighboring carbon atoms) of carbon also contribute to the existence of diverse allotropes [5] (Figure 2.1). Carbon allotropes are joined by strong covalent bonds, but different arrangements of carbon atoms cause different physical properties. Carbon allotropes can be classified into amorphous and crystalline allotropes, as represented in Figure 2.1. Carbon allotropes are employed for various applications [7].

DOI: 10.1201/9781003456025-2

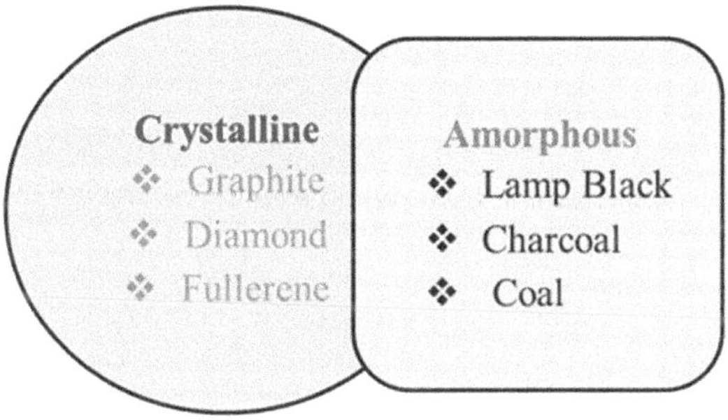

FIGURE 2.1 Classification of carbon allotropes.

2.2 IMPORTANT PROPERTIES

Under normal conditions, carbon allotropes are solid in nature. Graphite is the most thermodynamically stable form and is highly resistant to chemical reaction, as highly reactive oxygen needs a high temperature to react with graphite. The physical and chemical properties of some carbon allotropes are discussed here.

2.2.1 Graphite

It is the pure crystalline form of carbon. The structure of graphite is made of a flat two-dimensional layer of carbon atoms that are arranged hexagonally. In each layer of carbon, one carbon atom is linked to the other three carbon atoms through a covalent bond, and hybridization of carbon atom is sp2. The fourth bond is pi bond. The delocalization and mobilization of pi electrons make graphite electrically conductive in nature. Graphite has two forms α form and β form: The important property of graphite is depicted next [8–12].

- Greyish black color
- Solid material
- Opaque substance
- Lighter than diamond, smooth and slippery to touch
- Due to the presence of free electrons, able to conduct electricity
- Non-inflammable
- Soft due to weak Vander Waals forces

2.2.2 Diamond

Diamond is another allotrope of carbon that differs from graphite. It is the naturally occurring hardest carbon substance in which carbon atoms are arranged in a cubic crystal structure. In the structure of a diamond, each carbon atom is together by four carbon atoms, which are held by covalent bonds. The carbon atoms are sp3 hybridized, and the carbon bond length is consequently a three-dimensional network of strong covalent connections that arises in diamond. Diamond crystal structure is isometric, which reflects that carbon atoms are united and means that carbon atoms are bonded in basically the same way in all directions. Due to its wide application in the industry, demand for the material is continuously rising. It showed the highest thermal conductivity and poor electrical conductivity. The significant property of diamond is listed next [13–15].

- A very high melting point
- Strong carbon-carbon covalent bonds
- A very hard material
- Insoluble in water and organic solvent
- High reflectivity
- Corrosion resistance
- Optical transparency

2.2.3 Fullerene

Fullerenes are a distinctive allotrope of carbon, consisting of molecules formed by carbon atoms linked through single and double bonds. These molecules exhibit polyhedral structures, commonly referred to as "buckyballs" or "buckytubes," depending on their shape. Structurally, they resemble graphite with interconnected hexagonal rings, but they also incorporate pentagonal and occasionally heptagonal rings, preventing the formation of a planar sheet and resulting in their characteristic spherical or tubular forms. Common examples of fullerenes include C60 (Buckminsterfullerene), C70, and higher-order variants such as C80 and C90, with their properties and applications influenced by the number of carbon atoms in the structure. Fullerenes possess remarkable properties, including high tensile strength, excellent electrical conductivity, superior ductility, high thermal conductivity, chemical stability, and solubility in organic solvents. These attributes make them versatile materials for diverse applications. In the biomedical field, fullerenes are utilized in drug delivery systems and as antioxidants due to their ability to encapsulate active molecules. In electronics, their semi-conductive properties enable their use in diodes, transistors, and solar cells, where they mimic photosynthetic processes to enhance energy efficiency. Fullerenes are also used in coatings for surface protection and as additives to improve durability and thermal resistance. Additionally, they find applications in aerospace and automotive industries, where their integration into metals enhances strength and fluidity, making them ideal for lightweight structural components. With their unique structural and electronic characteristics, fullerenes hold significant promise in advancing nanotechnology, material science, and energy storage systems [16, 17].

2.3 SYNTHESIS METHODS

2.3.1 Graphite

Graphite is a native element mineral found in metamorphic and igneous rocks [18]. It is an extremely soft and layered material, is resistant to heat, and is chemically inert with other materials. The obtaining of graphite from natural and synthetic routes has received considerable attention owing to the valuable applications of graphite. The nature produced graphite from the metamorphic rocks consequently reduction of sedimentary rock in which carbon atoms are subjected to extreme heat and pressure in the earth crust and upper mantle [19]. The separation of natural graphite involved multistep methods, such as mining, large-scale beneficiation, and purification [20]. This is a complex and unproductive procedure for the production of graphite. The purification step required a huge amount of hydrofluoric acid (HF) for the removal of mineral impurities, and rigorous purification is essential for obtaining battery-grade graphite for lithium-ion batteries [21]. The separation of natural graphite is time-consuming and not considered eco-friendly [22]. Hence, the synthetic method is a convenient way for the synthesis of graphite. The graphite preparation method generally involves carbonizing carbon sources (biomass and organic matter) at very high temperatures and graphitizing amorphous carbon [23]. The graphite gained from petroleum coke at high temperature (above 2800°C) over a long period of time [24, 25]. This method faced the drawback of requiring a high amount of energy for the arrangement of carbon atoms into graphitized lattices. The synthesis of graphite by extracting Fe using Cl_2 has been devised [26]. The temperature below 500°C produced amorphous or disordered carbon. In the range of 600–1100°C, well-ordered graphite microcrystals were formed, and at 1130°C, the growth of large graphite crystals were found from the liquid phase like the creation of kish graphite by precipitation of carbon at high temperatures from supersaturated molten iron. The green and robust method for the synthesis of graphite is reported from the carbon dioxide (CO_2) in the absence of transition metal catalyst at ultralow temperatures [27].

2.3.2 Diamond

Diamond is a solid form of pure carbon. It has naturally 99.95% carbon, and the remaining 0.05% are the other impurities (not an inherent part of diamond). Hence, it is one of the expensive materials. Some trace elements can influence its color or crystal shape. The carbon allotrope graphite possesses only carbon, but its formation process and crystal structure are very different from diamond. Diamond is the most wonderful and unique gift to mankind from nature. It is an amazing miracle of nature consisting of billions of carbon atoms. Similar to graphite, diamond is also excavated worldwide [28]. It has been found that since 1870, most of the world's gem-quality diamonds have been quarried in Africa. In recent decades, the mining of diamonds has spread to many parts of the globe, including India. The diamond production has been full-fledged in Russia and Canada. Both countries produced almost half of the world's diamonds, but the core of natural gem-quality diamond production still remains in Africa. Diamond is found deep below the earth's surface under intense heat and high pressure [29]. Under these ambitious conditions, carbon crystallizes to form diamonds. Intense heat and high pressure need to be eliminated from the formed diamond. To facilitate this, a part of the mantle melts into magma. The magma comes out rapidly to the surface of the earth and transports diamonds into earth's deep crust. When the magma cools down, it converts into hard material known as kimberlite in the vertical structures called kimberlite pipes [30–32]. It is the most important source of natural diamonds. However, mining for natural diamonds frequently has a noteworthy environmental effect, such as deforestation, soil erosion, and disruption of local ecosystems, and the cost of natural diamonds is usually more expensive than their synthetic counterparts due to mining, transportation, and processing costs. The synthetic diamonds are produced by two different methods: (1) high-temperature, high-pressure process (HTHP process) and (2) chemical vapor deposition (CVD) method. The first method [28, 33–38], HTHP, is the mimic version of natural diamond production. In the HTHP method, raw materials along with a "seed diamond" are enclosed in a capsule, which is then subjected to intense heat and pressure. Over time, the diamond grows on the seed. In contrast, the Chemical Vapor Deposition (CVD) method uses methane gas (as a carbon source) and hydrogen gas to synthesize diamonds. Methane and hydrogen are dissociated in a high-temperature plasma, allowing carbon atoms from the methane to recombine and form diamonds. Hydrogen atoms play a crucial role by etching away any undesired graphite formed during the process, resulting in high-quality diamonds. The CVD method is simpler and more efficient compared to HTHP processes, enabling the production of a substantial number of diamonds on a large scale. The quality and purity of the gas used can be adjusted to produce diamonds with tailored properties, and impurities can be deliberately introduced to achieve specific characteristics. Two prominent variations of the CVD method include the Microwave Plasma CVD and the Hot Filament CVD techniques. In the Microwave Plasma CVD method, microwaves generate plasma within a reaction chamber, facilitating the dissociation of gaseous raw materials, which then form diamonds. This method allows precise control over crystal quality and purity, producing high-quality diamonds with excellent consistency. The Hot Filament CVD method, on the other hand, uses electrically heated filaments to reach extremely high temperatures, producing electrons that dissociate the gaseous molecules, ultimately leading to diamond formation. By using multiple filaments in a single chamber, this method enables the production of a large volume of diamonds. While natural diamonds take millions of years to form within the Earth, synthetic methods like these provide a controlled and efficient alternative for diamond production tailored to specific needs.

2.3.3 Fullerene

Fullerenes are the allotrope of carbon whose molecule consists of carbon atoms linked by single and double bonds. They are polyhedral carbon cages and perhaps the most prominent ball-shaped molecules in organic chemistry. They are one of the allotropes of the carbon atom, which are not evolved by nature. Since the discovery of C_{60} fullerene in 1985, it has gained significant position in chemistry due to its diverse applications such as drug delivery systems, sensors, catalysts, and many others [39–42]. The first synthesis of C_{60} fullerene was carried out by vaporizing graphite by laser irradiation, which produced a remarkably stable cluster of carbon atoms made up of 60 carbon atoms [43]. But this method bore many drawbacks, such as needing harsh reaction conditions (1300°C and pressure of 1 kbar), producing a mixture of C_{60} and C_{70}, and a lesser amount of higher fullerenes, such as C84, and requiring excessive efforts to purify the samples. But quantities of fullerene C60 were advanced prepared by arc heating of graphite and laser irradiation of polyaromatic

FIGURE 2.2 Characterization techniques for carbon allotropes.

hydrocarbon (PAHs). There are four main methods to synthesize fullerenes: (1) laser irradiation of carbon, (2) laser irradiation of PAHs, (3) electrical arc heating of graphite, and (4) resistive arc heating of graphite [44]. The structure of carbon allotropes is characterized by various techniques, as summarized in Figure 2.2.

2.4 APPLICATIONS

2.4.1 Graphite

Graphite is a promising material and has found applications in various fields [18, 45–47]. It is generally used in pencil tips, arc lamp electrodes, as a lubricant, coating, metallurgy, paint production, and moderatorand refractory component in nuclear reactors. Due to high thermal stability and chemical inertness, graphite is the perfect material in the refractory industry. It is used in the production of "Mag-Carbon" (Magnesia-Carbon) refractory bricks for lining steel-making furnaces [48]. Graphite carbon nitride-based materials are used as photo-catalysts for the deactivation of bacteria [23]. The natural flake graphite-based electrochemical coating exhibited many electrothermal applications [49]. Graphite is used as a fabrication material on the polymer substrate as a strain-resistive sensor. Graphite-based nanomaterials have found utility in drug delivery. In short, possessing significant properties such as small thermal expansion, high conductivity, high thermal resistance, excellent electrical conductivity, and brilliant lubricity makes graphite a valuable material in widespread industries like machinery, metallurgy, petrochemicals, light industry, electronics, electrical appliances, military defenses, and aerospace.

2.4.2 Diamond

Diamond, a unique allotrope of carbon, is widely used in jewelry due to its unparalleled beauty and durability. Often referred to as a "forever material," diamonds are cherished for their brilliance, making them a popular choice for engagement rings, earrings, and other fine jewelry pieces. Beyond aesthetics, the exceptional strength and distinctive properties of diamonds make them versatile for a broad range of applications [14, 50–52]. The majority of mined diamonds lack the quality required for use as gemstones, with approximately 80% of rough diamonds being allocated for industrial purposes. Due to their exceptional hardness, scoring a perfect 10 on the Mohs Hardness Scale, and remarkable durability, diamonds are extensively utilized in applications such as cutting, drilling, and polishing. These properties make them particularly valuable in the mining industry and military operations. Tiny diamond particles are embedded into tools like saw blades, drill bits, and grinding wheels, significantly enhancing their ability to cut through tough materials. As a result, diamonds play a crucial role in industrial cutting and drilling processes [53]. Diamonds, even in small sizes, are strategically embedded at the edges and tips of tools to enhance cutting and grinding efficiency. Diamond powder, often processed into a paste, is widely used for polishing and ultra-fine grinding applications. Due to their unmatched hardness and resistance to scratching, diamonds are ideal for engraving hard stones such as granite and quartz [54]. Diamond membranes, known for their durability, heat and abrasion resistance, and transparency, are utilized in advanced applications such as covering openings in lasers, vacuum chambers, and X-ray machines. Industrial-grade diamonds also contribute to superior sound quality by improving the performance of high-quality speakers, as their stiffness allows thin diamond domes to vibrate rapidly without deformation, preserving sound fidelity. Similarly, diamonds are employed in record player needles and DJ equipment for their precision and durability [55]. In addition, diamonds exhibit excellent thermal conductivity, making them suitable for use in heat sinks—devices designed to dissipate heat from electronic components such as processors. Their ability to withstand extreme temperature variations ensures consistent performance [56].Beyond industrial and electronic applications, diamonds have found a place in the beauty industry. Diamond-infused products, such as facial treatments and exfoliators, leverage diamond dust to reduce wrinkles and enhance skin appearance, albeit at a premium cost due to the high value of the material. This versatility underscores the widespread utility of diamonds across diverse fields [57].

2.4.3 Fullerene

Different types of fullerenes showed prominent applications in diverse fields. They have been the focus of attention in the different fields of science since their discovery. They are used as conductors, adsorbents for the removal of impurities, lubricants, and in cosmetic-related products, and find a place in the biomedical field as well as surface coating [17, 58, 59]. The introduction of fullerene in the metals (aluminum, duralumin, etc.) alters the properties of metals. The strength of metal increases the fluidity. Therefore, their applications find aircraft, automotive, and lightweight structures for spaceships [60–62]. Fullerene has been employed as an additive in fire-retardant paint. Introduction of fullerene, paint under the influence of temperature during a fire swells, a sufficient dense foam of layer is formed that several times enhanced the heating time to the critical temperature of the protected structure [63, 64]. Fullerenes have been employed in the production of Technos C60 oil, which offers a triple-layer protection mechanism. In the medical field, water-soluble fullerenes have demonstrated significant activity against Human Immunodeficiency Viruses (HIV) and exhibit strong antioxidant properties due to their high number of double bonds, making them effective as powerful free radical scavengers. Additionally, the molecular crystal structure of fullerenes exhibits semiconducting properties with a band gap of 1.5 eV, comparable to that of conventional semiconductors. This characteristic enables their application in electronic devices such as diodes, transistors, photocells, and other semiconductor-based technologies [44]. Owing to its unique electronic properties, one of the most innovative applications of fullerene is its use in solar cells, where it mimics the photosynthetic processes found in plants to enhance energy conversion efficiency. Additionally, fullerenes play a significant role in nanomedicine [59]. Their hollow structure allows them to encapsulate beneficial molecules, which can then be specifically directed to targeted sites, such as bacteria or cancer cells. Once at the intended location, the fullerene spheres dissolve, releasing their contents precisely where needed, ensuring that the effectiveness of the active substances remains intact. This targeted delivery system underscores the potential of fullerenes in advanced medical treatments and drug delivery systems.

2.5 CONCLUSION AND FUTURE PERSPECTIVE

The present chapter highlights the general physical properties of three carbon allotropes, some synthetic methods, and common synthetic applications. Carbon is able to form many allotropes due to its valency. Carbon atom possesses unique allotropes like graphite, diamond, fullerene, and many more. The carbon atom itself, connected with other carbon atoms in a three-dimensional arrangement, resulted in various allotropes. Each allotrope exhibits unique physical and chemical features that make it usable in each aspect of human life. The allotropes of carbon are promising materials for science and technology. Graphite and diamond are found in nature and can be synthesized, while various types of fullerenes can be prepared by a chemical route. Advanced carbon allotrope-based nanomaterials are potential flourishes in material-based technology. Advanced carbon allotropes will be forthcoming in subsequent scientific revolutions in electronics, biomedical, energy, and electrical engineering from a technological standpoint.

REFERENCES

[1] L. Fan, D. Yang, D. Li, A review on metastable silicon allotropes, *Materials.* 14 (2021) 3964. https://doi.org/10.3390/ma14143964.

[2] D.S. Sabirov, I.S. Shepelevich, Information entropy of oxygen allotropes. A still open discussion about the closed form of ozone, *Comput. Theor. Chem.* 1073 (2015) 61–66. https://doi.org/10.1016/j.comptc.2015.09.016.

[3] Z. Zhao, F. Tian, X. Dong, Q. Li, Q. Wang, H. Wang, X. Zhong, B. Xu, D. Yu, J. He, H.-T. Wang, Y. Ma, Y. Tian, Tetragonal allotrope of group 14 elements, *J. Am. Chem. Soc.* 134 (2012) 12362–12365. https://doi.org/10.1021/ja304380p.

[4] O.O. Kurakevych, Y. Le Godec, W.A. Crichton, T.A. Strobel, Silicon allotropy and chemistry at extreme conditions, *Energy Procedia.* 92 (2016) 839–844. https://doi.org/10.1016/j.egypro.2016.07.084.

[5] S. Nasir, M. Hussein, Z. Zainal, N. Yusof, Carbon-based nanomaterials/allotropes: A glimpse of their synthesis, properties and some applications, *Materials.* 11 (2018) 295. https://doi.org/10.3390/ma11020295.

[6] S.K. Tiwari, V. Kumar, A. Huczko, R. Oraon, A.D. Adhikari, G.C. Nayak, Magical allotropes of carbon: Prospects and applications, *Crit. Rev. Solid State Mater. Sci.* 41 (2016) 257–317. https://doi.org/10.1080/10408436.2015.1127206.
[7] W. Han, Z. Li, Y. Li, X. Fan, F. Zhang, G. Zhang, W. Peng, The Promoting role of different carbon allotropes cocatalysts for semiconductors in photo catalytic energy generation and pollutants degradation, *Front. Chem.* 5 (2017) 84. https://doi.org/10.3389/fchem.2017.00084.
[8] R. Sengupta, M. Bhattacharya, S. Bandyopadhyay, A.K. Bhowmick, A review on the mechanical and electrical properties of graphite and modified graphite reinforced polymer composites, *Prog. Polym. Sci.* 36 (2011) 638–670. https://doi.org/10.1016/j.progpolymsci.2010.11.003.
[9] E. Solfiti, F. Berto, A review on thermophysical properties of flexible graphite, *Procedia Struct. Integr.* 26 (2020) 187–198. https://doi.org/10.1016/j.prostr.2020.06.022.
[10] S. Kumar, Himanshi, J. Prakash, A. Verma, Suman, R. Jasrotia, A. Kandwal, R. Verma, S. Kumar Godara, M.A.M. Khan, S.M. Alshehri, J. Ahmed, A review on properties and environmental applications of graphene and its derivative-based composites, *Catalysts.* 13 (2023) 111. https://doi.org/10.3390/catal13010111.
[11] M.J. Mochane, T.E. Motaung, S.V. Motloung, Morphology, flammability, and properties of graphite reinforced polymer composites. Systematic review, *Polym. Compos.* 39 (2018) E1487–E1499. https://doi.org/10.1002/pc.24379.
[12] S.K. Khatkar, N.M. Suri, S. Kant, Pankaj, A Review on Mechanical and tribological properties of graphite reinforced self-lubricating hybrid metal matrix composites, *Rev. Adv. Mater. Sci.* 56 (2018) 1–20. https://doi.org/10.1515/rams-2018-0036.
[13] H. Yang, Y. Ma, Y. Dai, Progress of structural and electronic properties of diamond: A mini review, *Funct. Diam.* 1 (2021) 150–159. https://doi.org/10.1080/26941112.2021.1956287.
[14] N. Yang, S. Yu, J.V. Macpherson, Y. Einaga, H. Zhao, G. Zhao, G.M. Swain, X. Jiang, Conductive diamond: Synthesis, properties, and electrochemical applications, *Chem. Soc. Rev.* 48 (2019) 157–204. https://doi.org/10.1039/C7CS00757D.
[15] J.E. Field, The mechanical and strength properties of diamond, *Rep. Prog. Phys.* 75 (2012) 126505. https://doi.org/10.1088/0034-4885/75/12/126505.
[16] J. Yadav, Fullerene: Properties, synthesis and application, *J. Phys.* 6 (2017) 1–6.
[17] P.J.F. Harris, Fullerene polymers: A brief review, *C.* 6 (2020) 71. https://doi.org/10.3390/c6040071.
[18] A.D. Jara, A. Betemariam, G. Woldetinsae, J.Y. Kim, Purification, application and current market trend of natural graphite: A review, *Int. J. Min. Sci. Technol.* 29 (2019) 671–689. https://doi.org/10.1016/j.ijmst.2019.04.003.
[19] N. Vasumathi, T.V. Vijaya Kumar, S. Ratchambigai, S. Subba Rao, G. Bhaskar Raju, Flotation studies on low grade graphite ore from eastern India, *Int. J. Min. Sci. Technol.* 25 (2015) 415–420. https://doi.org/10.1016/j.ijmst.2015.03.014.
[20] J. Li, S. Hou, J. Su, K. Li, L. Wei, L. Ma, W. Shen, F. Kang, Z. Huang, Beneficiation of ultra-large flake graphite and the preparation of flexible graphite sheets from it, *New Carbon Mater.* 34 (2019) 205–210. https://doi.org/10.1016/S1872-5805(19)60012-0.
[21] K. Zaghib, X. Song, A. Guerfi, R. Rioux, K. Kinoshita, Purification process of natural graphite as anode for Li-ion batteries: Chemical versus thermal, *J. Power Sources.* 119–121 (2003) 8–15. https://doi.org/10.1016/S0378-7753(03)00116-2.
[22] F.A. Al-Sairafi, C. Jiang, Z. Zhong, B. Saleh, Study on purification of flake graphite by heat activation and hydrofluoric acid, *Adv. Mater. Process. Technol.* 8 (2022) 4564–4578. https://doi.org/10.1080/2374068X.2022.2079172.

[23] Y.W. Yap, N. Mahmed, M.N. Norizan, S.Z. Abd Rahim, M.N. Ahmad Salimi, K. Abdul Razak, I.S. Mohamad, M.M.A.-B. Abdullah, M.Y. Mohamad Yunus, Recent advances in synthesis of graphite from agricultural bio-waste material: A review, *Materials*. 16 (2023) 3601. https://doi.org/10.3390/ma16093601.

[24] J.M. Wrogemann, O. Fromm, F. Deckwirth, K. Beltrop, A. Heckmann, M. Winter, T. Placke, Impact of degree of graphitization, surface properties and particle size distribution on electrochemical performance of carbon anodes for potassium-ion batteries, *Batter. Supercaps*. 5 (2022). https://doi.org/10.1002/batt.202200045.

[25] K. Shen, Z.-H. Huang, K. Hu, W. Shen, S. Yu, J. Yang, G. Yang, F. Kang, Advantages of natural microcrystalline graphite filler over petroleum coke in isotropic graphite preparation, *Carbon*. 90 (2015) 197–206. https://doi.org/10.1016/j.carbon.2015.03.068.

[26] S. Dimovski, A. Nikitin, H. Ye, Y. Gogotsi, Synthesis of graphite by chlorination of iron carbide at moderate temperatures, *J. Mater. Chem*. 14 (2004) 238. https://doi.org/10.1039/b311938f.

[27] C. Liang, Y. Chen, M. Wu, K. Wang, W. Zhang, Y. Gan, H. Huang, J. Chen, Y. Xia, J. Zhang, S. Zheng, H. Pan, Green synthesis of graphite from CO2 without graphitization process of amorphous carbon, *Nat. Commun*. 12 (2021) 119. https://doi.org/10.1038/s41467-020-20380-0.

[28] V. Zhdanov, P. Smirnov, L. Andrzejewski, J. Bondareva, S. Evlashin, Comparative analysis of labor input required to produce one carat at different methods of synthesis and mining of diamonds, *Heliyon*. 8 (2022) e11519. https://doi.org/10.1016/j.heliyon.2022.e11519.

[29] B.V. Derjaguin, D.B. Fedoseev, The synthesis of diamond at low pressure, *Sci. Am*. 233 (1975) 102–109.

[30] A. Baruah, A.K. Gupta, N. Mandal, R.N. Singh, Rapid ascent conditions of diamond-bearing kimberlitic magmas: Findings from high pressure-temperature experiments and finite element modeling, *Tectonophysics*. 594 (2013) 13–26. https://doi.org/10.1016/j.tecto.2013.03.006.

[31] D.R.M. Pattison, A.A. Levinson, Are euhedral microdiamonds formed during ascent and decompression of kimberlite magma? Implications for use of microdiamonds in diamond grade estimation, *Appl. Geochem*. 10 (1995) 725–738. https://doi.org/10.1016/0883-2927(95)00037-2.

[32] A. Dongre, S. Tappe, Kimberlite and carbonatite dykes within the Premier diatreme root (Cullinan Diamond Mine, South Africa): New insights to mineralogical-genetic classifications and magma CO2 degassing, *Lithos*. 338–339 (2019) 155–173. https://doi.org/10.1016/j.lithos.2019.04.020.

[33] O.A. Ivanov, A.B. Muchnikov, V.V. Chernov, S.A. Bogdanov, A.L. Vikharev, J.E. Butler, Experimental study of hydrogen plasma etching of (100) single crystal diamond in a MPACVD reactor, *Mater. Lett*. 151 (2015) 115–118. https://doi.org/10.1016/j.matlet.2015.03.073.

[34] A.V. Khomich, V.G. Ralchenko, A.V. Vlasov, R.A. Khmelnitskiy, I.I. Vlasov, V.I. Konov, Effect of high temperature annealing on optical and thermal properties of CVD diamond, *Diam. Relat. Mater*. 10 (2001) 546–551. https://doi.org/10.1016/S0925-9635(00)00517-3

[35] Q. Liang, C. Yan, Y. Meng, J. Lai, S. Krasnicki, H. Mao, R.J. Hemley, Recent advances in high-growth rate single-crystal CVD diamond, *Diam. Relat. Mater*. 18 (2009) 698–703. https://doi.org/10.1016/j.diamond.2008.12.002.

[36] J. Lyu, S. Wang, B. Wang, K. Nishimura, N. Jiang, Synthesis of multiple single crystal diamonds by DC-GD-CVD, *Surf. Eng*. 35 (2019) 91–95. https://doi.org/10.1080/02670844.2018.1447270.

[37] M. Schwander, K. Partes, A review of diamond synthesis by CVD processes, *Diam. Relat. Mater.* 20 (2011) 1287–1301. https://doi.org/10.1016/j.diamond.2011.08.005.

[38] S. Eaton-Magana, J.E. Shigley, C.M. Breeding, Observations on HPHT-grown synthetic diamonds: A review, *Gems Gemol.* 53 (2017) 262–284. https://doi.org/10.5741/GEMS.53.3.262.

[39] M. Ajrin, A. Akther, Review on fullerene: A cutting edge trend in drug delivery, *Int. J. Pharm. Sci. Rev. Res.* 60 (2020) 84–89.

[40] R. Ganesamoorthy, G. Sathiyan, P. Sakthivel, Review: Fullerene based acceptors for efficient bulk heterojunction organic solar cell applications, *Sol. Energy Mater. Sol. Cells.* 161 (2017) 102–148. https://doi.org/10.1016/j.solmat.2016.11.024.

[41] H. Kaur, J. Kaur, R. Kumar, A review on all boron fullerene (B40): A promising material for sensing and device applications, *Mater. Today Proc.* 48 (2022) 1095–1102. https://doi.org/10.1016/j.matpr.2021.07.465.

[42] K. Aschberger, H.J. Johnston, V. Stone, R.J. Aitken, C.L. Tran, S.M. Hankin, S.A.K. Peters, F.M. Christensen, Review of fullerene toxicity and exposure—appraisal of a human health risk assessment, based on open literature, *Regul. Toxicol. Pharmacol.* 58 (2010) 455–473. https://doi.org/10.1016/j.yrtph.2010.08.017.

[43] M. Mojica, J.A. Alonso, F. Méndez, Synthesis of fullerenes: Synthesis of fullerenes, *J. Phys. Org. Chem.* 26 (2013) 526–539. https://doi.org/10.1002/poc.3121.

[44] A. Nimibofa, E.A. Newton, A.Y. Cyprain, W. Donbebe, Fullerenes: Synthesis and applications, *J. Mater. Sci. Res.* 7 (2018) 22. https://doi.org/10.5539/jmsr.v7n3p22.

[45] D. Barrett, M. Dentith, Geophysical exploration for graphite at Uley, South Australia, *Geophys. Sign, South Aust. Mineral Depos.* (2019) 47–58. https://doi.org/10.1071/asegspec12_04

[46] O. Baltag, G. Rosu, Applications of Graphite Materials in the field of electromagnetic compatibility, in: C. Miron, P. Mele, S. Kaneko, T. Endo (Eds.), *Carbon-Related Materials*, Springer International Publishing, Cham, 2020: pp. 19–44. https://doi.org/10.1007/978-3-030-44230-9_2.

[47] D.K. Das, P.C. Mishra, S. Singh, S. Pattanaik, Fabrication and heat treatment of ceramic-reinforced aluminium matrix composites—a review, *Int. J. Mech. Mater. Eng.* 9 (2014) 6. https://doi.org/10.1186/s40712-014-0006-7.

[48] I.O. Otunniyi, Z.V. Theko, B.L.E. Mokoena, B. Maramba, Major determinant of service life in magnesia-graphite slagline refractory lining in secondary steelmaking ladle furnace, *IOP Conf. Ser. Mater. Sci. Eng.* 655 (2019) 012003. https://doi.org/10.1088/1757-899X/655/1/012003.

[49] Z. Lei, S. Duan, X. Wu, C. Gao, L. Wang, X. Min, Z. Huang, M. Fang, R. Luo, B. Luo, High-performance electrothermal coatings based on natural flake graphite for multifunctional electrothermal applications, *J. Mater. Res. Technol.* 23 (2023) 1013–1026. https://doi.org/10.1016/j.jmrt.2023.01.071.

[50] O.A. Shenderova, A.I. Shames, N.A. Nunn, M.D. Torelli, I. Vlasov, A. Zaitsev, Review article: Synthesis, properties, and applications of fluorescent diamond particles, *J. Vac. Sci. Technol. B.* 37 (2019) 030802. https://doi.org/10.1116/1.5089898.

[51] R.K. Roy, K.-R. Lee, Biomedical applications of diamond-like carbon coatings: A review, *J. Biomed. Mater. Res. B Appl. Biomater.* 83B (2007) 72–84. https://doi.org/10.1002/jbm.b.30768.

[52] M. Shellaiah, K.W. Sun, A review on potential applications of diamond nanomaterials, *Sci. Adv. Today* 2 (2016) 25248.

[53] P.S. Martins, J.R. Gonçalves Carneiro, E.C.T. Ba, V.F. Vieira, Study on roughness and form errors linked with tool wear in the drilling process of an Al-Si alloy under high cutting speed using coated diamond-like carbon high-speed steel drill bits, *J. Manuf. Process.* 62 (2021) 711–719. https://doi.org/10.1016/j.jmapro.2021.01.006.

[54] A. Kubota, S. Nagae, S. Motoyama, High-precision mechanical polishing method for diamond substrate using micron-sized diamond abrasive grains, *Diam. Relat. Mater.* 101 (2020) 107644. https://doi.org/10.1016/j.diamond.2019.107644.

[55] K. Mershad, B. Said, DIAMOND: A tool for monitoring the participation of students in online lectures, *Educ. Inf. Technol.* 27 (2022) 4955–4985. https://doi.org/10.1007/s10639-021-10801-y.

[56] Q. Yang, J. Zhao, Y. Huang, X. Zhu, W. Fu, C. Li, J. Miao, A diamond made micro-channel heat sink for high-density heat flux dissipation, *Appl. Therm. Eng.* 158 (2019) 113804. https://doi.org/10.1016/j.applthermaleng.2019.113804.

[57] L. Lawson, Women working in the Thai coloured gemstone industry: Insights from entrepreneurial ecosystems, *Extr. Ind. Soc.* 6 (2019) 1066–1074. https://doi.org/10.1016/j.exis.2019.08.002.

[58] W. Hou, G. Shi, S. Wu, J. Mo, L. Shen, X. Zhang, Y. Zhu, Application of Fullerenes as photosensitizers for antimicrobial photodynamic inactivation: A review, *Front. Microbiol.* 13 (2022) 957698. https://doi.org/10.3389/fmicb.2022.957698.

[59] P. Bhakta, B. Barthunia, Fullerene and its applications: A review, *J. Indian Acad. Oral Med. Radiol.* 32 (2020) 159. https://doi.org/10.4103/jiaomr.jiaomr_191_19.

[60] M.M. Harussani, S.M. Sapuan, G. Nadeem, T. Rafin, W. Kirubaanand, Recent applications of carbon-based composites in defence industry: A review, *Def. Technol.* 18 (2022) 1281–1300. https://doi.org/10.1016/j.dt.2022.03.006.

[61] O. Gohardani, M.C. Elola, C. Elizetxea, Potential and prospective implementation of carbon nanotubes on next generation aircraft and space vehicles: A review of current and expected applications in aerospace sciences, *Prog. Aerosp. Sci.* 70 (2014) 42–68. https://doi.org/10.1016/j.paerosci.2014.05.002.

[62] M. Shafique, X. Luo, Nanotechnology in transportation vehicles: An overview of its applications, environmental, health and safety concerns, *Materials.* 12 (2019) 2493. https://doi.org/10.3390/ma12152493.

[63] Y. Yang, J.L. Díaz Palencia, N. Wang, Y. Jiang, D.-Y. Wang, Nanocarbon-based flame retardant polymer nanocomposites, *Molecules.* 26 (2021) 4670. https://doi.org/10.3390/molecules26154670.

[64] E.D. Weil, Fire-protective and flame-retardant coatings—A State-of-the-art review, *J. Fire Sci.* 29 (2011) 259–296. https://doi.org/10.1177/0734904110395469.

3 Carbon quantum dots (CQDs) and their composites as anti-corrosive materials

Halil İbrahim Çetintaş and Burak Tüzün

3.1 INTRODUCTION

As a significant global problem for many industries, such as the construction, manufacturing, and transportation sectors, corrosion can be briefly described as the deterioration of the material over time due to exposure to aggressive environmental conditions. It generally results in undesired equipment failures and costly repairs. According to NACE (National Association of Corrosion Engineers), at US$2.5 trillion, the annual cost of corrosion worldwide is over 3.4% of the world's GDP [1]. Moreover, it leads to various environmental hazards, which pose a great risk to all living beings. Since many pollutants accelerate corrosion and corrosion byproducts also pollute the environment, it is accepted that corrosion and pollution are interconnected processes [2]. Due to all these reasons, it is highly crucial to prevent corrosion in order to eliminate the potential costs and hazards resulting from this undesirable phenomenon.

Various corrosion prevention methods have been developed to enhance corrosion resistance, including alloying and dealloying, coating, galvanizing, painting, and the application of anti-corrosive materials, also called corrosion inhibitors [3]. Among these methods, anti-corrosive materials are the most extensively used technique in practice due to their low cost, simplicity, and efficiency. These materials can be divided into two categories: inorganic anti-corrosive materials and organic anti-corrosive materials [4]. Inorganic anti-corrosive materials, such as chromate, molybdate, nitrite, and phosphate, generally prevent corrosion for a long time, even in high temperatures, by forming a film layer via a reaction of anodic and cathodic parts; however, these materials can be reactive, toxic, and carcinogenic [5–7]. Organic anti-corrosive materials such as imidazoles, aldehydes, alcohols, amines, alkyl phenones, and other organic materials are adsorbed onto the surface of the material physically, chemically, or both ways, and form a thin protective layer between the material and the aggressive environment. Nonetheless, compared to inorganic anti-corrosive materials, even though organic anti-corrosive materials show low toxicity, they are more expensive, and because organic materials alone are not effective enough

 DOI: 10.1201/9781003456025-3

for preventing corrosion, a mixture of intensifiers, surfactants, and some solvents in appropriate amounts is usually needed [8–10]. Hence, researchers have recently focused on the development and fabrication of green, cost-effective, efficient, and multifunctional anti-corrosive materials.

The rapid pace of technological change and advancements in nanoscience have created new opportunities across a variety of research fields, including corrosion engineering. Especially in the past few years, carbon-based nanomaterials, including carbon black, carbon dots, carbon nanorods, carbon nanotubes, fullerene, graphene, and graphene oxide, have attracted a great deal of attention as anti-corrosive materials. As a cost-effective and high-yielding alternative to most other carbon-based nanomaterials, carbon quantum dots (CQDs) are extremely well-liked as corrosion inhibitors [11–13].

CQDs are novel members of the carbon-based nanomaterials family with a zero-dimensional structure that is generally less than 10 nm in size [14, 15]. They demonstrate numerous unique features such as high photoluminescence activity, good biocompatibility, low toxicity, high quantum yield, strong adsorption ability, low cost, high solubility, high thermal stability, and facile synthesis (Figure 3.1) [16, 17].

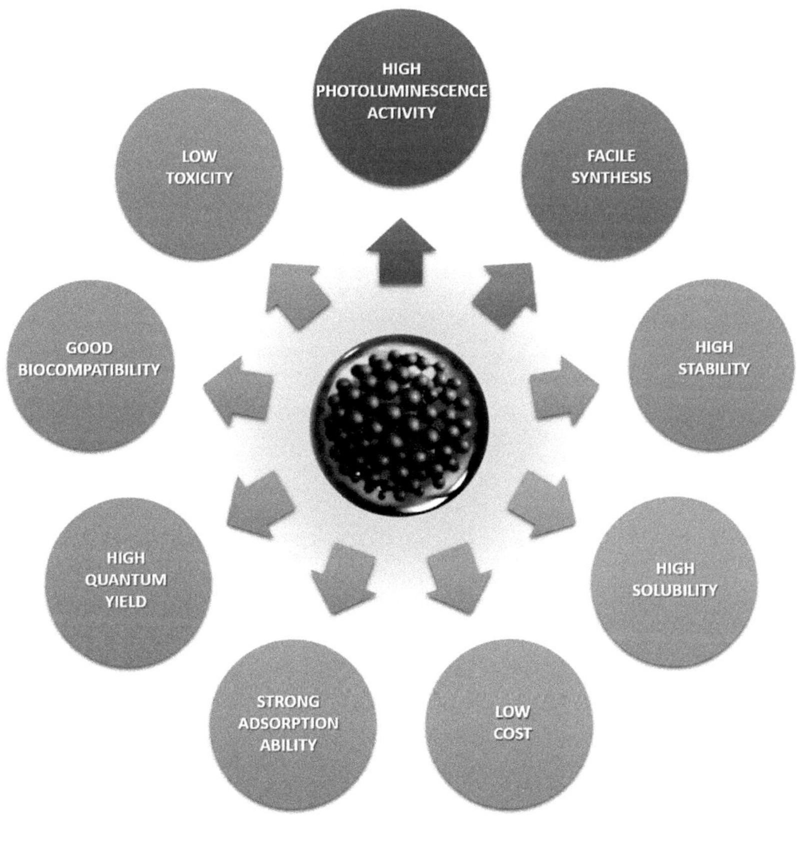

FIGURE 3.1 Properties of CQDs.

Also, their surface properties can be easily modified by grafting various functional groups, which provide additional features, power, and functionality to CQDs [18, 19]. Due to these unique qualities, CQDs are widely used for a range of biomedical applications, including bioimaging, biosensing, drug delivery, gene delivery, photodynamic therapy, photothermal therapy, pharmaceutical formulations, and industrial applications, such as catalysis, optoelectronics, and recently as anti-corrosive materials [20–24].

Even though the researchers are conducting extensive research about CQDs as anti-corrosive materials and the number of research studies is gradually increasing, it is still unclear what kind of influence the structures of CQDs have on corrosion protection performance, and further experimental studies are needed to shed light on this influence and to develop new CQD-based anti-corrosive materials [18]. This chapter discusses the synthesis, properties, and applications of CQDs and represents a comprehensive overview of the current studies about CQDs as anti-corrosive materials.

3.1.1 Carbon quantum dots

In 2004, Xu *et al.* published a report describing a preparative method for the purification of single-walled carbon nanotubes, and they accidentally isolated two novel classes of carbon nanomaterials. While one of these carbon-based nanomaterials was in a tubular shape, the other was a mixture of fluorescent carbon nanoparticles, called "carbon dots" (CDs) [25]. This study is accepted as the discovery of CQDs. Only two years after this report, Sun *et al.* conducted a study on fluorescent semiconductor carbon dots and produced nanoscale carbon particles in aggregates of various sizes. After a series of treatments, they obtained spherical nanoparticles around 5 nm in diameter and named them quantum-sized carbon dots [26]. This study is considered by the researchers as the origin of the name CQDs. There is actually a conceptual confusion in the literature regarding the term "CQDs." While some researchers consider CDs and CQDs to be the same concept [27–29], others classify CQDs as a distinct group of nanomaterials under the umbrella of CDs [30, 31].

CQDs are zero-dimensional quasi-spherical nanostructures with a diameter of less than 10 nm [32]. These carbon-based nanomaterials usually exist in sp^2 hybridization states, while in some cases, sp^3 hybridization states were also reported in the literature [33]. CQDs can be amorphous or crystalline, and their interplanar spacings are 0.21 and 0.34 nm, corresponding to (100) and (002) graphite crystallographic planes, respectively [34]. CQDs can be synthesized from several precursors via different techniques, and various surface functional groups can be incorporated during the synthesis [35]. These surface moieties give CQDs various additional features and make them more active and useful for many applications.

3.1.2 Synthesis

Synthetic approaches for CQD production are divided into two main categories as top-down and bottom-up. Top-down approaches refer to reducing macroscale materials to synthesize carbon nanoparticles by using various precursors [36]. These

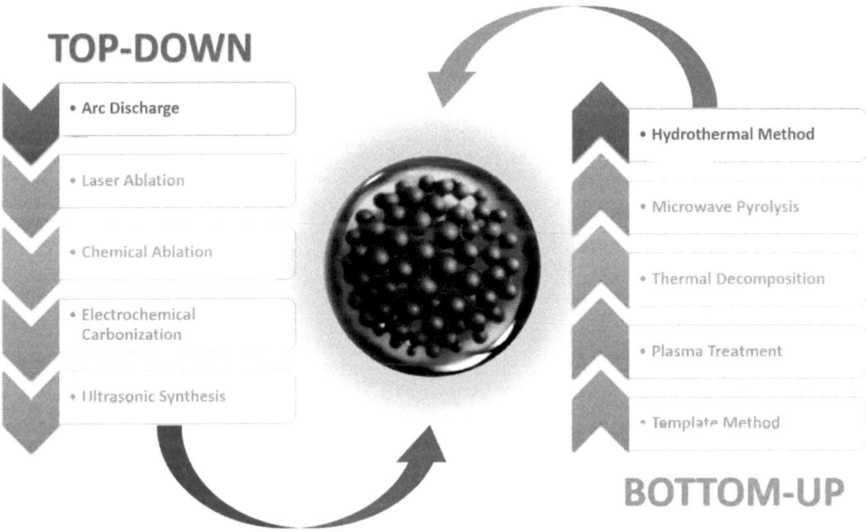

FIGURE 3.2 Top-down and bottom-up approaches to CQD synthesis [20].

precursors can be graphite, graphene, fullerene, carbon nanotubes, amorphous carbon, fiber, or even candle soot. Top-down approaches are more time-consuming and complex, and usually produce CQDs with lower emission quantum yields; however, it is mostly easier to control the structure and purity of the materials when compared to bottom-up methods [37, 38]. On the other hand, bottom-up approaches refer to producing CQDs of the desired size from smaller precursors or their combinations by using chemical reactions such as polymerization and carbonization [39]. These precursors can be glutamic acid, cysteine, glucose, glycerol, citric acid with L-aspartic acid, or glucose with ammonia [40–45]. Bottom-up approaches are simple, cost-effective, environmentally friendly, and allow for mass production [46]. With bottom-up methods, CQDs can even be produced from renewable sources, including waste, nutshells, fruit peels, or vegetable peels [47]. Also, in a study where a side-to-side comparison was made between top-down and bottom-up approaches, it was found that CDs synthesized from both approaches showed similar optical and morphological features except for several significant differences, such as surface structures and emission characteristics [48]. Top-down and bottom-up approaches to CQD synthesis are given in Figure 3.2.

3.1.2.1 Top-down approaches

3.1.2.1.1 Arc discharge

Arc discharge was the first method used for synthesizing CQDs. In 2004, Xu *et al.* synthesized the first CQDs from arc-discharged soot accidentally [25]. In this method, an anodic electrode composed of bulk carbon precursors and a cathode are placed a few millimeters from each other in a sealed reactor. A high-voltage direct current is created under an inert gas in order to produce plasma, and the temperature in the reactor reaches almost 4000 K. The carbon vapor derived from the anode is deposited on the cathode to produce CQDs [14, 49].

3.1.2.1.2 Laser ablation

Laser ablation, also known as laser irradiation, is one of the physical vapor deposition techniques [50]. As this technique is a physical approach, it allows for the production of small materials with very high purity [51]. In this technique, a laser beam with very high energy concentrates at a certain point on the carbon precursors in a liquid or colloidal dispersion form, and an ionization resulting in the removal of the surface atoms occurs. Thus, ablation leads to the formation of CQDs [52, 53].

3.1.2.1.3 Chemical ablation

Chemical ablation is a technique in which tiny organic molecules are carbonized to nano-carbonaceous materials by powerful oxidizing acids such as sulfuric acid or nitric acid, then pulled into thin sheets by a controlled oxidation process. While this technique is quite advantageous as it does not require complicated equipment and allows for large amounts of CQD production, it can be considered a disadvantage, in that the synthesis process is carried out in harsh reaction conditions [20].

3.1.2.1.4 Electrochemical carbonization

Alcohols are generally used as solvents in various solvent-based reactions; however, electrochemical carbonization of these solvents results in high-purity small nanoparticles. In this method, as carbon-based precursors, alcohols are directly converted into CQDs, especially in alkaline environments, via low-cost and simple electrochemical processing [54, 55].

3.1.2.1.5 Ultrasonic synthesis

In ultrasonic synthesis, or sonochemical synthesis, carbon-based precursors such as graphene, glucose, or dopamine are dissolved in alkali or acidic media, and then the mixed solution is given an ultrasonic treatment for a certain period of time at various temperatures [42, 56, 57].

3.1.2.2 Bottom-up approaches

3.1.2.2.1 Hydrothermal method

In the hydrothermal method, an aqueous solution of a carbon-based precursor is prepared and transferred to a Teflon-lined hydrothermal autoclave reactor. This closed system is heated and kept at a high temperature (100–1000°C) to let the hydrothermal reaction occur under high pressure [58, 59]. This method has several advantages, such as simplicity, low cost, mild reaction conditions, and the absence of the requirement for strong acids or bases [60]. Also, various characteristics of the precursors, including carbon atom number, carbon chain length, or functional groups, have important effects on the chemical and structural features of CQDs in hydrothermal processes [61]. When non-aqueous solvents are used in this method, it is also referred to as the solvothermal method.

3.1.2.2.2 Microwave pyrolysis

Microwave pyrolysis, one of the most well-known and convenient bottom-up approaches, is a method that utilizes microwave irradiation in order to synthesize large amounts of CQDs in only a few minutes [62, 63]. In this low-cost technique, a clear solution of carbon-based precursors is prepared and placed in a microwave oven. High-frequency energy created by the microwave oven generates atomic-level heating in the polar molecules due to microwave irradiation. As a result, a rapid reaction occurs, and high-purity CQDs in large amounts are produced [64].

3.1.2.2.3 Thermal decomposition

Thermal decomposition is a conventional bottom-up technique in which heat is used to decompose chemical bonds in the carbon-based precursors for producing CQDs. It is such an effective technique that it allows for monodisperse CQDs with different particle sizes to be produced from the same precursor [65]. Thermal decomposition is an endothermic reaction that can be either reversible or irreversible. Large amounts of CQDs can be produced by this simple and low-cost technique [66].

3.1.2.2.4 Plasma treatment

Plasma treatment or plasma-induced nanomaterial synthesis has emerged as a highly promising technique for producing eco-friendly CQDs [67]. In this technique, a neutral gas is heated or exposed to a strong electromagnetic field, and plasma occurs. Thus, the partially ionized particle beams with high energy generated from plasma interact with the precursors and lead to CQD synthesis within a few minutes [68, 69]. Due to this technique, materials with low reactivity and temperature sensitivity can be used as precursors [70].

3.1.2.2.5 Template method

The template approach, which has many benefits for regulating the structure, size, and morphology of the materials, is a low-cost and eco-friendly technique for CQD synthesis. This technique does not require hazardous chemicals and includes the use of a template for CQDs formation [71, 72]. Both soft templates, such as water-oil emulsion, and rigid templates, such as silica, can be used for synthesis [73, 74]. This technique consists of two stages. In the first stage, CQDs are produced by selecting an appropriate template. In the second stage, the etching process is applied in order to remove the templates, and pure CQDs are obtained. While the template method presents various advantages, including size control and prevention of agglomeration, it is generally hard to remove the templates, and this situation may affect many properties negatively, especially the purity of CQDs [21].

3.2.3 Structure and properties

CQDs are monodisperse, quasi-spherical carbon-based nanoparticles with a tunable size of less than 10 nm [75]. Due to the nucleation process, which includes a progressive growth of the inner part called the core and a self-passivated outer part called the shell consisting of functional groups, they feature a core-shell structure [76]. With a diameter of 2–3 nm and lattice spacing of 0.21 and 0.34 nm, the cores are very small, and they can exist in either sp^2-hybridized graphitic crystal state or mixed sp^2/sp^3 amorphous state depending on the synthetic parameters, including precursor type, temperature, duration, and pH [21, 77]. The core structures of CQDs are characterized through various analysis techniques. To determine the size and morphology of CQDs, scanning electron microscopy (SEM), transmission electron microscopy (TEM), high-resolution transmission microscopy (HR-TEM), or atomic force microscopy (AFM) can be used [78, 79]. The crystal structures of CQDs are determined by X-ray diffraction (XRD), and the amorphous or crystalline nature of CQDs is characterized by the selected area electron diffraction (SAED) patterns [21]. Specific surface area determinations of CQDs are performed by Brunauer-Emmett-Teller (BET) analysis [80]. Chemical structures and the surface functional groups of CQDs are characterized by Fourier transform infrared (FTIR), nuclear magnetic resonance (NMR), and Raman spectroscopy [81]. Specific atomic units exist on the surface of CQDs are identified by X-ray photoelectron spectroscopy (XPS), and the elemental compositions are determined by energy-dispersive X-ray (EDX) spectroscopy [15, 82]. Also, the optical properties are characterized through ultraviolet-visible (UV-Vis) and photoluminescence (PL) spectroscopy [83].

One of the most notable properties of CQDs is their stable and bright PL with a high quantum yield [84]. PL properties of CQDs are easily adjustable by modifying the carbon core and surface chemistry due to their nanoscale size [85]. CQDs also possess various important properties, including low toxicity, biocompatibility, and eco-friendliness [86]. These nanoscale materials can be nontoxic even without surface modification [87]. They can also be produced from renewable materials and waste. In addition to their photochemical and physicochemical stability, CQDs are water-soluble materials [88]. As a cost-effective alternative to other nanomaterials, the modification of CQDs is relatively easy. Modifying and functionalizing CQDs, including surface passivation and doping, can further enhance their unique properties [89, 90].

3.2 APPLICATIONS

CQDs and their composites have been used for a variety of purposes in a wide spectrum of applications, including bioimaging, biosensing, drug/gene delivery, photodynamic therapy, energy storage, light-emitting diodes (LEDs), food safety, catalysis, metal detection, water treatment, anti-counterfeiting, and anti-corrosive materials (Figure 3.3).

The potential applications of CQDs in biolabeling and bioimaging are very important. Ou *et al.* conducted a study that outlines a hydrothermal method for synthesizing an N-CQD composite that can label bacteria for bioimaging. They proved that the N-CQDs were effective in staining both gram-positive and gram-negative bacteria, with optimal concentrations of 1000 ppm and 500 ppm, respectively. In this study, the N-CQDs exhibited superior light stability and photobleaching resistance compared to the commercial dye FM4–64 and emitted red and green fluorescence at different wavelengths [91].

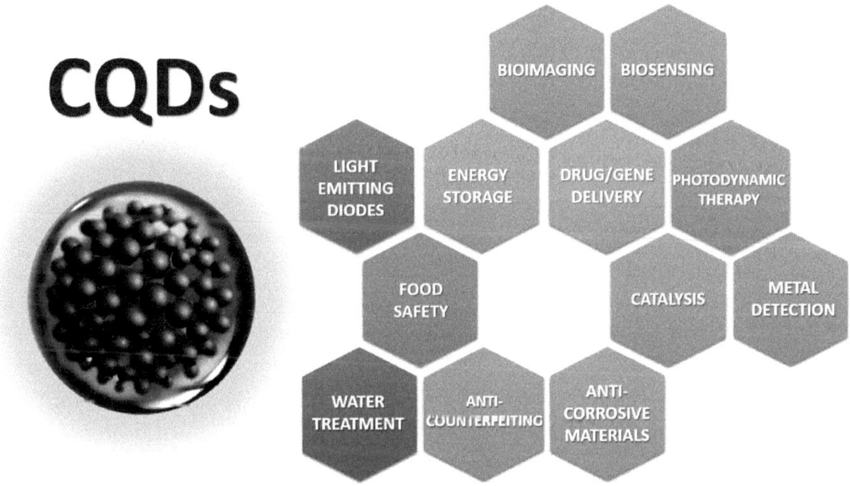

FIGURE 3.3 Applications of CQDs.

Because of their unique features, CQDs have been employed as effective biosensors for various biomolecules, such as glucose, nucleic acids, iron, and potassium [92]. Yu *et al.* fabricated a novel glucose biosensor that utilizes a fluorescent CQDs-GOD/CA complex-sensitive film for online detection of low-concentration glucose via optical fibers. The biosensor exhibited high sensitivity and repeatability, allowing real-time monitoring of glucose fluctuations over time. The change in fluorescence intensity was linearly proportional to glucose concentration in both the micromole and nanomole ranges [93].

Ting and his co-researchers developed a straightforward technique to produce CQDs from curcumin (CCM) as the precursor with desirable properties such as ultra-small size, rich hydrophilic groups, and positive potential. These CCM-CQDs exhibited potent antiviral activity against the porcine epidemic diarrhea virus (PEDV) and outperformed traditional CQDs. The antiviral mechanism involved multiple pathways, including blocking viral entry and inhibiting RNA synthesis and budding, as well as reducing reactive oxygen species accumulation and inducing the production of immune-related molecules. Based on these findings, they reported that CCM-CQDs have the potential to be a useful therapeutic option for various viral infections [94]. In another study, Hasanzadeh *et al.* developed a promising gene delivery vehicle by combining CQDs with poly-L-lysine (PLL) in a core-shell hybrid nanostructure. These nanoparticles had a narrow size distribution with an average diameter of less than 10 nm and exhibited excellent biocompatibility, water solubility, and tunable photoluminescence with a quantum yield of around 12%. The nanoparticles were also efficiently taken up by cells, with a cellular uptake of nearly 70%, and showed no significant toxicity to HEK-293T cells. The study showed novel CQDs/PLL core-shell nanoparticles had great potential as a non-toxic and effective gene delivery vector [95].

CQDs have been recently employed in photodynamic therapy as a potential treatment for different cancer types, including brain, lung, breast, and skin cancer [96]. In the research of He *et al.*, a modified $Bi_4O_5Br_2$ with CQDs was synthesized using a hydrolysis method at room temperature, resulting in an enhanced photo-killing effect of $Bi_4O_5Br_2$ against lung cancer (A549) cells. The incorporation of CQDs onto $Bi_4O_5Br_2$ was reported by these researchers to decrease the hydration ratio, increase the cellular uptake, and improve the production of photogenerated reactive oxygen species in comparison to the unmodified $Bi_4O_5Br_2$ [97].

The demand for cost-effective and eco-friendly energy solutions has led to significant advancements in energy storage and conversion techniques. CQDs are highly desirable for electrochemical applications because of their fast electron transfer and expansive surface area, leading to extensive research on their potential to address current energy challenges [98]. In 2022, Jo *et al.* carried out a research study about CQDs as energy storage devices. The researchers proposed a method to introduce P-doped CQDs (P-CQDs) into amorphous WO_3 (a-WO_3) at a low temperature, resulting in robust chemical linkages between them. The functional groups of P-CQDs formed strong hydrogen bonds with a-WO_3 in the form of quantum dot grafts, enabling them to withstand structural deformation following long and repetitive electrochemical cycles and bending cycles. Additionally, the functional groups and graphitic carbon of P-CQDs improved the Li-ion wettability and electrical conductivity, leading to excellent EC energy storage performance in P-CQDs/a-WO_3 electrodes [99].

CQDs, as novel carbon nanomaterials, have been extensively studied in LED applications due to their excellent properties [100]. The researchers, led by Mai, fabricated economical horticultural LEDs using red-emitting CQDs (R-CQDs) and reported that CQDs could be used as inexpensive, single-chip LED solutions for agricultural lighting [101]. In another study, Wen *et al.* integrated the photodynamic inactivation of bactericidal technology into food packaging materials by adding fluorescent CQDs derived from the turmeric plant into a chitosan matrix. They prepared several active composite food packaging films and discovered that the chitosan film containing CQDs had superior mechanical properties, UV protection, and hydrophobicity [102]. El-Shamy and his colleagues developed a new type of nanocomposite by incorporating CQD nanoparticles into polyvinyl alcohol (PVA). The resulting PVA/CQDs nanocomposites showed excellent performance in adsorbing and catalytically degrading large concentrations of methylene blue dye (30 mg/L) at room temperature and within a short period of time. This study suggested that PVA/CQDs have great potential for applications in catalytic water decomposition and desalination [103]. Chaudhary *et al.* synthesized highly fluorescent N, S co-doped CQDs (NS-CQDs), with an average size of 1.27 nm, from banana juice using a simple, one-step hydrothermal method that did not require any surface passivation, oxidizing agents, or inorganic salts, and used them as a fluorescent probe in order to detect the presence of Cu (II) in water. This study proved that green-synthesized NS-CQDs had shown considerable selectivity for heavy metals, such as Cu (II) [104]. Also, Zahmatkesh and his co-researchers synthesized CQDs, derived from peanut shells, and silver nanoparticles (Ag NPs), and integrated them into a polycarbonate copolymer membrane, resulting in a new low-cost microfiltration membrane system. Their study demonstrated that incorporating CQDs-Ag NPs into the membrane increased its strength and selectivity but reduced its pore size and permeability [105]. Mopan persimmon-based CQDs (MP-CQDs) were reported more recently by Ma and his group as an anti-counterfeiting material. They used a local persimmon fruit as a carbon source and prepared MP-CQDs via a hydrothermal method. By mixing MP-CQDs with sugar, they created an ink entitled MPI that they used to paint anti-counterfeiting patterns in combination with commercial fluorescent ink. Their study showed that MP-CQDs derived from biomass had great potential as effective materials for anti-counterfeiting applications [106]. Here, we will examine the topic of CQDS and their composites as anti-corrosive materials in detail under a separate heading.

3.3 CARBON QUANTUM DOTS AND THEIR COMPOSITES AS ANTI-CORROSIVE MATERIALS

Preventing corrosion is crucial for both economic and safety reasons, and advanced technologies must be developed to protect materials from this undesirable phenomenon. Corrosion inhibitors are widely recognized as the most effective way to achieve this, and currently, many different types of corrosion inhibitors have been used. CQDs are promising materials for corrosion prevention as they are highly effective and environmentally friendly compared to other options [107].

3.3.1 Applications in acidic media

Acidic corrosion is the process of materials being worn away or gradually destroyed by acidic compounds, and it is caused by electrochemical processes [108]. The majority of the research studies on the effectiveness of CQDS and their composites as anti-corrosive materials in acidic media were carried out on carbon steel in hydrochloric acid (HCl) solutions. In this regard, Lv *et al.* synthesized N-doped CQDs with luminescent properties from citric acid as a carbon source, and diethylenetriamine and urotropine as precursors by the microwave method. The effectiveness of CQDs in inhibiting corrosion of N80 carbon steel in 1 M hydrochloric acid was examined through static weight loss and electrochemical analysis. Their findings indicated that CQDs exhibited promising corrosion inhibition properties at 60°C, with an inhibition efficiency of up to 81.2% when the dosage was 600 ppm [109]. In another study, Ye *et al.* synthesized functionalized CQDs as anti-corrosive materials that were environmentally friendly and effective by conjugating imidazole and citric acid to CQDs and examined the corrosion inhibition behavior of these CQDs for Q235 carbon steel in HCl solution. According to the electrochemical tests they performed, CQDs showed a great ability to inhibit the corrosion of Q235 steel in HCl solution. Moreover, when the concentration of CQDs exceeded 100 mg/L, their inhibition efficiency on carbon steel in a 1 M HCl solution was over 90%. They attributed this remarkable performance to the adsorption film formed by CQDs that covered the surface of the steel [110]. To examine the anti-corrosive performance of CQDs for mild steel in HCl solution, in a recent study, Zhao *et al.* reported a new type of anti-corrosive material that was synthesized through a one-step process of self-polymerization and carbonization of 3,4-dihydroxy-L-phenylalanine (L-DOPA).

These nitrogen-doped polymer CQDs (NPDs) were in a water-dispersible form and had a 0D structure with an average diameter of 2.06 nm. The study showed that NPDs could effectively prevent the corrosion of mild steel in extremely acidic conditions with 97.95% inhibition efficiency by forming a passivation film on the metal surface and strong covalent bonding with iron [111].

It is well known that sulfuric acid (H_2SO_4), used for the removal of undesirable scale and rust in several industrial processes, triggers corrosion of pure metals and alloys [112]. The effect of this strong acid on carbon steel manifests itself as generalized and localized corrosion [113]. To evaluate the corrosion prevention effect of CQDs on Q235 carbon steel in H_2SO_4 solution, Xu and his colleagues synthesized nitrogen-doped CQDs (N-CQDs) with an average size of 2.5 ± 0.8 nm. They obtained these N-CQDs through a hydrothermal reaction of aminoguanidine and citric acid. At a concentration of 200 mg/L, the results demonstrated a maximum inhibition efficiency of 95.3%. They also reported that the quantum chemical calculations confirmed the positive effect of CQD nitrogen doping on anti-corrosive performance, and N-CQDs demonstrated a higher corrosion prevention compared to most other similar types of anti-corrosive materials at the same concentration [17].

Studies have demonstrated that CQDs and their composites are highly effective not only on carbon steel but also on other metals and alloys as anti-corrosive materials. N and S co-doped CQDs (N, S-CDs) synthesized by Cen *et al.* were first used as anti-corrosive materials for an aluminum alloy. The corrosion mitigation performance of N, S-CDs on 5052 aluminum alloy in HCl solution was investigated using weight loss analysis, surface characterization, and electrochemical measurements. The findings indicated that N, S-CDs can provide effective protection to aluminum, with mitigation efficiencies reaching as high as 85% even at a low concentration of 5 mg/L. They also stated that adhesion of N, S-CDs onto the aluminum surface was facilitated by the diffusion and agglomeration effects of nanomaterials, and the interaction arising from coordination bonds between aluminum ions and N, S-CDs led to the formation of a protective film on the surface of aluminum [114]. In another work, Zhang *et al.* fabricated N-doped CQDs (N-CDs) from citric acid and $_{L}$-serine amino acids via a solvothermal synthesis route. In their study investigating the anti-corrosive effect of CQDs for copper in H_2SO_4, they indicated that the N-CDs were effective corrosion inhibitors, with a high suppression efficiency of up to 98.5% after being immersed for 24 hours in a 0.5 M H_2SO_4 environment. They also reported that N-CDs interacted with the copper substrate through both chemical and physical adsorption, and the protective effect remained strong even at elevated temperatures [115].

CQDs and their composites are efficient anti-corrosive materials for not only metals and alloys but also metalloids such as silicon. In a recent report, Roy *et al.* synthesized chlorophyll(a)/CQDs bionanocomposite-decorated Si nanowires as photocathodes and researched their stability in H_2SO_4 media. As a result of their study, they reported that these newly developed bionanocomposite materials protected Si surfaces from acidic corrosion, and photocathodes exhibited impressive long-term stability when tested under acidic conditions for a period of three hours [116]. Table 3.1 summarizes the literature studies where CQDs and their composites were used as anti-corrosion materials in acidic environments.

TABLE 3.1
Some Examples of Recent Studies Focused on CQDs and Their Composites as Anti-corrosive Materials in Acidic Environments

Name	Precursors	Method	Substrate	Media	Exposure	Efficiency (%)	Year
N-doped CQDs (cdh-CDs)	Diethylenetriamine, urotropine, and citric acid	Microwave	N80 carbon steel	1 M HCl	–	81.20	2019 [109]
CDs	Aminosalicylic and thiourea	Hydrothermal	N80 carbon steel	1 M HCl	6 h	90.00	2023 [117]
Functionalized CQDs (CA-CDs)	N-(3-aminopropyl) imidazole and citric acid	Thermal decomposition	Q235 carbon steel	1 M HCl	24 h	94.00	2019 [110]
N-doped CQDs (N-CDs)	n-butylamine and methacrylic acid	Hydrothermal	Q235 carbon steel	1M HCl	24 h	94.96	2020 [118]
N-doped CQDs (N-CDs)	L-histidine and citric acid	Hydrothermal	Q235 carbon steel	0.1 M HCl	48 h	96.13	2020 [119]
N-doped CQDs (N-CDs)	Ammonium citrate	Thermal decomposition	Q235 carbon steel	1M HCl	24 h	97.43	2020 [120]
N, S co-doped CQDs (N, S-CDs)	Ammonium citrate and methionine	Hydrothermal	Q235 carbon steel	1M HCl	24 h	98.00	2021 [121]
N-doped CQDs (N-CQDs)	Aminoguanidine and citric acid	Hydrothermal	Q235 carbon steel	0.5 M H_2SO_4	60 h	95.30	2021 [17]
N-doped CQDs (NCQDs)	Folic acid and o-phenylenediamine	Hydrothermal	Q235 carbon steel	1M HCl	24 h	95.40	2021 [122]
N-doped CQDs (N-CDs)	L-histidine and 4-amino salicylic acid	Hydrothermal	Q235 carbon steel	1M HCl	24 h	93.00	2021 [123]
Water-soluble CDs	Anhydrous ethanol	Electrochemical Carbonization	Q235 carbon steel	1M HCl	72 h	98 00	2022 [124]
Schiff base CQDs (Sc-CDs)	o-phenylenediamine and p-benzoquinone	Hydrothermal	Q235 carbon steel	1M HCl	96 h	96.03	2022 [125]
Biomass CQDs (BCQDs)	Grapefruit peels	Hydrothermal	Q235 carbon steel	1M HCl	24 h	94.10	2023 [19]
High-yield CQDs	Acetaldehyde	Chemical Ablation	Q235 carbon steel	0.5 M HCl	72 h	95.30	2023 [126]
N-doped functionalized CQDs (N-CDs)	Urea and citric acid	Hydrothermal	Medium-carbon steel	0.5 M H_2SO_4	2 h	97.80	2021 [127]
S, N co-doped (CD1) and N doped (CD2)	Pyromelletic acid and thiourea	Solvothermal	Mild steel	15% HCl	–	96.40 90.00	2020 [128]
N-doped polymer nanodots (NPDs)	3,4-dihydroxy-L-phenylalanine	Hydrothermal	Mild steel	1M HCl	6 h	97.95	2022 [111]
N and S co-doped CQDs (N, S-CDs)	Aminosalicylic acid and thiourea	Hydrothermal	5052 aluminum alloy	0.1M HCl	12 h	85.90	2019 [114]
N-doped CQDs (N-CDs)	L-serine and citric acid	Solvothermal	Pure copper	0.5 M H_2SO_4	24 h	98.50	2021 [115]
CQDs	Pharmaceutical waste	Thermal decomposition	Silicon nanowires	0.5 M H_2SO_4	3 h	-	2020 [116]

3.3.2 Applications in saline media

In terms of chemistry, seawater is essentially a solution consisting of various salts dissolved in water, and its average salinity level is approximately 3.5%. Metals are typically more susceptible to corrosion when exposed to saline water compared to regular freshwater. If the saline water has a high concentration of sodium and chlorides, usually found in several thousand parts per million, it has a greater tendency to corrode many types of metals, especially steel and low-alloy steel. Moreover, it may also affect materials like aluminum alloys, copper alloys, and certain nickel-base alloys. The extent of corrosion may vary depending on the conditions, such as the concentration of salts, the presence of dissolved gases, the pH level, the temperature, and the tendency to form mineral deposits [129, 130]. CQDs, like in other environments, are green, low-cost, and effective anti-corrosive materials to prevent corrosion in saline environments.

In a research study, Li *et al.* synthesized N-doped CQDs (N-CQDs), ranging in size from 4 to 8 nm, using disodium ethylenediaminetetraacetate, sulfamic acid, and urotropine salt, and investigated the protective properties of these N-CQDs in inhibiting the corrosion of N80 carbon steel immersed in a saturated CO_2 3 wt% NaCl solution. The researchers concluded that N-CQDs displayed effective corrosion protection for carbon steel in a saline environment, achieving a protection efficiency of 83.7% at a concentration of 600 mg/L. Furthermore, they noted that higher concentrations of N-CQDs offered enhanced protection against corrosion for carbon steel in the presence of saline solutions [131]. Likewise, Wang *et al.* obtained N-CQDs through a solvothermal method by utilizing 4-aminosalicylic acid (ASA) as a precursor and examined their corrosion protection efficiency for waterborne epoxy resin coatings of carbon steel. They formulated waterborne epoxy resin coatings with different concentrations of N-CQDs (0.5 wt%, 1 wt%, and 2 wt%) and applied them to carbon steel electrodes. When they investigated the anti-corrosive performance of these coatings, both with and without N-CQDs, in a 3.5 wt% NaCl solution, the results indicated that the corrosion resistance of Q235 steel was significantly improved when composite coatings containing 0.5–1 wt% N-CQDs were used, surpassing the corrosion resistance offered by the pure waterborne epoxy resin coating [132].

Due to their remarkable attributes, including low density, high specific strength, stiffness, and excellent electromagnetic shielding capabilities, magnesium alloys have gained recognition as eco-friendly materials in automobile, electronics, and aerospace manufacturing. However, the corrosion resistance of magnesium alloys remains a critical concern, substantially affecting their long-term durability. In the past few decades, extensive research efforts have been dedicated to proposing diverse strategies aimed at improving the corrosion resistance of magnesium alloys. Among various methods, the surface coating of magnesium alloys with CQDs has been identified as an effective approach for enhancing their corrosion resistance [133, 134]. In this context, a composite coating of N-doped CQDs and polydopamine (PDA) was produced and coated on magnesium alloy by Zhang and his colleagues. The researchers investigating the anti-corrosive performance of CQDs for magnesium alloy in NaCl solution encountered noteworthy results. The findings indicated that the corrosion performance of the N-CQD coatings improved as the particle sizes increased. Moreover, they observed that after immersion in deionized water for 7 h, an obvious self-healing performance on the surface of the N-CQDs/PDA composite coating occurred, and the cross-scratches on the coating were filled with the PDA [134].

Copper is another metal that exhibits excellent properties. It offers exceptional electrical and thermal conductivity and a broad spectrum of controllable mechanical properties. These advantageous qualities make copper and its alloys highly beneficial for diverse applications where seawater is involved, such as distribution systems, treatment units, condensers, and heat exchangers [135, 136]. While copper is generally resistant to corrosion, it is more susceptible to corrosion in seawater due to the higher concentration of ions [137]. In order to minimize the corrosion of copper in saline environments, Zhang *et al.* developed high-N-doped CQDs from melamine and fructose through a solvothermal method and examined the anti-corrosive effects of the N-doped CQDs on copper in a 3 wt% NaCl solution. At the end of their research, they reported a significantly high anti-corrosive efficiency, reaching 96.1% at 200 mg/L, due to the protective film formed by N-CQDs on copper surfaces via the coordination between Cu^{2+} ions and the N-CQDs. Also, they stated that the increase in the N/C ratio has a significant impact on enhancing the performance of nitrogen-doped CQDs for corrosion prevention [138]. Table 3.2 lists the literature studies where CQDs and their composites were used as anti-corrosion materials in acidic environments.

3.3.3 Applications in alkaline media

Metal corrosion can happen in various environments, including freshwater, seawater, salt solutions, and alkaline or basic media [149]. Indeed, in some cases, it has also been reported that the corrosion rate can increase in alkaline media compared to acidic media [150]. Therefore, the prevention of corrosion in alkaline environments has become an area of interest for researchers. However, due to the relatively lower significance of corrosion in alkaline environments compared to acidic and saline environments, the number of literature studies focusing on the use of CQDs and their composites as anti-corrosive materials is still quite limited.

Recently, semiconductor CdS/multiwalled carbon nanotube quantum dots with a crystallite size of 2–3 nm have been successfully synthesized from *Opuntia ficus-indica* fruit sap using an ultrasound-assisted microwave method by Kandasamy and co-workers. The researchers conducted corrosion performance tests in three different solutions, including KOH, and reported that when they coated the zinc plate with the synthesized material, it provided effective corrosion protection [151]. In another study, Hu *et al.* synthesized solid-state CQDs from citric acid via a hydrothermal approach and incorporated them into Silicate-1 templates in order to utilize them for criminal investigations. Since latent fingerprints are commonly discovered on various contaminated surfaces during the investigation of criminal incidents, they investigated the corrosion resistance of the composite in different media, such as strong acids, alkalis, and organic solvents. They reported that the as-synthesized composite demonstrated

TABLE 3.2
Some Examples of Recent Studies Focused on CQDs and Their Composites as Anti-Corrosive Materials in Saline Environments

Name	Precursors	Method	Substrate	Media	Exposure	Efficiency (%)	Year
N-doped CQDs (N-CQDs)	Disodium EDTA, sulfamic acid, and urotropine salt	Microwave Pyrolysis	N80 carbon steel	CO_2– 3% NaCl	72 h	83.70	2020 [131]
N-doped CQDs (N-CDs)	4-aminosalicylic acid	Solvothermal	Q235 carbon steel	3.5% NaCl	70 days	–	2019 [132]
Functionalized carbon nanodots (CDs)	Garlic paste	Chemical Ablation	Q235 carbon steel	5% NaCl	30 days	99.99	2019 [139]
N-doped CQDs (N-CDs)	Methacrylic acid and ethyl(methyl)amine	Hydrothermal	Q235 carbon steel	3.5% NaCl	24 h	88.96	2020 [140]
CQDs	Citric acid	Microwave Pyrolysis	Q235 carbon steel	3.5% NaCl	9 h	–	2020 [141]
N-doped CQDs (N-CQDs)	Ethylenediamine and citric acid monohydrate	Hydrothermal	Q235 carbon steel	3.5% NaCl	90 days	80.80	2021 [142]
N-doped CQDs (N-CQDs)	L-tartaric acid and o-phenylenediamine	Hydrothermal	Q235 carbon steel	3.5% NaCl	30 min	–	2022 [143]
N and S co-doped CQDs (N, S-CDs)	Aminosalicylic acid and thiourea	Hydrothermal	Medium-carbon steel	CO^2– 3.5% NaCl	12 h	93.00	2019 [144]
CQDs	Ascorbic acid	Hydrothermal	403 stainless steel	0.5M NaCl	–	–	2019 [145]
Carbon nanodots (CDs)	Chitosan	Hydrothermal	BIS 2062 carbon steel	3.5% NaCl	–	–	2020 [146]
N-doped CQDs (N-CDs)	Benzotriazole and citric acid	Hydrothermal	Magnesium alloy	5% NaCl	4 h	–	2022 [134]
CDs pigments	Urea and citric acid	Hydrothermal	Molybdenum oxide	3.5% NaCl	72 h	–	2020 [147]
N-doped CQDs (N-CQDs)	Pre-baked carbon anode	Electrochemical	Pure copper	3.5% NaCl	24 h	96.32	2021 [148]
High-N-doped CQDs (N-doped CDs)	Melamine and fructose	Solvothermal	Pure copper	3% NaCl	24 h	96.10	2022 [138]

extra-high corrosion resistance in common organic solvents and various pH solutions ranging from 2 to 14 [152].

3.3.4 Lubricant applications

Lubricant applications of CQDs and their composites are an emerging field that shows great promise for enhancing the performance of lubricants. In the field of lubrication, CQDs can be used as additives to traditional lubricants to improve their tribological properties. These tiny particles can significantly lower friction, leading to lower energy consumption and reduced wear [153]. Wang *et al.* produced polyethylene glycol (PEG)-derived CQDs nanofluids through a slow thermal oxidation process and utilized them as lubricants in a diamond-like carbon film/bearing steel interface. The results showed that the PEG-CQDs nanofluids could effectively inhibit the tribo-oxidation and corrosion wear of the steel counterface. They concluded that the CQD nanofluids were promising candidates with excellent properties as lubricant additives [154].

These versatile nanomaterials also show outstanding self-healing abilities when they are used as lubricant additives. In a research study, for the first time, Wang and his colleagues synthesized eco-friendly menthol-modified CQDs (CDs-Menth) and evaluated their lubrication characteristics when used as additives to polyalphaolefin (PAO4) lubricant. In addition to remarkable anti-wear, friction-reducing, and load-carrying properties, CDs-Menth showed an amazing self-healing ability. Additionally, they reported that CDs-Menth have the potential to protect the friction interfaces from potential oxidation and corrosion [155]. Furthermore, Hu and his colleagues conducted research exploring the anti-corrosive and self-healing properties of CQDs. The researchers synthesized water-soluble CQDs from ammonium citrate via a thermal decomposition method. As-prepared CQDs exhibited excellent dispersibility in deionized water and demonstrated superior lubrication properties. They reported that CQDs also displayed notable anti-corrosive properties, with a corrosion inhibition efficiency of up to 60% compared to deionized water, in addition to their self-healing abilities [153].

3.3.5 Microbial corrosion applications

Microbial-induced corrosion (MIC) poses a significant challenge, and the prevention of MIC is crucial in various industrial applications, including pipelines and marine environments [156]. Multiple approaches have been employed to mitigate and minimize MIC. Generally, a comprehensive strategy that involves physical cleaning, mechanical modifications, and the utilization of chemical methods such as biocides is necessary to achieve effective biocorrosion control [157]. However, these processes often come with a significant financial burden, and the use of elevated levels of biocides can lead to adverse effects on various organisms and even humans, including general toxicity, bioaccumulation, and disruption of the endocrine system [158].

CQDs are eco-friendly and effective candidates for MIC solutions, and the studies in this field are still in their early stages. Kalajahi *et al.* made major contributions to MIC prevention by combining the CQDs with anti-bacterial metal nanoparticles. The research group synthesized Cu/CQDs via a hydrothermal method and examined the corrosion behavior of X60 steel in the presence of sulfate-reducing bacteria (SRB). While pitting corrosion and the formation of holes on the surface of X60 steel in SRB medium occurred, uniform corrosion was observed in the presence of Cu/CQDs. As a result of this study, they reported that the inhibitory effect of Cu/CQDs against SRB was higher compared to Cu nanoparticles in the same amount, and Cu/CQDs are green biocides to prevent MIC, even in low concentrations [157].

3.4 CONCLUSION

Corrosion is a significant problem in various sectors, ranging from industrial applications to infrastructure. It diminishes the durability of materials, compromises structural integrity, and leads to costly repairs or replacements. Moreover, corrosion-induced material loss can reduce energy efficiency and pose safety risks. Therefore, the prevention and control of corrosion are of paramount importance for sustainability, economy, and safety considerations. As novel carbon-based anti-corrosive materials, CQDs offer several advantages in corrosion prevention due to their unique properties, such as extra-small size, high surface area, tunable surface chemistry, eco-friendliness, low cost, and facile synthesis. They can effectively prevent corrosion by forming a protective layer or interacting with corrosive species. Additionally, CQDs present excellent stability, compatibility with different coatings and materials, and environmentally friendly features. Research and development in the field of CQDs and their composites are still ongoing, and further studies are needed to fully explore their potential in corrosion applications. However, the initial findings and advancements in this area suggest that CQDs have the potential to revolutionize corrosion prevention technology by offering significant characteristics, including high anti-corrosive performance, eco-friendliness, low cost, and facile synthesis in various industrial applications.

REFERENCES

[1] Ganjoo, R., Sharma, S., Verma, C., Quraishi, M. A., & Kumar, A. (2023). Heteropolysaccharides in sustainable corrosion inhibition: 4E (Energy, Economy, Ecology, and Effectivity) dimensions. *International Journal of Biological Macromolecules*, 123571.

[2] Guo, L., Verma, C., & Zhang, D. (Eds.). (2022). *Eco-Friendly Corrosion Inhibitors: Principles, Designing and Applications*. Elsevier.

[3] Usmani, M. A., Khan, I., Hasnat, A., & Moheman, A. (2023). Ionic liquids as green and efficient corrosion-protective materials for metals and alloys. In *Advanced Applications of Ionic Liquids* (pp. 185–196). Elsevier.

[4] Xu, Z., Wu, Y., Zhang, Z., Wang, Y., Hu, J., Ma, Y., . . . & Shi, C. (2023). A review on the research progress of LDHs as corrosion inhibitors for reinforced concrete. *Journal of Building Engineering*, 106303.

[5] Pradhan, B. (2022). A study on effectiveness of inorganic and organic corrosion inhibitors on rebar corrosion in concrete: A review. *Materials Today: Proceedings*, *65*(2022), 1360–1366.

[6] Calle, L. M., & Li, W. (2014). Microencapsulated indicators and inhibitors for corrosion detection and control. In *Handbook of Smart Coatings for Materials Protection* (pp. 370–422). Woodhead Publishing.
[7] De Damborenea, J., Conde, A., & Arenas, M. A. (2014). Corrosion inhibition with rare earth metal compounds in aqueous solutions. In *Rare Earth-Based Corrosion Inhibitors* (pp. 84–116). Woodhead Publishing.
[8] Quraishi, M. A., & Chauhan, D. S. (2021). Drugs as environmentally sustainable corrosion inhibitors. In *Sustainable Corrosion Inhibitors II: Synthesis, Design, and Practical Applications* (pp. 1–17). American Chemical Society.
[9] Verma, C., Hussain, C. M., & Ebenso, E. E. (Eds.). (2021). *Organic Corrosion Inhibitors: Synthesis, Characterization, Mechanism, and Applications*. John Wiley & Sons.
[10] Finšgar, M., & Jackson, J. (2014). Application of corrosion inhibitors for steels in acidic media for the oil and gas industry: A review. *Corrosion Science*, *86*, 17–41.
[11] Sheetal, Batra, R., Singh, A. K., Singh, M., Thakur, S., Pani, B., & Kaya, S. (2023). Advancement of corrosion inhibitor system through N-heterocyclic compounds: A review. *Corrosion Engineering, Science and Technology*, *58*(1), 73–101.
[12] Aslam, J., Verma, C., Verma, D. K., & Aslam, R. (Eds.). (2022). *Carbon Allotropes: Nanostructured Anti-Corrosive Materials*. Walter de Gruyter GmbH & Co KG.
[13] Liu, Y., Roy, S., Sarkar, S., Xu, J., Zhao, Y., & Zhang, J. (2021). A review of carbon dots and their composite materials for electrochemical energy technologies. *Carbon Energy*, *3*(5), 795–826.
[14] Wang, X., Feng, Y., Dong, P., & Huang, J. (2019). A mini review on carbon quantum dots: Preparation, properties, and electrocatalytic application. *Frontiers in Chemistry*, *7*, 671.
[15] Wang, Y., & Hu, A. (2014). Carbon quantum dots: Synthesis, properties and applications. *Journal of Materials Chemistry C*, *2*(34), 6921–6939.
[16] Zeng, S., Zhang, F., Liu, Y., Ouyang, S., Ye, Y. W., & Chen, H. (2022). Synthesis of Ce, N co-doped carbon dots as green and effective corrosion inhibitor for copper in acid environment. *Journal of the Taiwan Institute of Chemical Engineers*, *141*, 104608.
[17] Xu, X., Wei, H., Liu, M., Shen, G., Li, Q., Hussain, G., . . . &Ostrikov, K. K. (2021). Nitrogen-doped carbon quantum dots for effective corrosion inhibition of Q235 steel in concentrated sulphuric acid solution. *Materials Today Communications*, *29*, 102872.
[18] Guo, L., Zhu, M., Shi, W., Wang, K., Leng, S., & Brahmia, A. (2023). Evaluation of the corrosion inhibition behavior of N, S co-doped memory-type carbon dots for mild steel in HCl solution. *Materials Letters*, *330*, 133403.
[19] Zheng, S., Feng, L., Hu, Z., Li, J., Zhu, H., & Ma, X. (2023). Study on the corrosion inhibition of biomass carbon quantum dot self-aggregation on Q235 steel in hydrochloric acid. *Arabian Journal of Chemistry*, *16*(4), 104605.
[20] Azam, N., Najabat Ali, M., & Javaid Khan, T. (2021). Carbon quantum dots for biomedical applications: Review and analysis. *Frontiers in Materials*, *8*, 700403.
[21] Yadav, P. K., Chandra, S., Kumar, V., Kumar, D., & Hasan, S. H. (2023). Carbon quantum dots: Synthesis, structure, properties, and catalytic applications for organic synthesis. *Catalysts*, *13*(2), 422.
[22] Mirtchev, P., Henderson, E. J., Soheilnia, N., Yip, C. M., & Ozin, G. A. (2012). Solution phase synthesis of carbon quantum dots as sensitizers for nanocrystalline TiO 2 solar cells. *Journal of Materials Chemistry*, *22*(4), 1265–1269.
[23] Cui, M., Ren, S., Xue, Q., Zhao, H., & Wang, L. (2017). Carbon dots as new eco-friendly and effective corrosion inhibitor. *Journal of Alloys and Compounds*, *726*, 680–692.
[24] Verma, C., Alfantazi, A., & Quraishi, M. A. (2021). Quantum dots as ecofriendly and aqueous phase substitutes of carbon family for traditional corrosion inhibitors: A perspective. *Journal of Molecular Liquids*, *343*, 117648.

[25] Xu, X., Ray, R., Gu, Y., Ploehn, H. J., Gearheart, L., Raker, K., & Scrivens, W. A. (2004). Electrophoretic analysis and purification of fluorescent single-walled carbon nanotube fragments. *Journal of the American Chemical Society, 126*(40), 12736–12737.
[26] Sun, Y. P., Zhou, B., Lin, Y., Wang, W., Fernando, K. S., Pathak, P., . . . & Xie, S. Y. (2006). Quantum-sized carbon dots for bright and colorful photoluminescence. *Journal of the American Chemical Society, 128*(24), 7756–7757.
[27] Wang, R., Lu, K. Q., Tang, Z. R., & Xu, Y. J. (2017). Recent progress in carbon quantum dots: Synthesis, properties and applications in photocatalysis. *Journal of Materials Chemistry A, 5*(8), 3717–3734.
[28] Wang, L., Li, W., Yin, L., Liu, Y., Guo, H., Lai, J., . . . & Wu, M. (2020). Full-color fluorescent carbon quantum dots. *Science Advances, 6*(40), eabb6772.
[29] Zhao, D. L., & Chung, T. S. (2018). Applications of carbon quantum dots (CQDs) in membrane technologies: A review. *Water Research, 147*, 43–49.
[30] Molaei, M. J. (2020). Principles, mechanisms, and application of carbon quantum dots in sensors: A review. *Analytical Methods, 12*(10), 1266–1287.
[31] Liu, J., Li, R., & Yang, B. (2020). Carbon dots: A new type of carbon-based nanomaterial with wide applications. *ACS Central Science, 6*(12), 2179–2195.
[32] Janus, Ł., Radwan-Pragłowska, J., Piątkowski, M., & Bogdał, D. (2020). Facile synthesis of surface-modified carbon quantum dots (CQDs) for biosensing and bioimaging. *Materials, 13*(15), 3313.
[33] Molaei, M. J. (2019). Carbon quantum dots and their biomedical and therapeutic applications: A review. *RSC Advances, 9*(12), 6460–6481.
[34] Bhat, A. P., Dhoble, S. J., & Rewatkar, K. G. (2022). Medical applications of quantum dots. In *Graphene, Nanotubes and Quantum Dots-Based Nanotechnology* (pp. 803–836). Woodhead Publishing.
[35] Jain, B., Rawat, R., & Darko, D. A. (2022). Carbon allotropes: Properties and applications-state of the art. In *Carbon Allotropes: Nanostructured Anti-Corrosive Materials* (pp. 1–32). De Gruyter.
[36] Döring, A., Ushakova, E., & Rogach, A. L. (2022). Chiral carbon dots: Synthesis, optical properties, and emerging applications. *Light: Science & Applications, 11*(1), 75.
[37] Chu, K. W., Lee, S. L., Chang, C. J., & Liu, L. (2019). Recent progress of carbon dot precursors and photocatalysis applications. *Polymers, 11*(4), 689.
[38] Bruno, F., Sciortino, A., Buscarino, G., Soriano, M. L., Ríos, Á., Cannas, M., . . . & Agnello, S. (2021). A comparative study of top-down and bottom-up carbon nanodots and their interaction with mercury ions. *Nanomaterials, 11*(5), 1265.
[39] Sharma, A., & Das, J. (2019). Small molecules derived carbon dots: Synthesis and applications in sensing, catalysis, imaging, and biomedicine. *Journal of Nanobiotechnology, 17*(1), 1–24.
[40] Zhang, M., Wang, H., Wang, B., Ma, Y., Huang, H., Liu, Y., . . . & Kang, Z. (2019). Maltase decorated by chiral carbon dots with inhibited enzyme activity for glucose level control. *Small, 15*(48), 1901512.
[41] Hu, L., Sun, Y., Zhou, Y., Bai, L., Zhang, Y., Han, M., . . . & Kang, Z. (2017). Nitrogen and sulfur co-doped chiral carbon quantum dots with independent photoluminescence and chirality. *Inorganic Chemistry Frontiers, 4*(6), 946–953.
[42] Li, H., He, X., Liu, Y., Huang, H., Lian, S., Lee, S. T., & Kang, Z. (2011). One-step ultrasonic synthesis of water-soluble carbon nanoparticles with excellent photoluminescent properties. *Carbon, 49*(2), 605–609.
[43] Rai, S., Singh, B. K., Bhartiya, P., Singh, A., Kumar, H., Dutta, P. K., & Mehrotra, G. K. (2017). Lignin derived reduced fluorescence carbon dots with theranostic approaches: Nano-drug-carrier and bioimaging. *Journal of Luminescence, 190*, 492–503.

[44] Gao, P., Xie, Z., & Zheng, M. (2020). Chiral carbon dots-based nanosensors for Sn (II) detection and lysine enantiomers recognition. *Sensors and Actuators B: Chemical, 319*, 128265.
[45] Ma, Z., Ming, H., Huang, H., Liu, Y., & Kang, Z. (2012). One-step ultrasonic synthesis of fluorescent N-doped carbon dots from glucose and their visible-light sensitive photocatalytic ability. *New Journal of Chemistry, 36*(4), 861 864.
[46] Crista, D. M., Esteves da Silva, J. C., & Pinto da Silva, L. (2020). Evaluation of different bottom-up routes for the fabrication of carbon dots. *Nanomaterials, 10*(7), 1316.
[47] Kolanowska, A., Dzido, G., Krzywiecki, M., Tomczyk, M. M., Łukowiec, D., Ruczka, S., & Boncel, S. (2022). Carbon quantum dots from amino acids revisited: Survey of renewable precursors toward high quantum-yield blue and green fluorescence. *ACS Omega, 7*(45), 41165–41176.
[48] Bruno, F., Sciortino, A., Buscarino, G., Soriano, M. L., Ríos, Á., Cannas, M., . . . & Agnello, S. (2021). A comparative study of top-down and bottom-up carbon nanodots and their interaction with mercury ions. *Nanomaterials, 11*(5), 1265.
[49] Yang, L. (2015). *Nanotechnology-Enhanced Orthopedic Materials: Fabrications, Applications and Future Trends*. Woodhead Publishing.
[50] Nair, G. B., &Dhoble, S. J. (2020). *The Fundamentals and Applications of Light-Emitting Diodes: The Revolution in the Lighting Industry*. Woodhead Publishing.
[51] Shorgar, N., Bhati, I., & Jhalora, P. (2023). Laser ablation synthesis of quantum dots. In *Quantum Dots* (pp. 53–75). Elsevier.
[52] Kim, M., Osone, S., Kim, T., Higashi, H., & Seto, T. (2017). Synthesis of nanoparticles by laser ablation: A review. *KONA Powder and Particle Journal, 34*, 80–90.
[53] Kang, S., Ryu, J. H., Lee, B., Jung, K. H., Shim, K. B., Han, H., & Kim, K. M. (2019). Laser wavelength modulated pulsed laser ablation for selective and efficient production of graphene quantum dots. *RSC Advances, 9*(24), 13658–13663.
[54] An, Q., Lin, Q., Huang, X., Zhou, R., Guo, X., Xu, W., . . . & Chang, H. T. (2021). Electrochemical synthesis of carbon dots with a Stokes shift of 309 nm for sensing of Fe3+ and ascorbic acid. *Dyes and Pigments, 185*, 108878.
[55] Deng, J., Lu, Q., Mi, N., Li, H., Liu, M., Xu, M., . . . & Yao, S. (2014). Electrochemical synthesis of carbon nanodots directly from alcohols. *Chemistry—A European Journal, 20*(17), 4993–4999.
[56] Zhuo, S., Shao, M., & Lee, S. T. (2012). Upconversion and downconversion fluorescent graphene quantum dots: Ultrasonic preparation and photocatalysis. *ACS Nano, 6*(2), 1059–1064.
[57] Lu, M., & Zhou, L. (2019). One-step sonochemical synthesis of versatile nitrogen-doped carbon quantum dots for sensitive detection of Fe2+ ions and temperature in vitro. *Materials Science and Engineering: C, 101*, 352–359.
[58] Singh, H., Bamrah, A., Khatri, M., & Bhardwaj, N. (2020). One-pot hydrothermal synthesis and characterization of carbon quantum dots (CQDs). *Materials Today: Proceedings, 28*, 1891–1894.
[59] Ye, H., Liu, B., Wang, J., Zhou, C., Xiong, Z., & Zhao, L. (2022). A hydrothermal method to generate carbon quantum dots from waste bones and their detection of laundry powder. *Molecules, 27*(19), 6479.
[60] Qin, J. L., Li, S. J., Yuan, N. N., Chen, Y., Lei, B. F., Hu, C. F., & Liu, Y. L. (2020). Rapid hydrothermal method for the preparation of carbon quantum dots in the presence of hydrogen peroxide. *Micro & Nano Letters, 15*(7), 465–468.
[61] Shen, T., Wang, Q., Guo, Z., Kuang, J., & Cao, W. (2018). Hydrothermal synthesis of carbon quantum dots using different precursors and their combination with TiO2 for enhanced photocatalytic activity. *Ceramics International, 44*(10), 11828–11834.

[62] Chae, A., Choi, Y., Jo, S., Paoprasert, P., Park, S. Y., & In, I. (2017). Microwave-assisted synthesis of fluorescent carbon quantum dots from an A 2/B 3 monomer set. *RSC Advances*, *7*(21), 12663–12669.
[63] Yu, T., Wang, H., Guo, C., Zhai, Y., Yang, J., & Yuan, J. (2018). A rapid microwave synthesis of green-emissive carbon dots with solid-state fluorescence and pH-sensitive properties. *Royal Society Open Science*, *5*(7), 180245.
[64] Magesh, V., Sundramoorthy, A. K., & Ganapathy, D. (2022). Recent advances on synthesis and potential applications of carbon quantum dots. *Frontiers in Materials*, 383.
[65] Nagpal, R., & Gusain, M. (2022). Synthesis methods of quantum dots. In *Graphene, Nanotubes and Quantum Dots-Based Nanotechnology* (pp. 599–630). Woodhead Publishing.
[66] Khayal, A., Dawane, V., Amin, M. A., Tirth, V., Yadav, V. K., Algahtani, A., . . . & Jeon, B. H. (2021). Advances in the methods for the synthesis of carbon dots and their emerging applications. *Polymers*, *13*(18), 3190.
[67] Huang, X., Li, Y., Zhong, X., Rider, A. E., & Ostrikov, K. (2015). Fast microplasma synthesis of blue luminescent carbon quantum dots at ambient conditions. *Plasma Processes and Polymers*, *12*(1), 59–65.
[68] Li, C. X., Yu, C., Wang, C. F., & Chen, S. (2013). Facile plasma-induced fabrication of fluorescent carbon dots toward high-performance white LEDs. *Journal of Materials Science*, *48*, 6307–6311.
[69] Qin, X., Liu, J., Zhang, Q., Chen, W., Zhong, X., & He, J. (2021). Synthesis of yellow-fluorescent carbon nano-dots by microplasma for imaging and photocatalytic inactivation of cancer cells. *Nanoscale Research Letters*, *16*, 1–9.
[70] Lin, L., & Wang, Q. (2015). Microplasma: A new generation of technology for functional nanomaterial synthesis. *Plasma Chemistry and Plasma Processing*, *35*, 925–962.
[71] Gulati, S., Baul, A., Amar, A., Wadhwa, R., Kumar, S., & Varma, R. S. (2023). Eco-friendly and sustainable pathways to photoluminescent carbon quantum dots (CQDs). *Nanomaterials*, *13*(3), 554.
[72] Mo, F., Chen, Q., & Zhang, X. (2022). Synthesis of hollow calcium carbonate microspheres by a template method for DOX loading and release with carbon dots photosensitivity. *Materials*, *15*(24), 8768.
[73] Kwon, W., Lee, G., Do, S., Joo, T., & Rhee, S. W. (2014). Size-controlled soft-template synthesis of carbon nanodots toward versatile photoactive materials. *Small*, *10*(3), 506–513.
[74] Bourlinos, A. B., Stassinopoulos, A., Anglos, D., Zboril, R., Georgakilas, V., & Giannelis, E. P. (2008). Photoluminescent carbogenic dots. *Chemistry of Materials*, *20*(14), 4539–4541.
[75] Simsek, E. B. (2022). Carbon-based metal-free catalysts for photocatalytic reactions. In *Carbon-Based Metal Free Catalysts* (pp. 151–194). Elsevier.
[76] Kumar, P., Dua, S., Kaur, R., Kumar, M., & Bhatt, G. (2022). A review on advancements in carbon quantum dots and their application in photovoltaics. *RSC Advances*, *12*(8), 4714–4759.
[77] Martindale, B. C., Hutton, G. A., Caputo, C. A., Prantl, S., Godin, R., Durrant, J. R., & Reisner, E. (2017). Enhancing light absorption and charge transfer efficiency in carbon dots through graphitization and core nitrogen doping. *Angewandte Chemie*, *129*(23), 6559–6563.
[78] Wang, Y., Liang, Z., Su, Z., Zhang, K., Ren, J., Sun, R., & Wang, X. (2018). All-biomass fluorescent hydrogels based on biomass carbon dots and alginate/nanocellulose for biosensing. *ACS Applied Bio Materials*, *1*(5), 1398–1407.

[79] Li, H., Kang, Z., Liu, Y., & Lee, S. T. (2012). Carbon nanodots: Synthesis, properties and applications. *Journal of Materials Chemistry*, *22*(46), 24230–24253.
[80] Chen, G., Wu, S., Hui, L., Zhao, Y., Ye, J., Tan, Z., . . . & Zhu, Y. (2016). Assembling carbon quantum dots to a layered carbon for high-density supercapacitor electrodes. *Scientific Reports*, *6*(1), 1–9.
[81] Ahmed, H. B., & Emam, H. E. (2020). Environmentally exploitable biocide/fluorescent metal marker carbon quantum dots. *RSC Advances*, *10*(70), 42916–42929.
[82] Mohammadi, M., Khazaei, A., Rezaei, A., Huajun, Z., & Xuwei, S. (2019). Ionic-liquid-modified carbon quantum dots as a support for the immobilization of tungstate ions (WO42–): Heterogeneous nano catalysts for the oxidation of alcohols in water. *ACS Sustainable Chemistry & Engineering*, *7*(5), 5283–5291.
[83] Leong, K. H., Chin, Y. H., Sim, L. C., Tan, B., Dai, C., & Saravanan, P. (2022). Physical properties of quantum dots. In *Graphene, Nanotubes and Quantum Dots-Based Nanotechnology* (pp. 687–709). Woodhead Publishing.
[84] Lai, C., Lin, S., Huang, X., & Jin, Y. (2021). Synthesis and properties of carbon quantum dots and their research progress in cancer treatment. *Dyes and Pigments*, *196*, 109766.
[85] Vibhute, A., Patil, T., Gambhir, R., & Tiwari, A. P. (2022). Fluorescent carbon quantum dots: Synthesis methods, functionalization and biomedical applications. *Applied Surface Science Advances*, *11*, 100311.
[86] Purkait, M. K., Haldar, D., & Debnath, B. (2023). *Technological Advancements in Product Valorization of Tea Waste*. Elsevier.
[87] Ray, S. C., & Jana, N. R. (2017). Toxicology and biosafety of carbon nanomaterials. In Ray, S. C., & Jana, N. R. (Eds.), *Carbon Nanomaterials for Biological and Medical Applications* (1st ed., pp. 205–229). Elsevier.
[88] Nasrollahzadeh, M., Sajadi, M. S., Atarod, M., Sajjadi, M., & Isaabadi, Z. (2019). *An Introduction to Green Nanotechnology*. Academic Press.
[89] Pourmadadi, M., Rahmani, E., Rajabzadeh-Khosroshahi, M., Samadi, A., Behzadmehr, R., Rahdar, A., & Ferreira, L. F. R. (2023). Properties and application of carbon quantum dots (CQDs) in biosensors for disease detection: A comprehensive review. *Journal of Drug Delivery Science and Technology*, 104156.
[90] Dimos, K. (2016). Carbon quantum dots: Surface passivation and functionalization. *Current Organic Chemistry*, *20*(6), 682–695.
[91] Ou, S. F., Zheng, Y. Y., Lee, S. J., Chen, S. T., Wu, C. H., Hsieh, C. T., . . . & Hsueh, Y. H. (2021). N-doped carbon quantum dots as fluorescent bioimaging agents. *Crystals*, *11*(7), 789.
[92] Namdari, P., Negahdari, B., & Eatemadi, A. (2017). Synthesis, properties and biomedical applications of carbon-based quantum dots: An updated review. *Biomedicine & Pharmacotherapy*, *87*, 209–222.
[93] Yu, S., Ding, L., Lin, H., Wu, W., & Huang, J. (2019). A novel optical fiber glucose biosensor based on carbon quantum dots-glucose oxidase/cellulose acetate complex sensitive film. *Biosensors and Bioelectronics*, *146*, 111760.
[94] Ting, D., Dong, N., Fang, L., Lu, J., Bi, J., Xiao, S., & Han, H. (2018). Multisite inhibitors for enteric coronavirus: Antiviral cationic carbon dots based on curcumin. *ACS Applied Nano Materials*, *1*(10), 5451–5459.
[95] Hasanzadeh, A., Jahromi, M. A. M., Abdoli, A., Mohammad-Beigi, H., Fatahi, Y., Nourizadeh, H., . . . & Karimi, M. (2021). Photoluminescent carbon quantum dot/poly l-Lysine core-shell nanoparticles: A novel candidate for gene delivery. *Journal of Drug Delivery Science and Technology*, *61*, 102118.
[96] Karagianni, A., Tsierkezos, N. G., Prato, M., Terrones, M., & Kordatos, K. V. (2023). Application of carbon-based quantum dots in photodynamic therapy. *Carbon*, *203*, 273–310.

[97] He, B., Jin, H. Y., Wang, Y. W., Fan, C. M., Wang, Y. F., Zhang, X. C., . . . & Liu, J. W. (2022). Carbon quantum dots/Bi4O5Br2 photocatalyst with enhanced photodynamic therapy: Killing of lung cancer (A549) cells in vitro. *Rare Metals*, *41*(1), 132–143.
[98] Sikiru, S., Oladosu, T. L., Kolawole, S. Y., Mubarak, L. A., Soleimani, H., Afolabi, L. O., & Toyin, A. O. O. (2023). Advance and prospect of carbon quantum dots synthesis for energy conversion and storage application: A comprehensive review. *Journal of Energy Storage*, *60*, 106556.
[99] Jo, M. H., Kim, K. H., & Ahn, H. J. (2022). P-doped carbon quantum dot graft-functionalized amorphous WO3 for stable and flexible electrochromic energy-storage devices. *Chemical Engineering Journal*, *445*, 136826.
[100] Da, X., Han, Z., Yang, Z., Zhang, D., Hong, R., Tao, C., . . . & Huang, Y. (2022). Preparation of multicolor carbon dots with high fluorescence quantum yield and application in white LED. *Chemical Physics Letters*, *794*, 139497.
[101] Mai, X. D., Nguyen, S. H., Tran, D. L., Nguyen, V. Q., & Nguyen, V. H. (2023). Single-chip horticultural LEDs enabled by greenly synthesized red-emitting carbon quantum dots. *Materials Letters*, *341*, 134195.
[102] Wen, F., Li, P., Yan, H., & Su, W. (2023). Turmeric carbon quantum dots enhanced chitosan nanocomposite films based on photodynamic inactivation technology for antibacterial food packaging. *Carbohydrate Polymers*, *311*, 120784.
[103] El-Shamy, A. G., & Zayied, H. S. S. (2020). New polyvinyl alcohol/carbon quantum dots (PVA/CQDs) nanocomposite films: Structural, optical and catalysis properties. *Synthetic Metals*, *259*, 116218.
[104] Chaudhary, N., Gupta, P. K., Eremin, S., & Solanki, P. R. (2020). One-step green approach to synthesize highly fluorescent carbon quantum dots from banana juice for selective detection of copper ions. *Journal of Environmental Chemical Engineering*, *8*(3), 103720.
[105] Zahmatkesh, S., Ni, B. J., Klemeš, J. J., Bokhari, A., & Hajiaghaei-Keshteli, M. (2022). Carbon quantum dots-Ag nanoparticle membrane for preventing emerging contaminants in oil produced water. *Journal of Water Process Engineering*, *50*, 103309.
[106] Ma, H., Guan, L., Chen, M., Zhang, Y., Wu, Y., Liu, Z., . . . & Li, X. (2023). Synthesis and enhancement of carbon quantum dots from Mopan persimmons for Fe3+ sensing and anti-counterfeiting applications. *Chemical Engineering Journal*, *453*, 139906.
[107] Berdimurodov, E., Verma, D. K., Kholikov, A., Akbarov, K., & Guo, L. (2022). The recent development of carbon dots as powerful green corrosion inhibitors: A prospective review. *Journal of Molecular Liquids*, *349*, 118124.
[108] Viles, H. A. (1999). Acid corrosion (of stone and metal). In *Environmental Geology. Encyclopedia of Earth Science*. Springer.
[109] Lv, J., Fu, L., Zeng, B., Tang, M., & Li, J. (2019). Synthesis and acidizing corrosion inhibition performance of N-doped carbon quantum dots. *Russian Journal of Applied Chemistry*, *92*, 848–856.
[110] Ye, Y., Yang, D., & Chen, H. (2019). A green and effective corrosion inhibitor of functionalized carbon dots. *Journal of Materials Science & Technology*, *35*(10), 2243–2253.
[111] Zhao, H., Sun, T. Y., Huang, L. F., Wei, J., & Qiu, S. (2022). A green strategy for nitrogen-doped polymer nanodots with high oxygen and chloride corrosion resistance in extremely acidic condition. *Chemical Engineering Journal*, *437*, 135242.
[112] Raja, P. B., & Sethuraman, M. G. (2008). Inhibitive effect of black pepper extract on the sulphuric acid corrosion of mild steel. *Materials Letters*, *62*(17–18), 2977–2979.
[113] Panossian, Z., de Almeida, N. L., de Sousa, R. M. F., de Souza Pimenta, G., & Marques, L. B. S. (2012). Corrosion of carbon steel pipes and tanks by concentrated sulfuric acid: A review. *Corrosion Science*, *58*, 1–11.

[114] Cen, H., Zhang, X., Zhao, L., Chen, Z., & Guo, X. (2019). Carbon dots as effective corrosion inhibitor for 5052 aluminium alloy in 0.1 M HCl solution. *Corrosion Science*, *161*, 108197.

[115] Zhang, Y., Zhang, S., Tan, B., Guo, L., & Li, H. (2021). Solvothermal synthesis of functionalized carbon dots from amino acid as an eco-friendly corrosion inhibitor for copper in sulfuric acid solution. *Journal of Colloid and Interface Science*, *604*, 1–14.

[116] Roy, K., Ghosh, D., Sarkar, K., Devi, P., & Kumar, P. (2020). Chlorophyll (a)/carbon quantum dot bio-nanocomposite activated nano-structured silicon as an efficient photocathode for photoelectrochemical water splitting. *ACS Applied Materials & Interfaces*, *12*(33), 37218–37226.

[117] Cen, H., He, S., Gong, X., Wu, C., & Chen, Z. (2023). Effect of external parallel electric field on corrosion inhibition of carbon dots for carbon steel in 1 M HCl solution. *Journal of Materials Science*, 1–19.

[118] Ye, Y., Zou, Y., Jiang, Z., Yang, Q., Chen, L., Guo, S., & Chen, H. (2020). An effective corrosion inhibitor of N doped carbon dots for Q235 steel in 1 M HCl solution. *Journal of Alloys and Compounds*, *815*, 152338.

[119] Ye, Y., Zhang, D., Zou, Y., Zhao, H., & Chen, H. (2020). A feasible method to improve the protection ability of metal by functionalized carbon dots as environment-friendly corrosion inhibitor. *Journal of Cleaner Production*, *264*, 121682.

[120] Liu, Z., Ye, Y. W., & Chen, H. (2020). Corrosion inhibition behavior and mechanism of N-doped carbon dots for metal in acid environment. *Journal of Cleaner Production*, *270*, 122458.

[121] Luo, J., Cheng, X., Chen, X., Zhong, C. F., Xie, H., Ye, Y. W., . . . & Chen, H. (2021). The effect of N and S ratios in N, S co-doped carbon dot inhibitor on metal protection in 1 M HCl solution. *Journal of the Taiwan Institute of Chemical Engineers*, *127*, 387–398.

[122] Zhu, M., Guo, L., He, Z., Marzouki, R., Zhang, R., & Berdimurodov, E. (2022). Insights into the newly synthesized N-doped carbon dots for Q235 steel corrosion retardation in acidizing media: A detailed multidimensional study. *Journal of Colloid and Interface Science*, *608*, 2039–2049.

[123] Pan, L., Li, G., Wang, Z., Liu, D., Zhu, W., Tong, C., . . . & Hu, S. (2021). Carbon dots as environment-friendly and efficient corrosion inhibitors for Q235 steel in 1 M HCl. *Langmuir*, *37*(49), 14336–14344.

[124] Zhu, J., Zhu, M., Zhang, R., He, Z., Xiong, L., & Guo, L. (2022). Corrosion inhibition behavior of electrochemically synthesized carbon dots on Q235 carbon steel. *Journal of Adhesion Science and Technology*, 1–13.

[125] Long, W. J., Li, X. Q., Xu, P., Feng, G. L., & He, C. (2022). Facile and scalable preparation of carbon dots with Schiff base structures toward an efficient corrosion inhibitor. *Diamond and Related Materials*, *130*, 109401.

[126] Zhu, J., Zhu, M., He, Z., Xiong, L., Zhang, R., & Guo, L. (2023). Chemical oxidation synthesized high-yield carbon dots for acid corrosion inhibition of Q235 steel. *ChemistrySelect*, *8*(7), e202204621.

[127] Cao, S., Liu, D., Wang, T., Ma, A., Liu, C., Zhuang, X., . . . & Gui, J. (2021). Nitrogen-doped carbon dots as high-effective inhibitors for carbon steel in acidic medium. *Colloids and Surfaces A: Physicochemical and Engineering Aspects*, *616*, 126280.

[128] Saraswat, V., & Yadav, M. (2020). Carbon dots as green corrosion inhibitor for mild steel in HCl solution. *ChemistrySelect*, *5*(25), 7347–7357.

[129] Zakowski, K., Narozny, M., Szocinski, M., & Darowicki, K. (2014). Influence of water salinity on corrosion risk—the case of the southern Baltic Sea coast. *Environmental Monitoring and Assessment*, *186*, 4871–4879.

[130] Fink, F. W. (1963). Saline-water development poses problems in metal corrosion. *Groundwater*, *1*(4), 21–25.

[131] Li, J., Lv, J., Fu, L., Tang, M., & Wu, X. (2020). New ecofriendly nitrogen-doped carbon quantum dots as effective corrosion inhibitor for saturated CO 2 3% NaCl solution. *Russian Journal of Applied Chemistry*, *93*, 380–392.

[132] Wang, J., Du, P., Zhao, H., Pu, J., & Yu, C. (2019). Novel nitrogen doped carbon dots enhancing the anticorrosive performance of waterborne epoxy coatings. *Nanoscale Advances*, *1*(9), 3443–3451.

[133] Jiang, B. K., Chen, A. Y., Gu, J. F., Fan, J. T., Liu, Y., Wang, P., . . . & Wang, X. Y. (2020). Corrosion resistance enhancement of magnesium alloy by N-doped graphene quantum dots and polymethyltrimethoxysilane composite coating. *Carbon*, *157*, 537–548.

[134] Zhang, H. D., Chen, A. Y., Gan, B., Jiang, H., & Gu, L. J. (2022). Corrosion protection investigations of carbon dots and polydopamine composite coating on magnesium alloy. *Journal of Magnesium and Alloys*, *10*(5), 1358–1367.

[135] Nunez, L., Reguera, E., Corvo, F., Gonzalez, E., & Vazquez, C. (2005). Corrosion of copper in seawater and its aerosols in a tropical island. *Corrosion Science*, *47*(2), 461–484.

[136] Revie, R. W. (Ed.). (2011). *Uhlig's Corrosion Handbook* (Vol. 51). John Wiley & Sons.

[137] Xiao, L. U., Liu, Y. W., Zhao, H. T., Chen, P. A. N., & Wang, Z. Y. (2021). Corrosion behavior of copper in extremely harsh marine atmosphere in Nansha Islands, China. *Transactions of Nonferrous Metals Society of China*, *31*(3), 703–714.

[138] Zhang, T., Zhang, D., Wu, P., & Gao, L. (2022). Corrosion inhibition of high-nitrogen-doped CDs for copper in 3wt% NaCl solution. *Journal of the Taiwan Institute of Chemical Engineers*, *138*, 104462.

[139] Khatoon, H., Iqbal, S., & Ahmad, S. (2019). Influence of carbon nanodots encapsulated polycarbazole hybrid on the corrosion inhibition performance of polyurethane nanocomposite coatings. *New Journal of Chemistry*, *43*(26), 10278–10290.

[140] Ye, Y., Jiang, Z., Zou, Y., Chen, H., Guo, S., Yang, Q., & Chen, L. (2020). Evaluation of the inhibition behavior of carbon dots on carbon steel in HCl and NaCl solutions. *Journal of Materials Science & Technology*, *43*, 144–153.

[141] Feng, M., Liu, Y., Zhang, S., Liu, Y., Luo, N., & Wang, D. (2020). Carbon quantum dots (CQDs) modified TiO2 nanorods photoelectrode for enhanced photocathodic protection of Q235 carbon steel. *Corrosion Science*, *176*, 108919.

[142] Wang, X., Li, C., Zhang, M., Lin, D., Yuan, S., Xu, F., . . . & Wang, H. (2022). A novel waterborne epoxy coating with anti-corrosion performance under harsh oxygen environment. *Chemical Engineering Journal*, *430*, 133156.

[143] Di, Z., Zhao, Q., Xu, H., Wu, K., Yan, Y., Jin, X., . . . & Ran, B. (2022). N-doped carbon quantum dots with mechanical enhancement and fluorescence properties toward anticorrosion coating. *International Journal of Electrochemical Science*, *17*(2212101), 2.

[144] Cen, H., Chen, Z., & Guo, X. (2019). N, S co-doped carbon dots as effective corrosion inhibitor for carbon steel in CO2-saturated 3.5% NaCl solution. *Journal of the Taiwan Institute of Chemical Engineers*, *99*, 224–238.

[145] Guan, Z. C., Jin, P., Liu, Q., Wang, X., Chen, L. F., Xu, H., . . . & Du, R. G. (2019). Carbon quantum dots/Ag sensitized TiO2 nanotube film for applications in photocathodic protection. *Journal of Alloys and Compounds*, *797*, 912–921.

[146] Keerthana, A. K., & Ashraf, P. M. (2020). Carbon nanodots synthesized from chitosan and its application as a corrosion inhibitor in boat-building carbon steel BIS2062. *Applied Nanoscience*, *10*, 1061–1071.

[147] Liu, X. R., Sheng, X. X., Yuan, X. Y., Liu, J. K., Sun, X. W., & Yang, X. H. (2020). Research on correlation between corrosion resistance and photocatalytic activity of molybdenum zinc oxide modified by carbon quantum dots pigments. *Dyes and Pigments*, *175*, 108148.
[148] Zhou, Q., Yuan, G., Lin, M., Wang, P., Li, S., Tang, J., . . . & Zhang, Y. (2021). Large-scale electrochemical fabrication of nitrogen-doped carbon quantum dots and their application as corrosion inhibitor for copper. *Journal of Materials Science*, *56*, 12909–12919.
[149] Roberge, P. R. (2008). *Corrosion Engineering*. McGraw-Hill Education.
[150] Kumar, S. R., Krishnaa, S. D., Krishna, M. D., Gokulkumar, N. T., & Akilesh, A. R. (2021). Investigation on corrosion behaviour of aluminium 6061-T6 alloy in acidic, alkaline and salt medium. *Materials Today: Proceedings*, *45*, 1878–1881.
[151] Kandasamy, K., Surendhiran, S., Khadar, Y. S., & Rajasingh, P. (2021). Ultrasound-assisted microwave synthesis of CdS/MWCNTs QDs: A material for photocatalytic and corrosion inhibition activity. *Materials Today: Proceedings*, *47*, 757–762.
[152] Hu, Z., Dai, H., Zhou, W., Wei, J., Zhang, H., Ye, Z., . . . & Guo, R. (2021). Corrosion resistant solid-state carbon dots@ silicalite-1 composite for latent fingerprints detection. *Journal of Alloys and Compounds*, *889*, 161660.
[153] Hu, Y., Wang, Y., Wang, C., Ye, Y., Zhao, H., Li, J., . . . & Xue, Q. (2019). One-pot pyrolysis preparation of carbon dots as eco-friendly nano additives of water-based lubricants. *Carbon*, *152*, 511–520.
[154] Wang, F., Shang, L., Zhang, G., & Wang, Z. (2022). Polyethylene glycol derived carbon quantum dots nanofluids: An excellent lubricant for diamond-like carbon film/bearing steel contact. *Friction*, 1–12.
[155] Wang, B., Lian, J., Jing, M., & Lu, H. (2022). Synthesis of eco-friendly carbon dots as self-repairing additives of polyalphaolefin by means of a green solvation effect. *Langmuir*, *38*(50), 15756–15765.
[156] Tabish, M., Zarin, A., Malik, M. U., Khan, M. A., Zhao, J., & Yasin, G. (2022). Microbial-induced corrosion of metals with presence of nanoparticles. In *Biodegradation and Biodeterioration at the Nanoscale* (pp. 675–699). Elsevier.
[157] Kalajahi, S. T., Rasekh, B., Yazdian, F., Neshati, J., & Taghavi, L. (2020). Green mitigation of microbial corrosion by copper nanoparticles doped carbon quantum dots nanohybrid. *Environmental Science and Pollution Research*, *27*, 40537–40551.
[158] Liu, W. R., Yang, Y. Y., Liu, Y. S., Zhang, L. J., Zhao, J. L., Zhang, Q. Q., . . . & Ying, G. G. (2017). Biocides in wastewater treatment plants: Mass balance analysis and pollution load estimation. *Journal of Hazardous Materials*, *329*, 310–320.

4 Conducting polymers
Fundamental aspects, properties, synthesis, and structural elucidation

M. Rbaa, A. Bouabbadi, R. Hsissou, M. Oubaaqa, A. Bekhta, B. Tüzün, E. Berdimurodov, B. Lakhrissi, and A. Zarrouk

4.1 INTRODUCTION

Materials play a fundamental role in all areas of human activity and form the bedrock of our industrial society. They form the essential basis for the manufacture of objects essential to everyday life. The discovery of conductive polymers in the 1970s revolutionized polymer science and opened up new avenues for the development of advanced materials [1]. Conductive polymers are organic substances with high electrical conductivity. CPs are made up of consecutive single and double bonds in the conjugated carbon chain, which are widely delocalized and polarized. These materials exhibit remarkable optical, electronic, and electrochemical properties, making them particularly attractive for a wide range of applications, including energy storage, sensors, actuators, and electronic devices [2]. Conducting polymers can be classified into two main categories: conjugated polymers and inherently conducting polymers [3]. Conjugated polymers are made up of alternating single and double bonds, which create a continuous pathway for electron delocalization along the polymer backbone [4]. Inherently conducting polymers, on the other hand, contain conjugated and non-conjugated segments in their structures, resulting in a combination of electrical conductivity and mechanical strength [5]. The unique properties of conducting polymers arise from the delocalization of electrons along the polymer backbone, which leads to the formation of a highly conductive path. This process is facilitated by doping, which involves the addition of dopant molecules that introduce charge carriers into the polymer matrix [6]. The conductivity of conducting polymers can be modulated by changing the dopant concentration, temperature, or pH [7].

Over the past few decades, significant progress has been made in the synthesis, characterization, and application of conducting polymers [8]. These materials have been successfully used in various technological applications, including batteries, solar cells, electrochromic devices, sensors, and actuators [9]. Ongoing research is

DOI: 10.1201/9781003456025-4

aimed at developing new and improved conductive polymers with tailored properties for specific applications.

4.2 CHARACTERISTICS OF CONDUCTIVE POLYMERS

4.2.1 Polymers: definition and initial uses

Polymers are high molecular weight substances made up of long chains formed by the repetition of structural units called monomers. These are linked together by chemical bonds to form linear, branched, or cross-linked chains. Macromolecules are made up of repeated monomers, and their molecular weight is generally high, ranging from 5,000 to 250,000. Polymers have been recognized for their excellent electrical insulation properties since their first use. Thanks to these properties, they have found an important application, particularly in cable sheathing, where electrical insulators are sought after. Ease of implementation, flexibility, aesthetics, lightness, and chemical inertness are other major advantages.

4.2.2 Characteristics

Conducting polymers are a class of materials that exhibit electrical conductivity similar to that of metals but with the flexibility and other unique properties of polymers [10]. Here are some basic characteristics of conducting polymers:

Chemical structure: Conducting polymers consist of repeating units that are linked together to form long chains. These repeating units are typically composed of a conjugated backbone, which allows for the delocalization of electrons and thus electrical conductivity [11].

Doping: Doping consists of introducing chemical agents (dopants) into the polymer chain to modify the density of charge carriers (electrons) in the material, thereby significantly increasing the material's electrical performance [7].

Electrical conductivity: The electrical conductivity of conductive polymers can vary considerably, from insulating to highly conductive, depending on the chemical structure of the polymer and the level of doping applied.

Applications: Conductive polymers offer numerous potential applications, including as electronic and optoelectronic devices, sensors, and actuators. They can also be used in biomedical applications, such as drug delivery and tissue engineering.

Processing: Conductive polymers can take a variety of forms, such as thin films, fibers, and three-dimensional structures [12, 13]. They are therefore extremely adaptable to a wide range of uses.

Environmental impact: Conducting polymers have the potential to be more environmentally friendly than traditional electronic materials, as they can be synthesized using non-toxic starting materials and can be processed using environmentally friendly techniques. Overall, conducting polymers are a promising class of materials with unique properties and potential applications in a range of fields [14].

4.3 THE MAIN CLASSES OF CONDUCTING POLYMERS

4.3.1 Intrinsically conductive polymers (ICPs)

ICPs are organic polymers endowed with electrical conductivity, made possible by the presence of a π-conjugated system along their main chain. This conjugation promotes the delocalization of π-electrons, enabling charge transport. Their conductivity can be modulated by a doping process, giving them behavior ranging from semiconductor to metallic conductor. Among the most widely studied PCIs are polyaniline (PANI), polypyrrole (PPy), polythiophene (PT), and derivatives such as PEDOT. Thanks to their remarkable electrochemical, optical, and mechanical properties, these materials are finding numerous applications in fields such as organic electronics, sensors, energy storage, and biomedical technologies [15–17].

4.3.2 Conductive composite polymers (CPC)

CCPs, also known as extrinsic conducting polymers, consist of an insulating polymer matrix containing dispersed conductive particles. Unlike intrinsic conducting polymers, their conductivity depends primarily on the nature, concentration, and distribution of the conductive charges within the matrix.

4.4 SYNTHESES AND STRUCTURAL ELUCIDATION OF THE CONDUCTING POLYMERS

Conductive polymers can be synthesized using different methods, which can be categorized into several types:

- Chemical polymerization
- Electrochemical polymerization
- Photochemical polymerization
- Emulsion polymerization
- Metathesis polymerization
- Addition polymerization
- Solid-state polymerization
- Plasma polymerization
- Pyrolysis
- Soluble precursor polymerization

The monomers used to synthesize conductive polymers generally have a characteristic aromatic or polyconjugated structure that can be retained in the structure of the resulting polymer. For example, acetylene polymerization produces polyacetylene containing conjugated units. Similarly, the polymerization of benzene gives rise to poly(p-phenylene), made up of aromatic chains linked by covalent bonds. The extension of π-conjugation in the structure of these conductive polymers plays a key role in their electronic properties [18].

Chemical polymerization occurs when monomers react with each other through covalent bonds under the influence of an oxidizing agent. There are two chemical methods for synthesizing conductive polymers.

1) Oxidation of the monomer with iron (III) chloride (FeC13) or other transition metals
2) Nickel-catalyzed Grignard coupling of 2,5-dihaloheterocyclic compounds [19]

Electrochemical polymerization, also known as electropolymerization, involves the formation of a polymer film directly on the surface of a solid electrode and involves the oxidation of the monomer on the electrode, generating a cationic radical that triggers polymerization [20]. The first electrochemical synthesis of a conductive polymer was achieved by Letheby in 1862 with the production of polyaniline (PANi) [21]. Later, this method was subsequently taken up and developed by Szaways [22]. After this initial study, the field remained inactive for some time until research into the electrochemical polymerization of certain vinyl monomers appeared in the late 1940s [23].

The synthesis of conductive polymers by electrochemical means is carried out in a cell with one or two compartments, using a three-electrode system and a solution containing a supporting electrolyte [24]. During electrochemical polymerization, the monomer in solution dissolved in a suitable solvent in the presence of a supporting electrolyte undergoes oxidation or reduction at the surface of the working electrode. This process leads to the formation of a polymer film, which is deposited directly on the electrode surface [25]. Electrochemical polymerization can be performed using three distinct operating modes depending on the applied current-voltage system: potentiostatic, galvanostatic, or potentiodynamic. The choice of method depends mainly on the intended application. In general, the potentiostatic technique is used to obtain thin and homogeneous films, while the galvanostatic mode is preferred for the production of thicker films [24].

The standard triple electrode system consists of a working electrode, counter electrode, and reference electrode. Electrodes should not be oxidized as polymer films are formed by the oxidation process. Therefore, platinum (Pt), gold (Au), and tin oxide (SnO_2) substrates as working electrodes should be inert species, which are indium tin oxide (ITO) and stainless substrates. Pt, Au, and Ni are counter electrodes. Saturated calomel electrode and silver/silver chloride (Ag/AgCl) electrode are used as reference electrodes. The choice of supporting electrolyte is based on the criteria of solubility, dissociation, and nucleophilic properties. R_4NX type (R: Alkyl, aryl radical and X = Cl^-, Br^-, I^-, ClO_4^-, BF_4^-, PF_6^-, $CF_3SO_3^-$, $CH_3C_6H_4SO_3^-$) quaternary ammonium salts are easily soluble in aprotic solvents in anhydrous environments (acetonitrile (ACN), benzonitrile, etc.). These salts in solution are used as a supporting electrolyte.

4.5 REVIEW OF SOME SYNTHESIS METHODS FOR CONDUCTIVE POLYMERS

4.5.1 Chemical synthesis of polypyrrole as a conductive polymer

Polypyrrole (PPy) is a widely used conductive polymer in biomedical applications due to its high electrical conductivity, ease of preparation, and ease of surface

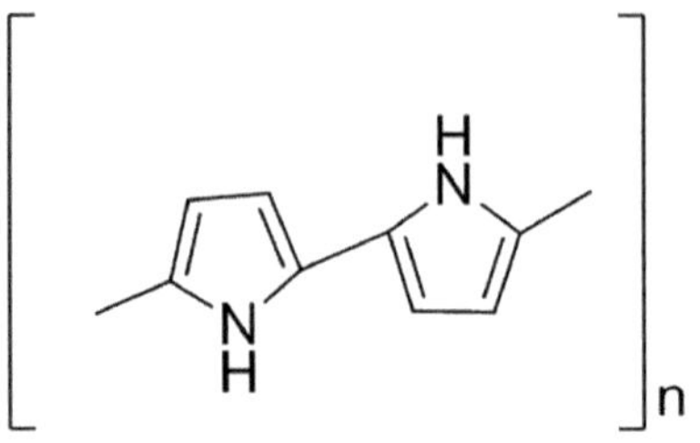

FIGURE 4.1 Polypyrrole chemical structure.

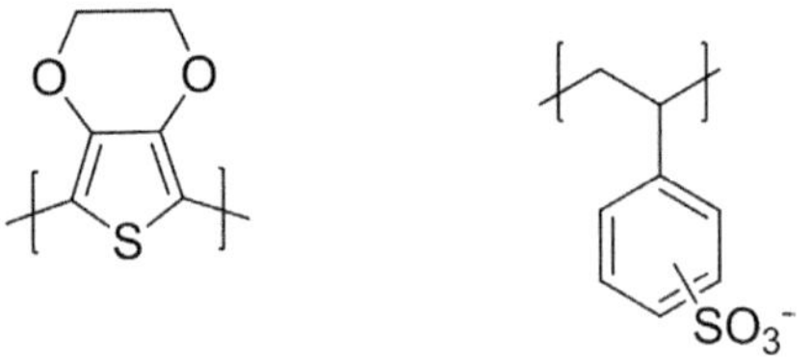

FIGURE 4.2 Poly(3,4-ethylenedioxythiophene) chemical structure.

modification. It is typically achieved by oxidative chemical polymerization of the pyrrole monomer using an oxidizing agent, such as ammonium persulfate. This reaction results in the formation of a conductive polymer on the electrode surface or in solution [26] (**Figure 4.1**).

4.5.2 Electrochemical synthesis of PEDOT as a conductive polymer

Electrochemical polymerization represents an efficient method for synthesizing conductive polymers by electrodeposition. Materials such as polyaniline or poly(3,4-ethylene-dioxythiophene) (PEDOT) can be produced by controlled oxidation of their respective monomers at the electrode surface. The process leads to the formation of an adherent polymer film with optimal conductive properties (**Figure 4.2**) [27, 28].

4.5.3 Synthesis of polymer conductors via plasma

Plasma polymerization is a gas-phase deposition technique used to synthesize conductive polymers such as polypyrrole and polythiophene [31]. This innovative process is based on the activation of a vaporized monomer by a cold plasma (ionized gas). It is carried out at low temperature (<100°C) and low pressure (0.1–10 Torr), making it compatible with thermosensitive substrates [29, 30] (**Figure 4.3**).

4.5.4 Radiation polymerization

Radiation polymerization is an advanced synthesis technique for producing conductive polymers. Based on the exposure of monomers to ionizing radiation

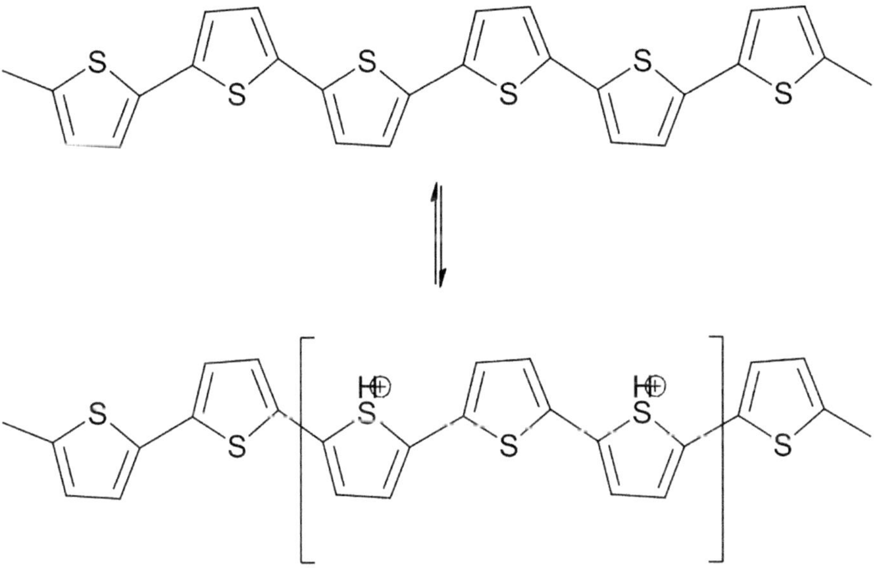

FIGURE 4.3 Mesomerism (dopage) in polythiophene chemical structure.

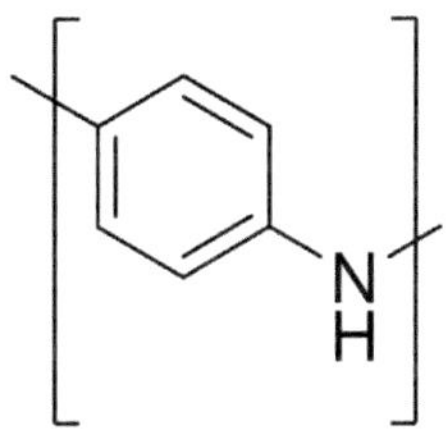

FIGURE 4.4 Template polymerization in polyaniline chemical structure.

(γ-rays, electron beams, or intense UV light), it enables the synthesis of conductive polymers without the need for chemical initiators [32–34]. Materials such as polypyrrole (**Figure 4.1**) and polythiophene can be obtained using this technique.

4.5.5 Template polymerization

Matrix polymerization is a process in which a monomer is polymerized within a matrix (solid, liquid, or gel), serving to control the growth and organization of the polymer chains. This matrix guides the formation of the polymer by influencing its structure and orientation. This process allows for the creation of composite materials or films with specific characteristics by exploiting the unique properties of the matrix and the polymer [35, 36]. An example of a conducting polymer synthesized through template polymerization is polyaniline [27] (**Figure 4.4**).

4.6 CONDUCTIVE POLYMERS: CURRENT USES AND PERSPECTIVES

Thanks to their remarkable physical and chemical properties, conductive polymers are used in a wide range of applications:

- **Energy storage:** Conducting polymers are widely used in the design of batteries, supercapacitors, and fuel cells due to their high conductivity and fast charge and discharge rates [37]. These materials have the potential to replace traditional electrode materials, resulting in lighter and more efficient energy storage devices [38, 39].
- **Electronic devices:** Thanks to their excellent optical and electrical properties, conductive polymers are used in the manufacture of devices such as OLEDs, organic field-effect transistors (OFETs), and sensors. Their compatibility with flexible substrates makes them the materials of choice for electronics development [40].
- **Biomedical applications:** Conductive polymers are used in biomedical applications such as biosensors, drug delivery systems, and tissue engineering. They are biocompatible and can be easily modified to incorporate various biomolecules, making them ideal for targeted drug delivery and tissue engineering applications [30].
- **Antistatic coatings**: Thanks to their biocompatibility and their electroactive properties, conducting polymers such as polypyrrole and PEDOT find innovative applications in the field of health. They are notably used for the design of sensitive biocaptures, devices for controlled delivery of drugs, as well as conductive supports in tissue engineering. Their aptitude to be functionalized with biomolecules allows to develop targeted and highly specific applications [41].
- **Corrosion protection**: Conductive polymers, when applied as coatings, create a protective barrier against moisture, oxygen, and chemicals. They are easy and efficient to apply, making them particularly suitable for harsh marine and industrial environments [42, 43].

4.7 CONCLUSION

In conclusion, conducting polymers represent a fascinating class of materials with unique properties and great potential in various applications. In this review, we have provided a comprehensive overview of the fundamental aspects, properties, synthesis, and structural elucidation of conducting polymers. The electronic and optical properties of these materials have been discussed, highlighting their importance in electronics, energy storage, and sensing applications. Various synthetic methods have been covered, including chemical oxidation, electrochemical polymerization, plasma polymerization, template polymerization, and solid-state polymerization, providing a range of options for the synthesis of conducting polymers. Structural elucidation techniques, such as X-ray diffraction, nuclear magnetic resonance spectroscopy, and infrared spectroscopy, have been presented, enabling the characterization and optimization of conducting polymer properties. Challenges and opportunities in the field of conducting polymers have also been discussed, emphasizing the need for continued research and development in this area. In the future, conducting polymers are expected to play a significant role in addressing global challenges related to energy, sustainability, and health.

REFERENCES

[1] D. Braun, H. Cherdron, M. Rehahn, H. Ritter, B. Voit, *Polymer synthesis: Theory and practice: Fundamentals, methods, experiments*, 2005. https://doi.org/10.1007/b138247.

[2] T.A. Skotheim, J.R. Reynolds, *Conjugated polymers: Theory, synthesis, properties, and characterization*, 2007. CRC Press.

[3] Y.Z. Long, M.M. Li, C. Gu, M. Wan, J.L. Duvail, Z. Liu, Z. Fan, Recent advances in synthesis, physical properties and applications of conducting polymer nanotubes and nanofibers, *Prog Polym Sci.* 36 (2011). https://doi.org/10.1016/j.progpolymsci.2011.04.001.

[4] A.K. Agrawal, S.A. Jenekhe, Electrochemical properties and electronic structures of conjugated polyquinolines and polyanthrazolines, *Chem Mater.* 8 (1996). https://doi.org/10.1021/cm9504753.

[5] D.A. Zatale, S.T. Rathod, K.N. Pawar, Synthesis and characterization of conducting polymer, *Int J Sci Res Sci Technol.* (2021). https://doi.org/10.32628/ijsrst218328.

[6] M.M. Chehimi, M.L. Abel, E. Pigois-Landureau, M. Delamar, Characterization of conducting polymers by inverse gas chromatography Part II. Effect of dopant on the dispersive and specific properties of polypyrrole, *Synth Met.* 60 (1993). https://doi.org/10.1016/0379-6779(93)91278-A.

[7] M.Y. Zhang, Y. Song, D. Yang, Z. Qin, D. Guo, L.J. Bian, X.G. Sang, X. Sun, X.X. Liu, Redox poly-counterion doped conducting polymers for pseudocapacitive energy storage, *Adv Funct Mater.* 31 (2021). https://doi.org/10.1002/adfm.202006203.

[8] K. Yoshino, K. Kaneto, S. Takeda, Applications of conducting polymers as electronics and opto-electronics devices, *Synth Met.* 18 (1987). https://doi.org/10.1016/0379-6779(87)90972-6.

[9] J.E. Yoo, K.S. Lee, A. Garcia, J. Tarver, E.D. Gomez, K. Baldwin, Y. Sun, H. Meng, T.Q. Nguyen, Y.L. Loo, Directly patternable, highly conducting polymers for broad applications in organic electronics, *Proc Natl Acad Sci U S A.* 107 (2010). https://doi.org/10.1073/pnas.0913879107.

[10] A. Kumar, W.C. Wilisch, N.S. Lewis, The electrical properties of semiconductor/metal, semiconductor/liquid, and semiconductor/conducting polymer contacts, *Crit Rev Solid State Mater Sci.* 18 (1993). https://doi.org/10.1080/10408439308243731.

[11] L. Qin, Q. Tao, X. Liu, M. Fahlman, J. Halim, P.O.Å. Persson, J. Rosen, F. Zhang, Polymer-MXene composite films formed by MXene-facilitated electrochemical polymerization for flexible solid-state microsupercapacitors, *Nano Energy.* 60 (2019). https://doi.org/10.1016/j.nanoen.2019.04.002.

[12] J.E. Kuhlmann, S.S.Y. Liu, K. Dirnberger, M. Zharnikov, S. Ludwigs, Electrochemical characterization of redox probes confined in 3D conducting polymer networks, *Chem Eur J.* 27 (2021). https://doi.org/10.1002/chem.202103257.

[13] H. Yuk, B. Lu, S. Lin, K. Qu, J. Xu, J. Luo, X. Zhao, 3D printing of conducting polymers, *Nat Commun.* 11 (2020). https://doi.org/10.1038/s41467-020-15316-7.

[14] J. Yang, Y. Liu, S. Liu, L. Li, C. Zhang, T. Liu, Conducting polymer composites: Material synthesis and applications in electrochemical capacitive energy storage, *Mater Chem Front.* 1 (2017). https://doi.org/10.1039/c6qm00150e.

[15] S. Özbağ, *İletkenpolimerlerinkarakterizasyon* (2010) Ph.D. Thesis, Institute of Science, Harran University.

[16] S. Kee, N. Kim, B. Park, B.S. Kim, S. Hong, J.H. Lee, S. Jeong, A. Kim, S.Y. Jang, K. Lee, Highly deformable and see-through polymer light-emitting diodes with all-conducting-polymer electrodes, *Advanced Materials*. 30 (2018). https://doi.org/10.1002/adma.201703437.

[17] B. Devadas, T. Imae, Effect of carbon dots on conducting polymers for energy storage applications, *ACS Sustain Chem Eng*. 6 (2018). https://doi.org/10.1021/acssuschemeng.7b01858.

[18] E. Turaç, *Tiyofentüreviyeni tip iletkenpolimerlerinsentezvekarakterizasyonu* (2010) Ph.D. Thesis, Niğde University, Institute of Science, Department of Chemistry, Niğde.

[19] R.D. McCullough, Chemistry of conducting polythiophenes, *Adv Mater*. 10 (1998), 93–116.

[20] S. Özbağ, *İletkenpolimerlerinkarakterizasyonu*, Institute of Science, Harran University, 2010.

[21] H. Letheby, XXIX. On the production of a blue substance by the electrolysis of sulphate of aniline, *J Chem Sci*. 15 (1862), 161.

[22] E.C. Szaways, Electrochemical preparation of polyaniline, *J Chem Soc*. 77 (1900), 207.

[23] E.A. Rembold, *The Political Economy of Food Aid: The Case of Public Law 480 in India* (1974), Ph D. Dissertation, Ohio State University, Columbus.

[24] D. Kumar, R.C. Sharma, Advances in conductive polymers, *Eur Poly J*. 34 (1998), 1053–1060.

[25] E. Turaç, *Tiyofentüreviyeni tip iletkenpolimerlerinsentezvekarakterizasyonu* (2010), Ph.D. Thesis, Niğde University, Institute of Science, Department of Chemistry, Niğde.

[26] R.M. Apetrei, G. Carac, A. Ramanaviciene, G. Bahrim, C. Tanase, A. Ramanavicius, Cell-assisted synthesis of conducting polymer—polypyrrole—for the improvement of electric charge transfer through fungal cell wall, *Colloids Surf B Biointerfaces*. 175 (2019). https://doi.org/10.1016/j.colsurfb.2018.12.024.

[27] F. Saouti, S. Belaaouad, A. Cherqaoui, Y. Naimi, Polyaniline thin film prepared by electrochemical polymerization method, *Biointerface Res Appl Chem*. 12 (2022). https://doi.org/10.33263/BRIAC124.55235533.

[28] R. Kandpal, M. Shahadat, S.W. Ali, S.Z. Ahammad, Alternating current based electrochemically polymerized polyaniline modified flexible conducting paper as an exoelectrogenic biofilm platform for a microbial fuel cell, *J Water Process Eng*. 46 (2022). https://doi.org/10.1016/j.jwpe.2021.102459.

[29] J.Y. Kim, H.J. Jang, G.T. Bae, C.S. Park, E.Y. Jung, H.S. Tae, Improvement of nanostructured polythiophene film uniformity using a cruciform electrode and substrate rotation in atmospheric pressure plasma polymerization, *Nanomaterials*. 12 (2022). https://doi.org/10.3390/nano12010032.

[30] J. Carneiro de Oliveira, M. de Meireles Brioude, A. Airoudj, F. Bally-Le Gall, V. Roucoules, Plasma polymerization in the design of new materials: Looking through the lens of maleic anhydride plasma polymers, *Mater Today Chem*. 23 (2022). https://doi.org/10.1016/j.mtchem.2021.100646.

[31] E. Sainz-García, M. López, R. Múgica-Vidal, B. Rojo-Bezares, C. Lozano, A. González-Marcos, P. Toledano, I. Muro-Fraguas, A. Sainz-García, Y. Sáenz, F. Alba-Elías, Promotion of biofilm production via atmospheric-pressure plasma-polymerization for biomedical applications, *Appl Surf Sci*. 581 (2022). https://doi.org/10.1016/j.apsusc.2021.152350.

[32] S.S. Metwally, R.S. Hassan, E.H. El-Masry, E.H. Borai, Gamma-induced radiation polymerization of kaolin composite for sorption of lanthanum, europium and uranium ions from low-grade monazite leachate, *J Radio Anal Nucl Chem*. 315 (2018). https://doi.org/10.1007/s10967-017-5638-7.

[33] A. Ashfaq, M.C. Clochard, X. Coqueret, C. Dispenza, M.S. Driscoll, P. Ulański, M. Al-Sheikhly, Polymerization reactions and modifications of polymers by ionizing radiation, *Polymers (Basel).* 12 (2020). https://doi.org/10.3390/polym12122877.

[34] J.P. Fouassier, J. Lalevée, *Photoinitiators for polymer synthesis: Scope, reactivity and efficiency*, 2012. https://doi.org/10.1002/9783527648245.

[35] R. Saito, Combination of template polymerization and atom transfer radical polymerization: Strategy for synthesis of specifically structural polymers, *Polymer (Guildf).* 49 (2008). https://doi.org/10.1016/j.polymer.2008.03.011.

[36] S. Połowiński, Template polymerisation and co-polymerisation, *Prog Polym Sci.* 27 (2002). https://doi.org/10.1016/S0079-6700(01)00035-1.

[37] A.M. Bryan, L.M. Santino, Y. Lu, S. Acharya, J.M. D'Arcy, Conducting polymers for pseudocapacitive energy storage, *Chem Mater.* 28 (2016). https://doi.org/10.1021/acs.chemmater.6b01762.

[38] Y. Shi, L. Peng, G. Yu, Nanostructured conducting polymer hydrogels for energy storage applications, *Nanoscale.* 7 (2015). https://doi.org/10.1039/c5nr03403e.

[39] J. Wang, J. Wang, Z. Kong, K. Lv, C. Teng, Y. Zhu, Conducting-polymer-based materials for electrochemical energy conversion and storage, *Adv Mater.* 29 (2017). https://doi.org/10.1002/adma.201703044.

[40] L. Pan, H. Qiu, C. Dou, Y. Li, L. Pu, J. Xu, Y. Shi, Conducting polymernano structures: Template synthesis and applications in energy storage, *Int J Mol Sci.* 11 (2010). https://doi.org/10.3390/ijms11072636.

[41] H. Yoon, Current trends in sensors based on conducting polymer nanomaterials, *Nanomaterials.* 3 (2013). https://doi.org/10.3390/nano3030524.

[42] K. Auepattana-Aumrung, T. Phakkeeree, D. Crespy, Polymer-corrosion inhibitor conjugates as additives for anticorrosion application, *Prog Org Coat.* 163 (2022). https://doi.org/10.1016/j.porgcoat.2021.106639.

[43] R. Teijido, L. Ruiz-Rubio, A.G. Echaide, J.L. Vilas-Vilela, S. Lanceros-Mendez, Q. Zhang, State of the art and current trends on layered inorganic-polymer nanocomposite coatings for anticorrosion and multi-functional applications, *Prog Org Coat.* 163 (2022). https://doi.org/10.1016/j.porgcoat.2021.106684.

5 Polyaniline, polypyrrole, and polythiophene as anticorrosive materials

Shokoh Parham, Hadi Sarvari, and David J. Edwards

5.1 INTRODUCTION

Thermodynamically, metals and their alloys are the most susceptible to corrosion. Corrosion is the breakdown or corrosion of a material caused by its interaction with its environment. The application of paint or organic coating to the metal substrate is arguably the most widespread method for avoiding corrosion. In an ideal circumstance, organic coatings would provide long-lasting corrosion protection. However, this quality is dependent upon the coating's ability to preserve an exposed metal surface from developing defects during its service life [1].

Corrosion prevention is a global concern for the electrical and metallurgical industries [2]. Numerous industries, including the construction and oil industries as well as the healthcare sector, have a vested interest in the study of metal corrosion suppression [3–5]. Throughout the years, ways of preventing metal corrosion have evolved. Conductive polymer coatings are used in various contexts. Electrodeposited conductive polymers (CPs) on metal electrodes have been a key area of electrochemistry research over the past several decades. Traditional coatings, such as paints, contain no hazardous or lethal chemicals, whereas the CPs do. In addition, CPs provide physical and electronic barrier effects as well as electromagnetic interference shielding, enhancing the protection behavior [2]; this is in contrast to traditional coatings, which only provide physical barriers against corrosive conditions. In recent years, shape memory materials and intrinsically conductive polymers (CPs) have been regarded as possible materials for corrosion protection of metals because of the design and development of alternative self-healing organic coatings [1].

The corrosion of materials continues to pique the scientific community's curiosity. The researchers in the field of corrosion inhibition are unrelenting in their pursuit of new and more efficient ways for shielding metals from corrosion. The addition of corrosion inhibitors to corrosive environment is a shift from standard suppression approaches. Preventing corrosion of materials has been an important topic of study and recognition for centuries [6].

DOI: 10.1201/9781003456025-5

Various anticorrosive materials have been developed [7]. Chemicals such as conductive polymers can be used to prevent corrosion. The creation of conductive polymers as an alternative to metal conductors and semiconductors is a noteworthy accomplishment. Many researchers have been attempting to develop a polymer with precisely tunable mechanical, thermal, optical, and electrical properties. A significant amount of research has been conducted on conductive polymers such as polyacetylene, polyaniline, polypyrrole, polythiophene, poly(p-phenylene), and poly(phenylene vinylene) [8]. Numerous studies [9–15] have investigated conductive polymers and other anticorrosive compounds. This chapter elucidates the corrosion resistance of polyaniline, polypyrrole, and polythiophene.

5.2 CORROSION INHIBITORS

Carbon steel, copper (Cu), magnesium, and nickel (Ni) are among the metals and alloys subject to corrosion in industrial environments [16]. The addition of corrosion inhibitors (in extremely small quantities) to a corrosive environment retards the corrosion of a designated metal. They react with the corroding surface to generate a protective layer and can be thought of as catalysts or substances that slow corrosion [17]. When an anticorrosive coating is placed on a metal surface, the likelihood of corrosion is reduced since the coating shelters the metal from its corrosive environment. The use of anticorrosive coating can protect metal surfaces against rust and corrosion. Anticorrosion coatings containing corrosion inhibitors have enabled the protection of metals from oxidation and corrosion in recent years [16].

Conductive polymers are an example of a type of corrosion inhibitor [18, 19]. MacDiarmid (1985) stated that using conductive polymers can prevent corrosion [20, 21]. The majority of conductive polymers (CPs) are polyheterocycles, polyanilines, and poly(phenylene vinylene)s. CPs have the potential for chemical and electrochemical synthesis. Electrochemical generation by anodic oxidation has been demonstrated for the majority of CPs, allowing for the formation of conductive coatings. Electrochemical doping, chemical doping via charge transfer, photodoping, doping by acid-base chemistry (only polyaniline is doped using this method), and charge injection at a metal-semiconducting polymer interface are all methods for converting CPs from an insulating to a conducting state [21]. New corrosion inhibitors using conductive polymers have been the subject of extensive research in numerous industries ranging from construction and medicine to petroleum and energy [22–28]. The adoption of conducting polymer as an alternative coating material for steel could [29]. Due to their effectiveness in protecting metals from corrosion, conductive polymers have garnered considerable interest. Polyaniline, polypyrrole, and other common conducting polymers are used to protect metals by coating them. Various components and preparation techniques have been used to generate composite conducting polymers. According to scientific studies, anodic protection and the application of corrosion inhibitors are the two most effective methods for delaying the corrosion of metals. It has been demonstrated and analyzed that coatings on CPs dramatically prevent corrosion. New scientific coatings or methods for preserving precious metals will certainly dominate future research [30].

5.3 CONDUCTING POLYMERS AS ANTICORROSIVE MATERIALS

Almost until 30 years ago, it was believed that all carbon-based polymers were insulators. In the past, the notion that polymers could be made to transmit electricity was scoffed at. Due to this feature, plastics are widely used in the electrical industry. They were also used as insulation. The discovery of a new class of polymers known as inherently conductive or electroactive polymers is fast transforming this limited perspective. Similar to the plastic industry in the 1930s and 1950s, this class is in its infancy, but its potential applications are vast [31, 32].

Conducting polymers (CPs), also known as conductive polymers, organic polymeric conductors, or conjugated conductive polymers, are polymeric materials that are intrinsically conductive without conducting additions. As insulating materials, polymers have maintained their unique status due to their electrical insulating properties. Hideki Shirakawa, Alan Heeger, and Alan MacDiarmid created semiconducting polyacetylene [33] almost half a century ago. Since then, additional CP servicing diverse demands within the industrial community has been synthesized. The rigid backbone of conjugated conducting polymers consists of a highly conjugated system of conjugated segments coupled with atoms to provide lone pairs (such as S, N), which are responsible for an electrical charge through the constant overlap of π-orbitals along the polymer structure. The doping process or conjugation of π electrons in the polymer backbone determines the electrical properties of this kind. Due to their unique characteristics, including technological supremacy and cheaper production costs compared to other materials [34], they have garnered considerable attention. Figure 5.1 illustrates several types of conducting polymers [35].

Due to their valuable mechanical, optical, and electrical properties as well as their high economic worth and extraordinary environmental resilience, conducting polymers (CPs) have drawn a great deal of interest. Anticorrosive materials, electrostatic materials, conducting adhesives, electromagnetic interference shielding, artificial nerves, aircraft structures, diodes, transistors, energy storage, actuators, drug delivery, neural interfaces, and protein purification are just a few of the many applications for conducting polymers [36].

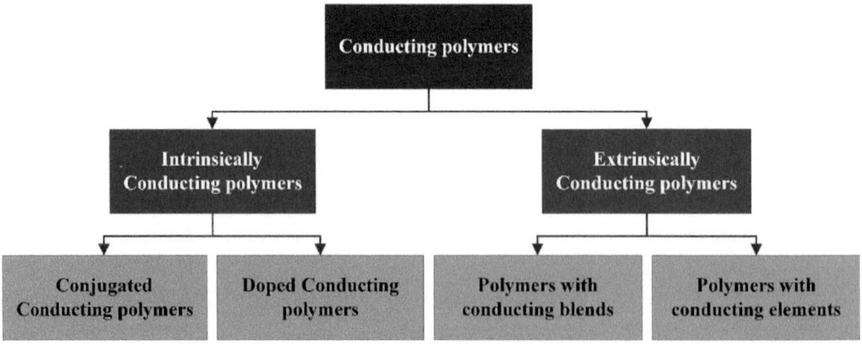

FIGURE 5.1 Conducting polymer classification.

A considerable amount of work has been conducted on polyaniline (PANI), polythiophenes (PTh), polypyrrole (PPy), polyacetylene (PA), polyorthotoluidine (POT), polyparaphenylene vinylene (PPV), and polyparaphenylene (PPPh) and its derivatives. Conductive polymers are characterized by their low cost, low weight, high conductivity, high corrosion resistance, and high photochromic and electrochromic capabilities and features. Both their structure and functional groups are adaptable, allowing them to be adapted to specific demands. Conducting polymers have low stability, restricted processability, and restricted solubility. To counteract these disadvantages, new blends, hybrid conducting polymers (HCPs), interpenetrating polymer networks (IPNs), and co-polymers are being developed [37].

All of the major industries that use corrosive materials acknowledge the relevance of corrosion, but the chemical sector in particular does so. Corrosion control methods are the most effective means of addressing this issue. Typically, a layer of conductive polymers is applied to the medium to provide corrosion resistance for the base material, which might comprise toxic and environmentally damaging compounds [7]. Because of these substantial investments, new protective coating materials and technologies [38–46] using carbon-based compounds such as conductive polymers have been created. According to Olad *et al.*, the ternary hybrid nanocomposite Polyaniline/ZnO/PVC was produced by in situ chemical polymerization. The ternary hybrid material exhibited superior anticorrosion properties compared to pure PANI coatings. The barrier and adhesion properties of coatings can be enhanced by adding ZnO and PVC [38]. As detailed by Ozkazanc *et al.*, metal ions ($NiCl_2$, $ZnCl_4$, $FeCl_2$, and $CuCl_2$) were deposited on a PPy-coated Cu electrode. PPy/Ni and PPy/Zn coatings demonstrated superior corrosion resistance compared to the competition [39]. A review study seeks to comprehend the properties, manufacture, and technological influence of epoxy/polyaniline nanocomposites. This article summarizes recent advances in the design and properties of epoxy, polyaniline, and epoxy/polyaniline nanocomposites. Epoxy/polyaniline nanocomposites have demonstrated efficacy as antibacterial agents, corrosion inhibitors, radiation shielding materials, and supercapacitor components [40]. Alam *et al.*, who synthesized eco-friendly anticorrosive coating by combining PANI with soya oil alkyd, discovered that PANI loading of only 2.5 wt% improved the corrosion resistance of MS. At the metal coating interface, a compact iron/dopant combination may have formed, rendering that region inert. Before the PANI can undergo a continuous charge transfer reaction from emeraldine salt form (ES) to emeraldine base (EB) at the metal-coating interface, this layer serves as a protective layer [41]. Rajasekharan *et al.* determined the anticorrosion characteristics of organic acid-doped PANI coatings (oxalic acid, citric acid, and tartaric acid), significantly enhancing the corrosion resistance of PANI layer with tartaric acid. This is because the tartrate ion is significantly larger and heavier than the oxalate and citrate ions. In addition, the tetrahedral structure of tartrate anions inhibits polymer chain release under severe circumstances [42]. In another publication, coating technologies such as conductive polymer coatings, conventional coatings, nanopolymer composite, and hydrophobic coatings are examined in detail. Focus is placed on the most popular and promising coatings including conducting polymers and nanopolymer composites. This article discusses corrosion protection and its affecting factors and the advantages and disadvantages of various coating processes. Finally, suggestions for future research and applications are provided. The coating is one of the most prevalent and effective anticorrosive strategies, and corrosion is the leading cause of damage to steel products. Different types of coatings use distinct substances, application techniques, and levels of success [43]. Corrosion of steel components is a significant problem in the construction industry. Consequently, numerous studies [44–46] have studied the viability of using conductive polymers as protective steel materials. Figure 5.2 depicts the diverse applications of conducting polymers [47], and Figure 5.3 depicts many forms of conducting polymer-based protective coatings [7].

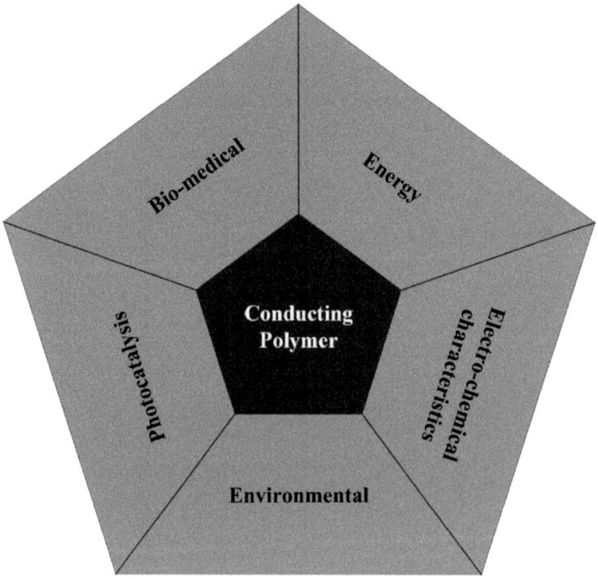

FIGURE 5.2 Applications of conducting polymers in different industries.

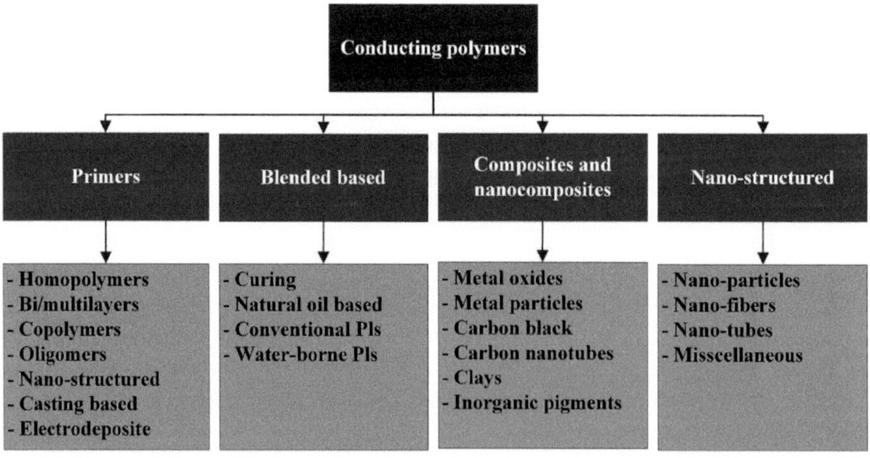

FIGURE 5.3 Several forms of conducting polymer-based protective coatings.

5.3.1 POLYANILINE AS ANTICORROSIVE MATERIALS

Due to its peculiar conduction mechanism and high oxygen and water resistance, the polyaniline family of conjugated polymers is exceptional among classes of conductive polymers. Polyaniline, a conductive polymer, has existed for quite some time. Letheby synthesized it for the first time in 1862 using anodic oxidation of aniline in sulfuric acid to produce four different octamers [8].

Due to their resistance to corrosion, conductive polymers are a new form of metal compounds that have been the focus of substantial research. Polyaniline (PANI) is the best material for this application due to its great environmental stability and simple manufacture. DeBerry (1985) and Ahmad and MacDiarmid (1987) [48, 49] showed a decade ago that CPs and polyaniline, in particular, may inhibit corrosion. The electrodeposition of PANI in a highly acidic environment increased the corrosion resistance of passivated steel, according to the research. The mechanism underlying the prevalent idea that PANI can protect carbon steel from corrosion is now under investigation. Several hypotheses have been proposed on the mechanism through which CPs and PANI in particular give corrosion protection. PANI contributes to the formation of an electric field on the surface of the metal, which restricts the rate at which electrons can flow from the metal to the oxidant. PANI facilitates metal oxide protective coatings [31, 50].

Numerous investigations [51–62] have demonstrated the anticorrosive qualities of polyaniline-based compounds. Polyaniline nanofibers could replace conventional polyaniline aggregates in mild steel corrosion prevention [35]. The electrical properties of ferroelectric copolymers are improved using conducting polyaniline nanofiber nanofillers. By applying in situ chemical oxidation polymerization, CeO_2-PANI conductive polymer capsules were made. After allowing the epoxy resin to dry, F51 epoxy-containing distributed capsules were sprayed over the steel. EIS investigation of corrosion resistance of coatings in 3.5 wt% NaCl solution. After 30 days, the coating with CeO_2-PANI capsules exhibited the highest corrosion resistance with an impedance modulus greater than $10^9 \Omega \cdot cm^2$ (10 times that of ordinary PANI-Epoxy coatings). Because of the synergistic impact of PANI and Ce^{3+} ions, a passive film (Fe_3O_4 + γ-Fe_2O_3 + CeO_2) was produced, which provided the boost. Despite PANI's capacity to polarize the anodic process, Ce^{3+} inhibited the cathodic reaction [51]. The efficacy of recently developed epoxy-based paint systems, including polyaniline as a pigment to prevent corrosion of mild steel reinforcement, has been explored. This article explains how polyaniline is manufactured and why it is used in epoxy-based coating systems. Electrochemical impedance spectroscopy, potential time studies, cathodic disbandment test, anodic polarization research, salt spray test, and chemical resistance test were used to investigate the corrosion resistance of epoxy polyaniline system coated on mild steel. The corrosion resistance of an epoxy polyaniline coating system applied to concrete-encased reinforcing bars was also evaluated by accelerated time to crack tests [52]. In another study, it was discovered that by copolymerizing 2-aminobenzenesulfonic acid with aniline, a processable and well-dispersed common organic solvent material was created. On a steel substrate, SPANI/epoxy composite coatings were applied before anticorrosive testing in 3.5 wt% NaCl solution using electrochemical impedance spectroscopy and polarization curves.

Unlike plain epoxy coatings, which lose their protective efficiency after only 80 days of immersion, composite coatings, including SPANI, demonstrated exceptional protective performance for the whole 120 days. To prove that a composite coating comprising SPANI successfully passivated a steel substrate, the content and morphology of rust under the coatings were examined. By creating a layer of Fe_2O_3 and Fe_3O_4, self-doped SPANI improved the corrosion resistance of composite coatings [53].

Polyaniline was chemically produced using N-methyl-2-pyrrolidone (NMP) solution and then cast onto iron samples in one investigation. Various electrochemical studies, including corrosion potential and corrosion current, were conducted on iron samples coated with polyaniline under various circumstances. The findings revealed that polyaniline could provide corrosion prevention even in highly severe environments. The protective qualities of the conducting polymer were compared to those of a more common polymer, such as polyvinyl chloride (PVC). Experiments have demonstrated that polyaniline derived from emeraldine is more shielding than other forms [54]. In one study, polyaniline (PANi), poly(2-toluidine) (PT), and poly(aniline-co-2-toluidine) (co-PT) were evaluated as corrosion inhibitors for 304-stainless steel (SS) in a 0.5 M HCl solution. Under cyclic voltammetry, coating SS substrates with PANi, PT, and co-PT necessitated an acetonitrile solution, a supporting electrolyte of tetrabutylammonium perchlorate (TBAP), and an acid of hydrochloric acid ($HClO_4$). These coatings were studied using spectroscopy techniques such as CV, FTIR, and UV-vis. The corrosion protective coating qualities of PANi, PT, and co-PT on SS were compared using open circuit potential-time (E_{ocp}-time), electrochemical impedance spectroscopy (EIS), and potentiodynamic polarization studies. This study demonstrates that PANi, PT, and co-PT are all effective corrosion-protective coatings for SS, with each coating offering over 80% protection after 48 h in a corrosive test solution [55]. Kraljic and colleagues [56] demonstrated that electrosynthesis of a polyaniline layer on low-alloy steel using H_2SO_4 and H_3PO_4 revealed its remarkable anticorrosion properties. El-Shazly *et al.* [57] investigate the viability of coating reinforced concrete with polyaniline (PANi) to increase its corrosion resistance in chloride and sulfate-ion-dominated corrosive environments. To generate a polyaniline layer, the galvanostatic approach was used. Using the potentiodynamic method, the corrosion resistance of PANi-coated steel in conjunction with stainless steel was investigated while it supported a layer of concrete containing different concentrations of corrosive components such as NaCl and H_2SO_4. Using a silver/silver chloride reference electrode, the corrosion current was measured. The results revealed that Cl^- corrodes steel more effectively than SO_4^{-2} [57].

When addressing corrosion resistance, conducting polymers, particularly polyaniline (PANI) and its derivatives, deserve consideration. Organic lightweight batteries, chemical sensors, and field-effect transistors could all benefit from PANI's outstanding environmental stability, straightforward manufacturing, and reversible acid/base doping/dedoping performance. PANI and its derivatives are suitable corrosion inhibitors due to the presence of π-electrons in the benzene rings and nitrogen on the conjugated backbone. According to professionals [53], metals like mild steel, stainless steel, aluminum, aluminum alloys, and copper are less likely to corrode when treated with polyaniline.

5.3.2 Polypyrrole as anticorrosive materials

The class of polymers known as polypyrrole (PPy) has been the focus of multiple synthetic PPy characterization studies [63]. Polypyrrole (PPy) possesses intrinsic anti-corrosion properties, making it one of the most promising polymers with corrosion resistance. Numerous protection techniques for Ppy-coated metals and alloys are available. These include ennobling or anodic passivation, cathodic protection, oxygen reduction mediation, barrier protection, and controlled inhibitor-release mechanisms [64]. PPy has garnered more attention than other CPs due to its better features (such as stability in the oxidized form, amazing redox capabilities, good optical and electrical properties, and the ability to provide high electrical conductivity) [2]. Consequently, PPy is commonly regarded as one of the best anticorrosive coating choices among conductive polymers [65]. Numerous studies [65–72] have explored the anticorrosion capabilities of polypyrrole due to its ease of polymerization and mechanical stability.

This study compares and contrasts two possible mechanisms of employing conductive polymer coatings to prevent corrosion. The ennobling procedure depends on the polymer's high potential, which is believed to maintain the metal's passive state and obliterate oxygen reduction sites. Using a model sample arrangement, electrochemically generated polypyrrole coatings may prevent the corrosion of small pinhole-like defects. When submerged in a solution devoid of chloride in chloride-free solutions, the defect loses its inactive state within seconds, while polypyrrole can keep it inactive for hours. The extent of the problem has a significant impact on the safety that can be guaranteed. In cyclic environmental conditions, polypyrrole showed efficiency as a protective agent due to its self-recharging properties upon dissociation from the flaw. However, when a larger defect is present, the positive impact of the conducting polymer is lost, resulting in rapid delamination. Experiments conducted in $^{18}O_2$ atmosphere helped pinpoint the location of the oxygen reduction process. ToF-SIMS sputter profiles reveal that $^{18}O^-$ and $^{18}OH^-$ ions are produced somewhat above the polypyrrole layer during delamination. This prevents the obscuration of oxygen reduction sites by strongly conducting polymers such as pure polypyrrole [73]. In the galvanostatic electropolymerizing technique, conductive polypyrrole electrodes were produced by electropolymerizing aqueous polypyrrole solutions on mild steel electrodes. The electrochemical response of coated electrodes in both doped and undoped states was examined in comparison to mild steel electrodes. The corrosion resistance of the electrode coated with undoped polypyrrole improved dramatically [2]. A polypyrrole (PPy) film was deposited onto the surface of the Q235 steel using potentiostatic deposition with changing constant potentials. PPy-Q235 series samples have stronger corrosion resistance than Q235 series samples, and PPy-Q235 series samples without FeC_2O_4 in the film have greater corrosion resistance than PPy-Q235 series samples with FeC_2O_4 in. Regarding corrosion resistance, PPy-Q235-0.95 is superior to [74]. Munot *et al.* reported developing and painting a sample of low-carbon steel with conducting polypyrrole in a separate study. Using the linear polarization method, corrosion rates of both uncoated and painted low-carbon steel samples were found. Electrochemical tests were conducted on steel samples that were either uncoated or coated with conducting polypyrrole before being submerged in a solution designed to mimic the pores of carbonated concrete. As demonstrated by a positive shift in corrosion potential, the polypyrrole covering shields low-carbon steel from anodic corrosion. However, as the concentration of chloride ions in the carbonated pore solution increased, the corrosion resistance of the low-carbon steel samples coated with polypyrrole decreased [75]. Using the computational technique, the mechanical properties of polypyrrole (PPy)/polyaminobenzene sulfonic acid functionalized single-walled carbon nanotubes (CNT-PABS) and polypyrrole (PPy)/carboxylic acid functionalized single-walled carbon nanotubes (CNT-CA) composites are anticipated. The results unequivocally proved that CNT-PABS and CNT-CA were distributed sufficiently in composite coatings and contributed positively to the coatings' mechanical integrity. In addition, composite coatings offer higher anticorrosion performance compared to PPy [76]. This article examines how strengthening PPy network with additives can better protect coated steel substrates from corrosion using a single study evaluation. Either the PPy composites are electrodeposited directly onto the surface or combined with epoxy coatings. This article discusses the electrochemical processes and chemical procedures required to produce PPy composites. Experiments on corrosion are used to compare and study the corrosion behavior of steel substrates coated with pure PPy and various PPy composites [77]. Since then, research into the use of PPy as a conductive polymer in formulations for protective coatings with enhanced anticorrosion properties has blossomed. The corrosion resistance of both pure and composite PPy was enhanced. In addition to being environmentally safe, PPy has applications in the nautical, defense, construction, and building industries [64].

5.3.3 Polythiophene as anticorrosive materials

Due to its excellent environmental stability, higher thermal stability, and lower band gap energy, polythiophene polymer has garnered considerable interest from academics and industry. Due to their exceptional mechanical properties, ease of processability, electrical, semiconducting, and optical activities, polythiophene composites have garnered considerable interest [78]. Polythiophene is a promising material due to its high electrical conductivity, low cost, optical properties, and environmental resistance [79]. The unique electrochemical and physical features of polythiophene and its derivatives have been investigated in studies [80–87]. Enhanced polythiophene performance can improve technology, such as electronic devices, supercapacitors, lithium batteries, sensors, and corrosion prevention [80].

Using a one-of-a-kind anticorrosion pigment, an intelligent epoxy coating with bi-active/passive protective characteristics was produced as part of a research effort. To accomplish this, a graphene oxide (GO) structure with dual phytic acid and polythiophene functionality was meticulously developed. Thiophene monomer sheets polymerized well when treated with rGO-phytic acid (rGO-Phy). The polythiophene nanoparticles had appealing decorations. The epoxy covering was strengthened against corrosion by using nanocomposites. Incorporating rGO-Phytic acid-Polythiophene (rGO-Phy-Pth) nanocomposites into the epoxy composite coating confers self-healing performance in addition to the exceptional barrier property, resulting in approximately nine weeks of stable protection as demonstrated by the corrosion retardation performance of the applied coatings on the mild steel panels. The acquired exceptional self-healing potential is attributed to the formation of a film with increasing resistance up to

50 kΩ cm^2 at the damaged area of the GO sheets, whereas the blank sample exhibited a decreasing trend [81]. Another article demonstrated for the first time that electrodeposited superhydrophobic conducting polythiophene covering can effectively prevent corrosion on the steel substrate. This is accomplished by inhibiting water absorption onto the coating, hence preventing corrosive chemicals and corrosion products from diffusing through the coating, and by producing an anodic change in the corrosion potential as it galvanically connects to the metal substrate. Any metallic surface can be coated with superhydrophobic anticorrosion nanostructured conducting polymer in two simple stages [82]. Polythiophene (PTh) has a long history of use in conjugated polymer materials due to its resistance to high temperatures and harsh conditions. The remarkable optical, electronic, and magnetic properties of polythiophenes, as well as their potential applications in various fields, such as chemical and optical sensors, light-emitting diodes, transistors, photovoltaic devices, displays, solar cells, polymer electronic interconnects, molecular devices, corrosion protection, and DNA detection, have attracted considerable interest and contributed to their status as one of the better-characterized conducting polymers. PTh can be produced using chemical and electrochemical techniques, among others. This work employed electrochemical techniques to apply polythiophene layers via electro-polymerization to stainless steel, carbon steel, and aluminum in order to examine their corrosion-resistance capabilities. The electrochemical approach is gaining popularity for numerous reasons, including its low environmental impact [83]. As indicated in Table 5.1, polyaniline, polypyrrole, and polythiophene all have anticorrosive characteristics [88–107].

TABLE 5.1
Anticorrosive Polyaniline, Polypyrrole, and Polythiophene Compounds [88–107]

Materials	Conductive Polymer	Result	Ref.
Polypyrrole-SiO_2 composites	Polypyrrole	Using multilayered composite polypyrrole/SiO_2 coatings enhanced the corrosion resistance of stainless steel substrate.	[88]
The electrochemical deposition of polyaniline (PANI) on 316-L stainless steel (316LSS)	Polyaniline	Effective protection against the ingress of corrosive substances.	[89]
The PPy coatings by loading ZnO nanoparticles	Polypyrrole	The Al alloy was adequately protected against corrosion by the optimal coating.	[90]
Epoxy/DBSA (dodecylebenzene sulfonic acid) doped polyaniline-TiO_2 nanocomposite	Polyaniline	The anticorrosive performance of EP/DBSA-doped PANI-TiO_2 is superior to that of EP/DBSA-doped PANI.	[91]
Conducting composite polypyrrole/polyaniline coatings on type 304 stainless steel	Polypyrrole and polyaniline	The bilayer coatings could decrease alloy corrosion far more efficiently than Ppy coatings alone.	[92]
Polypyrrole/$PW_{12}O_{40}^{3-}$ coatings on carbon steel electrodes	Polypyrrole	The developed coatings offer excellent corrosion resistance in chloride-containing environments.	[93]

(*Continued*)

TABLE 5.1 *(Continued)*
Anticorrosive Polyaniline, Polypyrrole, and Polythiophene Compounds [88–107]

Materials	Conductive Polymer	Result	Ref.
Polyaniline/polyvinyl chloride blended coatings	Polyaniline	Superior corrosion resistance performance.	[94]
Bilayer coatings of polyaniline and poly(N-methylaniline) on mild steel	Polyaniline	Greater corrosion resistance is displayed.	[95]
Polythiophene-coated steel	Polythiophene	The determination of volume fraction, water uptake, and delamination area was used to bolster the corrosion resistance of conducting polythiophene coating.	[96]
Polypyrrole doped with Nickel Cobalt Oxide	Polypyrrole	Increase the resistance to corrosion of ferritic stainless steel.	[97]
Emeraldine base (EB) and salt (ES) (hydrofluoric acid and camphorsulfonic acid doped) forms of polyaniline (PANI) coatings on mild steel	Polyaniline	The ES (particularly camphorsulfonic acid-doped) forms of PANI coating displayed superior corrosion resistance, although all PANI coating forms could provide corrosion protection for mild steel.	[98]
Polythiophene (PTh) films	Polythiophene	After long immersion times (779 h), 0.3 mol thiophene in BFEE (Boron trifluoride diethyl etherate) potentiostatically synthesized at 1.3 V for 1000 s provided the best protection efficiency.	[99]
Polypyrrole coatings on mild steel	Polypyrrole	The best performance against mild steel corrosion during extended immersion time.	[100]
Polyaniline and polyaniline-thiokol rubber composites	Polyaniline	As a protective coating for mild steel, PANI/TR composite conductive polymer can significantly reduce corrosion.	[101]
Epoxy-silicone-polythiophene interpenetrating polymer network	Polythiophene	The corrosion resistance of IPN (interpenetrating polymer network) coatings with lower silicone concentration is exceptional.	[102]
Perfluorinated lubricant/ polypyrrole composite	Polypyrrole	The perfluorinated lubricant/polypyrrole composite material exhibited excellent corrosion inhibition properties.	[103]
Polyaniline/graphene oxide/ Zn-doped TiO_2 nanocomposite	Polyaniline	These coatings have exceptional corrosion-resistant qualities.	[104]
Polythiophene on nickel-coated mild steel	Polythiophene	Excellent corrosion protection against mild steel was obtained.	[105]
Polypyrrole-titanate nanotube composites	Polypyrrole	Stainless steel corrosion rates decreased in the presence of composite films.	[106]
Graphene/Polyaniline Hybrid Coating	Polyaniline	To prevent the corrosion of copper metal.	[107]

5.4 CONCLUSION AND OUTLOOK

Modern technologies necessitate inventive new materials. Metal corrosion control is a crucial activity with environmental, economic, aesthetic, and technological significance [108]. Numerous studies concentrate on the use of various materials to control metal corrosion in various areas, including medical, construction, and oil sectors [109–117]. In terms of corrosion protection, the employment of various materials, including conductive polymers, is particularly successful. In this chapter, we describe the incorporation of conductive polymers (Polyaniline, Polypyrrole, and Polythiophene) into various metals using various procedures and approaches. Several commercially used conductive polymers (Polyaniline, Polypyrrole, Polythiophene) and their evaluation and risks are discussed in order to find novel compounds to put in various metals in order to control corrosion. To replace environmentally dangerous substances, scientists are exploring natural alternatives. Depending on the necessity of controlling the corrosion of metals in various industries, natural anticorrosive materials, especially those derived from natural and cost-effective, nontoxic, and renewable anticorrosive agents, are anticipated to gain widespread favor in the near future. In the next years, the composite of conductive polymers with anticorrosive natural materials will be the most in-demand anticorrosive agents, and their synthesis and application methods to control the corrosion of various metals in various industries will be the most researched topics.

REFERENCES

1. Deshpande PP, Jadhav NG, Gelling VJ, Sazou D. Conducting polymers for corrosion protection: A review. *Journal of Coatings Technology and Research*. 2014;11:473–494.
2. Bazli L, Yusuf M, Farahani A, Kiamarzi M, Seyedhosseini Z, Nezhadmansari M, Aliasghari M, Iranpoor M. Application of composite conducting polymers for improving the corrosion behavior of various substrates: A review. *Journal of Composites and Compounds*. 2020;2(5):228–240.
3. Short NR, Dennis JK. Corrosion resistance of zinc-alloy coated steel in construction industry environments. *Transactions of the IMF*. 1997;75(2):47–52.
4. Brondel D, Edwards R, Hayman A, Hill D, Mehta S, Semerad T. Corrosion in the oil industry. *Oilfield Review; (Netherlands)*. 1994;6(2).
5. Corbett RA. Laboratory corrosion testing of medical implants. In *Proceedings of Materials and Processes for Medical Devices Conference 2004* (pp. 166–171). ASM International.
6. Malik MA, Hashim MA, Nabi F, Al-Thabaiti SA, Khan Z. Anti-corrosive ability of surfactants: A review. *International Journal of Electrochemical Science*. 2011;6(6):1927–1948.
7. Kaliyannan GV, Velusamy MK, Palaniappan SK, Anandraj MK, Rathanasamy R. Polymer coatings for corrosive protection. *Polymer Coatings: Technology and Applications*. 2020:371–395.
8. Kumar D, Sharma RC. Advances in conductive polymers. *European Polymer Journal*. 1998;34(8):1053–1060.
9. Abd El-Ghaffar MA, Abdelwahab NA, Fekry AM, Sanad MA, Sabaa MW, Soliman SM. Polyester-epoxy resin/conducting polymer/barium sulfate hybrid composite as a smart eco-friendly anti-corrosive powder coating. *Progress in Organic Coatings*. 2020;144:105664.

10. Al-Gamal AG, Farag AA, Elnaggar EM, Kabel KI. Comparative impact of doping nano-conducting polymer with carbon and carbon oxide composites in alkyd binder as anti-corrosive coatings. *Composite Interfaces*. 2018;25(11):959–980.
11. Nautiyal A, Ren MQ, Huang TS, Zhang X, Cook J, Bozack MJ, Farag R. High-performance engineered conducting polymer film towards antimicrobial/anticorrosion applications. *Engineered Science*. 2018;4(24):70–78.
12. Chen Q, Lei Y. Applications of conducting polymers for anti-corrosive in marine environments. In *Corrosion and Protection of Marine Engineering Materials: Applications of Conducting Polymers and Their Composites* (pp. 105). CRC Press, 2023.
13. Abd El-Ghaffar MA, Abdel-Wahab NA, Sanad MA, Sabaa MW. High performance anti-corrosive powder coatings based on phosphate pigments containing poly (o-aminophenol). *Progress in Organic Coatings*. 2015;78:42–48.
14. Jlassi K, Radwan AB, Sadasivuni KK, Mrlik M, Abdullah AM, Chehimi MM, Krupa I. Anti-corrosive and oil sensitive coatings based on epoxy/polyaniline/magnetite-clay composites through diazonium interfacial chemistry. *Scientific Reports*. 2018;8(1):13369.
15. Naresh U, Kumar NS, Basha DB, Naidu KC, Sarma MS, Kumar RJ, Pothu R, Boddula R. Conducting polymer coatings for corrosion resistance in electronic materials. *Polymer Coatings: Technology and Applications*. 2020:175–188.
16. Cui G, Bi Z, Zhang R, Liu J, Yu X, Li Z. A comprehensive review on graphene-based anti-corrosive coatings. *Chemical Engineering Journal*. 2019;373:104–121.
17. Page CL. Corrosion and protection of reinforcing steel in concrete. *Durability of Concrete and Cement Composites*. 2007:136–186.
18. Rohwerder M. Conducting polymers for corrosion protection: A review. *International Journal of Materials Research*. 2009;100(10):1331–1342.
19. Sitaram SP, Stoffer JO, O'Keefe TJ. Application of conducting polymers in corrosion protection. *Journal of Coatings Technology*. 1997;69:65–69.
20. MacDiarmid AG. *Short Course on Conductive Polymers*. SUNY, 1985.
21. Zarras P, Anderson N, Webber C, Irvin DJ, Irvin JA, Guenthner A, Stenger-Smith JD. Progress in using conductive polymers as corrosion-inhibiting coatings. *Radiation Physics and Chemistry*. 2003;68(3–4):387–394.
22. Umoren SA, Solomon MM. Protective polymeric films for industrial substrates: A critical review on past and recent applications with conducting polymers and polymer composites/nanocomposites. *Progress in Materials Science*. 2019;104:380–450.
23. Shukla SK, Quraishi MA, Prakash R. A self-doped conducting polymer "polyanthranilic acid": An efficient corrosion inhibitor for mild steel in acidic solution. *Corrosion Science*. 2008;50(10):2867–2872.
24. Rohwerder M, Michalik A. Conducting polymers for corrosion protection: What makes the difference between failure and success? *Electrochimica Acta*. 2007;53(3):1300–1313.
25. Spinks GM, Dominis AJ, Wallace GG, Tallman DE. Electroactive conducting polymers for corrosion control: Part 2. Ferrous metals. *Journal of Solid State Electrochemistry*. 2002;6:85–100.
26. Ahmad N, MacDiarmid AG. Inhibition of corrosion of steels with the exploitation of conducting polymers. *Synthetic Metals*. 1996;78(2):103–110.
27. Tallman DE, Spinks G, Dominis A, Wallace GG. Electroactive conducting polymers for corrosion control: Part 1. General introduction and a review of non-ferrous metals. *Journal of Solid State Electrochemistry*. 2002;6:73–84.
28. Patil RC, Radhakrishnan S. Conducting polymer based hybrid nano-composites for enhanced corrosion protective coatings. *Progress in Organic Coatings*. 2006;57(4):332–336.

29. Rahman FA, Basirun WJ, Johan MR, Ghazali N. Corrosion inhibition of particulate and tubular polyaniline in polyvinyl butyral for mild steel protection. *Materials Letters*. 2023:133898.
30. Kumar A. Role of conducting polymers in corrosion protection. *World Journal of Advanced Research and Reviews*. 2023;17(2):045–047.
31. Al-Dulaimi AA. *Evaluation of Polyaniline Composites and Nanostructures as Anticorrosive Pigments for Carbon Steel* (Doctoral dissertation, Universiti Teknologi Malaysia), 2010.
32. Umoren SA, Solomon MM, Saji VS. Chapter 19—Conducting polymers. In *Polymeric Materials in Corrosion Inhibition* (Umoren SA, Solomon MM, Saji VS, Eds., pp. 443–466). Elsevier, 2022.
33. Le T-H, Kim Y, Yoon H. Electrical and electrochemical properties of conducting polymers. *Polymers (Basel)*. 2017;9. https://doi.org/10.3390/polym9040150.
34. AL-Refai HH, Ganash AA, Hussein MA. Polythiophene and its derivatives-Based nanocomposites in electrochemical sensing: A mini review. *Materials Today Communications*. 2021;26:101935.
35. Rawat NK, Ahmad S. Unveiling nanoconducting polymers and composites for corrosion protection. In *Nanomaterials-Based Coatings* (pp. 373–395). Elsevier, 2019.
36. Das TK, Prusty S. Review on conducting polymers and their applications. *Polymer-Plastics Technology and Engineering*. 2012;51(14):1487–1500.
37. Iqbal S, Ahmad S. Recent development in hybrid conducting polymers: Synthesis, applications and future prospects. *Journal of Industrial and Engineering Chemistry*. 2018;60:53–84.
38. Olad A, Nosrati R. Preparation and corrosion resistance of nanostructured PVC/ZnO-polyaniline hybrid coating. *Progress in Organic Coatings*. 2013;76(1):113–118.
39. Ozkazanc H, Zor S. Electrochemical synthesis of polypyrrole (PPY) and PPY|metal composites on copper electrode and investigation of their anticorrosive properties. *Progress in Organic Coatings*. 2013;76(4):720–728.
40. Kausar A. Versatile epoxy/polyaniline and derived nanocomposite: From strategic design to advance application. *Materials Research Innovations*. 2021;25(6):321–330.
41. Alam J, Riaz U, Ahmad S. High performance corrosion resistant polyaniline/alkyd ecofriendly coatings. *Current Applied Physics*. 2009;9(1):80–86.
42. Rajasekharan V, Stalin T, Viswanathan S, Manisankar P. Electrochemical evaluation of anticorrosive performance of organic acid doped polyaniline based coatings. *International Journal of Electrochemical Science*. 2013;8:11327–11336.
43. Xu H, Zhang Y. A review on conducting polymers and nanopolymer composite coatings for steel corrosion protection. *Coatings*. 2019;9(12):807.
44. Fontana JJ, Reams W, Scannell W, Sprinkel M. Electrically conductive polymer concrete overlay as secondary anode for cathodic protection, 1990. https://trid.trb.org/View/1181524.
45. Fontana JJ, Elling D, Reams W. *Electrically Conductive Polymer Concrete Coatings*. Brookhaven National Lab. (BNL), Upton, NY (United States), 1988.
46. Cho LY, Guiotti LG, Liu AS. Corrosion performance of polypyrrole-bilayer coating on carbon steel. *Journal of Materials Science*. 2023:1–6.
47. Raza S, Li X, Soyekwo F, Liao D, Xiang Y, Liu C. A comprehensive overview of common conducting polymer-based nanocomposites; Recent advances in design and applications. *European Polymer Journal*. 2021;160:110773.
48. DeBerry DW. Modification of the electrochemical and corrosion behavior of stainless steels with an electroactive coating. *Journal of the Electrochemical Society*. 1985;132(5):1022.

49. Ahmad N, MacDiarmid AG. *Inhibition of Corrosion of Steels Using Conducting Polymers*, 1987. https://www.sid.ir/paper/629793/en.
50. McAndrew TP. Corrosion prevention with electrically conductive polymers. *Trends in Polymer Science*. 1997;1(5):7–12.
51. Fan W, Wang H, Wang C, Liu Z, Zhu Y, Li K. Epoxy coating capable of providing multi-component passive film for long-term anti-corrosive of steel. *Applied Surface Science*. 2020;521:146417.
52. Saravanan K, Sathiyanarayanan S, Muralidharan S, Azim SS, Venkatachari G. Performance evaluation of polyaniline pigmented epoxy coating for corrosion protection of steel in concrete environment. *Progress in Organic Coatings*. 2007;59(2):160–167.
53. Qiu S, Chen C, Zheng W, Li W, Zhao H, Wang L. Long-term corrosion protection of mild steel by epoxy coating containing self-doped polyaniline nanofiber. *Synthetic Metals*. 2017;229:39–46.
54. Mirmohseni A, Oladegaragoze A. Anti-corrosive properties of polyaniline coating on iron. *Synthetic Metals*. 2000;114(2):105–108.
55. Hür E, Bereket G, Şahin Y. Anti-corrosive properties of polyaniline, poly (2-toluidine), and poly (aniline-co-2-toluidine) coatings on stainless steel. *Current Applied Physics*. 2007;7(6):597–604.
56. Kraljić M, Mandić Z, Duić L. Inhibition of steel corrosion by polyaniline coatings. *Corrosion Science*. 2003;45(1):181–198.
57. El-Shazly AH, Wazzan AA, Radain TA. Investigation for the possibility of improving the reinforced concrete corrosion resistance using polyaniline coated steel. *International Journal of Electrochemical Science*. 2012;7:2416–2429.
58. Kalendová A, Veselý D, Stejskal J. Organic coatings containing polyaniline and inorganic pigments as corrosion inhibitors. *Progress in Organic Coatings*. 2008;62(1):105–116.
59. Zhao Y, Tian S, Lin D, Zhang Z, Li G. Functional anti-corrosive and anti-bacterial surface coatings based on cuprous oxide/polyaniline microcomposites. *Materials & Design*. 2022;216:110589.
60. Pour-Ali S, Dehghanian C, Kosari A. Corrosion protection of the reinforcing steels in chloride-laden concrete environment through epoxy/polyaniline-camphorsulfonate nanocomposite coating. *Corrosion Science*. 2015;90:239–247.
61. Mrada M, Dhouibia L. Corrosion prevention of mild steel in a simulated concrete medium by the electro-application of a polyaniline coating. *Journal of the Tunisian Chemical Society*. 2017;19:359–367.
62. Zhou S, Liang X, Gao J, Ye B, Li C, Qian J, Shu J, Ye Y. Applications of polyaniline-alkaline copper carbonate composites for anti-bio-attachment of concrete in marine environments. *Journal of Marine Science and Engineering*. 2023;11(2):419.
63. Kostić R, Raković D, Stepanyan SA, Davidova IE, Gribov LA. Vibrational spectroscopy of polypyrrole, theoretical study. *The Journal of Chemical Physics*. 1995;102(8):3104–3109.
64. Zadeh MK, Yeganeh M, Shoushtari MT, Esmaeilkhanian A. Corrosion performance of polypyrrole-coated metals: A review of perspectives and recent advances. *Synthetic Metals*. 2021;274:116723.
65. Pandey U, Singh AK, Sharma C. Development of anti-corrosive novel nickel-graphene oxide-polypyrrole composite coatings on mild steel employing electrodeposition technique. *Synthetic Metals*. 2022;290:117135.
66. Ferreira CA, Aeiyach S, Coulaud A, Lacaze PC. Appraisal of the polypyrrole/cataphoretic paint bilayer system as a protective coating for metals. *Journal of Applied Electrochemistry*. 1999;29(2):259–263.

67. Jlassi K, Sliem MH, Benslimane FM, Eltai NO, Abdullah AM. Design of hybrid clay/polypyrrole decorated with silver and zinc oxide nanoparticles for anticorrosive and antibacterial applications. *Progress in Organic Coatings*. 2020;149:105918.
68. Jiang L, Syed JA, Lu H, Meng X. In-situ electrodeposition of conductive polypyrrole-graphene oxide composite coating for corrosion protection of 304SS bipolar plates. *Journal of Alloys and Compounds*. 2019;770:35–47.
69. Munota H, Deshpandea P, Modherab C. Electrochemical investigations on semi-conducting properties of a passive film on conducting polypyrrole coated low carbon steel in a simulated concrete pore solution. *Portugaliae Electrochimica Acta*. 2019;37(1):43–50.
70. Munot H, Deshpande P, Modhera C. Effects of chloride ions on the corrosion resistance of conducting polypyrrole coated steel in simulated concrete pore solution. *UPB Scientific Bulletin*. 2017;79.
71. Martins JI, Reis TC, Bazzaoui M, Bazzaoui EA, Martins L. Polypyrrole coatings as a treatment for zinc-coated steel surfaces against corrosion. *Corrosion Science*. 2004;46(10):2361–2381.
72. Ladan M, Basirun WJ, Kazi SN, Rahman FA. Corrosion protection of AISI 1018 steel using Co-doped TiO2/polypyrrole nanocomposites in 3.5% NaCl solution. *Materials Chemistry and Physics*. 2017;192:361–373.
73. Michalik A, Rohwerder M. Conducting polymers for corrosion protection: A critical view. (1547–1559). De-Gruyter, 2005.
74. Zhang L, Liu S, Han H, Zhou Y, Hu S, He C, Yan Q. Studies on the formation process and anti-corrosive performance of polypyrrole film deposited on the surface of Q235 steel by an electrochemical method. *Surface and Coatings Technology*. 2018;341:95–102.
75. Munot H, Deshpande P, Modhera C. Conducting polypyrrole coated rebar in a carbonated concrete pore solution: Electrochemical investigations. *Portugaliae Electrochimica Acta*. 2018;36(5):365–375.
76. Ioniţă M, Prună A. Polypyrrole/carbon nanotube composites: Molecular modeling and experimental investigation as anti-corrosive coating. *Progress in Organic Coatings*. 2011;72(4):647–652.
77. Ashassi-Sorkhabi H, Kazempour A. Incorporation of organic/inorganic materials into polypyrrole matrix to reinforce its anticorrosive properties for the protection of steel alloys: A review. *Journal of Molecular Liquids*. 2020;309:113085.
78. Ramesan MT, Suhailath K. Role of nanoparticles on polymer composites. In *Micro and Nano Fibrillar Composites (MFCs and NFCs) from Polymer Blends* (pp. 301–326). Woodhead Publishing, 2017.
79. Kausar A. Chapter 6—Emerging hybrids derived from polythiophene and graphene, In *Conducting Polymer-Based Nanocomposites* (Kausar A, Ed., pp. 129–156). Elsevier, 2021.
80. Ai L, Liu Y, Zhang XY, Ouyang XH, Ge ZY. A facile and template-free method for preparation of polythiophene microspheres and their dispersion for waterborne corrosion protection coatings. *Synthetic Metals*. 2014;191:41–46.
81. Tehrani ME, Ramezanzadeh M, Bahlakeh G, Ramezanzadeh B. S, P-codoped rGO-phytic acid-polythiophene core-shell; synthesis, modeling, and dual active-passive anti-corrosive performance of epoxy nanocomposite. *Journal of Industrial and Engineering Chemistry*. 2021;103:102–117.
82. de Leon AC, Pernites RB, Advincula RC. Superhydrophobic colloidally textured polythiophene film as superior anticorrosion coating. *ACS Applied Materials & Interfaces*. 2012;4(6):3169–3176.

83. Adbulaziz SA, Alsammarraie AA. Anticorrosion behavior of deposited nanostructured polythiophene on stainless steel, carbon steel, and aluminum in sea water. *International Journal of Engineering Research & Science*. 2016;2(2).
84. Bouabdallaoui M, Aouzal Z, El Guerraf A, Jadi SB, Bazzaoui M, Wang R, Bazzaoui EA. Influence of polythiophene overoxidation on its physicochemical properties and corrosion protection performances. *Materials Today: Proceedings*. 2020;31:S69–S74.
85. Tüken T, Yazici B, Erbil M. The use of polythiophene for mild steel protection. *Progress in Organic Coatings*. 2004;51(3):205–212.
86. Lu Y, Wang S, Xiong C, Hu GH. Synthesis and characterization of a liquid-like polythiophene and its potential applications. *Synthetic Metals*. 2020;270:116603.
87. Xavier JR. Corrosion protection performance and interfacial interactions of polythiophene/silanes/MnO_2 nanocomposite coatings on magnesium alloy in marine environment. *International Journal of Polymer Analysis and Characterization*. 2021;26(4):309–329.
88. Grari O, Taouil AE, Dhouibi L, Buron CC, Lallemand F. Multilayered polypyrrole-SiO2 composite coatings for functionalization of stainless steel: Characterization and corrosion protection behavior. *Progress in Organic Coatings*. 2015;88:48–53.
89. Li P, Ding X, Yang Z, Chen M, Wang M, Wang X. Electrochemical synthesis and characterization of polyaniline-coated PEMFC metal bipolar plates with improved corrosion resistance. *Ionics*. 2018;24:1129–1137.
90. Fekri F, Shahidi Zandi M, Foroughi MM. Polypyrrole coatings for corrosion protection of Al alloy2024: Influence of electrodeposition methods, solvents, and ZnO nanoparticle concentrations. *Iranian Polymer Journal*. 2019;28:577–585.
91. Hosseini MG, Sefidi PY. Electrochemical impedance spectroscopy evaluation on the protective properties of epoxy/DBSAdoped polyaniline-TiO2 nanocomposite coated mild steel under cathodic polarization. *Surface and Coatings Technology*. 2017;331:66–76.
92. Ren YJ, Zeng CL. Effect of conducting composite polypyrrole/polyaniline coatings on the corrosion resistance of type 304 stainless steel for bipolar plates of proton-exchange membrane fuel cells. *Journal of Power Sources*. 2008;182(2):524–530.
93. Bonastre J, Garcés P, Huerta F, Quijada C, Andión LG, Cases F. Electrochemical study of polypyrrole/PW12O403−coatings on carbon steel electrodes as protection against corrosion in chloride aqueous solutions. *Corrosion Science*. 2006;48(5):1122–1136.
94. Bandeira RM, van Drunen J, de Souza Ferreira FA, Rodrigues-Filho UP, Tremiliosi-Filho G. Chemically synthesized polyaniline/polyvinyl chloride blended coatings for the corrosion protection of AA7075 aluminum alloy. *Corrosion Science*. 2018;139:35–46.
95. Narayanasamy B, Rajendran S. Electropolymerized bilayer coatings of polyaniline and poly (N-methylaniline) on mild steel and their corrosion protection performance. *Progress in Organic Coatings*. 2010;67(3):246–254.
96. Kousik G, Pitchumani S, Renganathan NG. Electrochemical characterization of polythiophene-coated steel. *Progress in Organic Coatings*. 2001;43(4):286–291.
97. Banu A, Marcu M. Anticorrosion behaviour of polypyrrole doped with nickel cobalt oxide used as a primer on ferritic stainless steel in different environments. *Revista de Chimie Bucharest*. 2014;65:210–214.
98. Zhang Y, Shao Y, Liu X, Shi C, Wang Y, Meng G, Zeng X, Yang Y. A study on corrosion protection of different polyaniline coatings for mild steel. *Progress in Organic Coatings*. 2017;111:240–247.
99. Dai Y, Zhu F, Zhang H, Ma H, Wang W, Lei J. Electrosynthesis and characterization of polythiophene and corrosion protection for stainless steel. *International Journal of Electrochemical Science*. 2016;11:4084–4091.

100. Arabzadeh H, Shahidi M, Foroughi MM. Electrodeposited polypyrrole coatings on mild steel: Modeling the EIS data with a new equivalent circuit and the influence of scan rate and cycle number on the corrosion protection. *Journal of Electroanalytical Chemistry*. 2017;807:162–173.
101. Ding K, Jia Z, Ma W, Tong R, Wang X. Polyaniline and polyaniline-thiokol rubber composite coatings for the corrosion protection of mild steel. *Materials Chemistry and Physics*. 2002;76(2):137–142.
102. Palraj S, Selvaraj M, Vidhya M, Rajagopal G. Synthesis and characterization of epoxy-silicone-polythiophene interpenetrating polymer network for corrosion protection of steel. *Progress in Organic Coatings*. 2012;75(4):356–363.
103. Zhu H, Hou J, Qiu R, Zhao J, Xu J. Perfluorinated lubricant/polypyrrole composite material: Preparation and corrosion inhibition application. *Journal of Applied Polymer Science*. 2014;131(9).
104. Ladan M, Basirun WJ, Kazi SN. Polyaniline/graphene oxide/Zn-doped TiO2 nanocomposite coatings for the corrosion protection of carbon steel. *Journal of Adhesion Science and Technology*. 2021;35(22):2483–2505.
105. Tüken T, Yazıcı B, Erbil ME. Electrochemical synthesis of polythiophene on nickel coated mild steel and corrosion performance. *Applied Surface Science*. 2005;239(3–4):398–409.
106. Herrasti P, Kulak AN, Bavykin DV, de Léon CP, Zekonyte J, Walsh FC. Electrodeposition of polypyrrole-titanate nanotube composites coatings and their corrosion resistance. *Electrochimica Acta*. 2011;56(3):1323–1328.
107. Kim S, Le TH, Park CS, Park G, Kim KH, Kim S, Kwon OS, Lim GT, Yoon H. A solution-processable, nanostructured, and conductive graphene/polyaniline hybrid coating for metal-corrosion protection and monitoring. *Scientific Reports*. 2017;7(1):15184.
108. Salunkhe BP, Rane SP. Treatise on conducting polymers for corrosion protection-Advanced approach. *Paint India*. 2016;61.
109. Eliaz N. Corrosion of metallic biomaterials: A review. *Materials*. 2019;12(3):407.
110. Child TF, Van Ooij WJ. Application of silane technology to prevent corrosion of metals and improve paint adhesion. *Transactions of the IMF*. 1999;77(2):64–70.
111. Bertolini L, Elsener B, Pedeferri P, Redaelli E, Polder RB. *Corrosion of Steel in Concrete: Prevention, Diagnosis, Repair*. John Wiley & Sons, 2013.
112. Hihara LH, Latanision RM. Corrosion of metal matrix composites. *International Materials Reviews*. 1994;39(6):245–264.
113. Yadla SV, Sridevi V, Lakshmi MV, Kumari SK. A review on corrosion of metals and protection. *International Journal of Engineering Science & Advanced Technology*. 2012;2(3):637–644.
114. Baldissera AF, Ferreira CA. Coatings based on electronic conducting polymers for corrosion protection of metals. *Progress in Organic Coatings*. 2012;75(3):241–247.
115. Tallman DE, Pae Y, Bierwagen GP. Conducting polymers and corrosion: Polyaniline on steel. *Corrosion*. 1999;55(08).
116. Tallman DE, Pae YQ, Bierwagen GP. Conducting polymers and corrosion: Part 2 polyaniline on aluminum alloys. *Corrosion*. 2000;56(04).
117. Ohtsuka T. Corrosion protection of steels by conducting polymer coating. *International Journal of Corrosion*. 2012;2012(1):915090. Wiley.

6 Composites, nanocomposites, and nanostructured conducting polymers as anti-corrosive materials

B Gpal Krishna, Sanjay Tiwari, Dhriti Sundar Ghosh, and M Jagannadha Rao

6.1 INTRODUCTION

Organic coatings are a successful method of preventing corrosion in metal-based products. Nevertheless, enough ions may eventually get through the coating even in the absence of imperfections to cause corrosion at the interface. However, coating imperfections are more harmful. These imperfections may result from production procedures or external harm, such as stone or scratch impact. The most popular method for increasing metallic materials' corrosion resistance is the application of protective coatings. Conductive polymers have attracted a lot of attention recently due to their potential uses in a wide range of fields, including biomedicine, corrosion protection, sensors, rechargeable batteries, catalysis, and microbial fuel cells, among many others. As a result, depending on the polymerization conditions (concentration, current, voltage), electrolyte (dopant) composition, and nature of the metallic substrate, either an exceptional protective capacity or a terrible corrosion assault can be attained (Balint et al., 2014; Rohwerder & Michalik, 2007; Ruhi et al., 2014).The bonding of the polymers to the substrate is one of the most important parameters because delamination, which is the main consequence of corrosion on the substrate, occurs when adhesion to the substrate is lost. This is because the efficacy of polymers as a barrier depends on the degree of attachment to the metal surface (Brock, 2004). Figure 6.1 illustrates the various synthesis techniques, conducting polymers' characteristics, and the numerous applications for them.

Corrosion-inhibiting pigments are incorporated into the organic coatings used to cover metallic surfaces to reduce corrosive assault. The most effective pigments are those that contain chromates (often in the form of strontium chromate), but because of their toxicity and carcinogenicity, their use must gradually be

 DOI: 10.1201/9781003456025-6

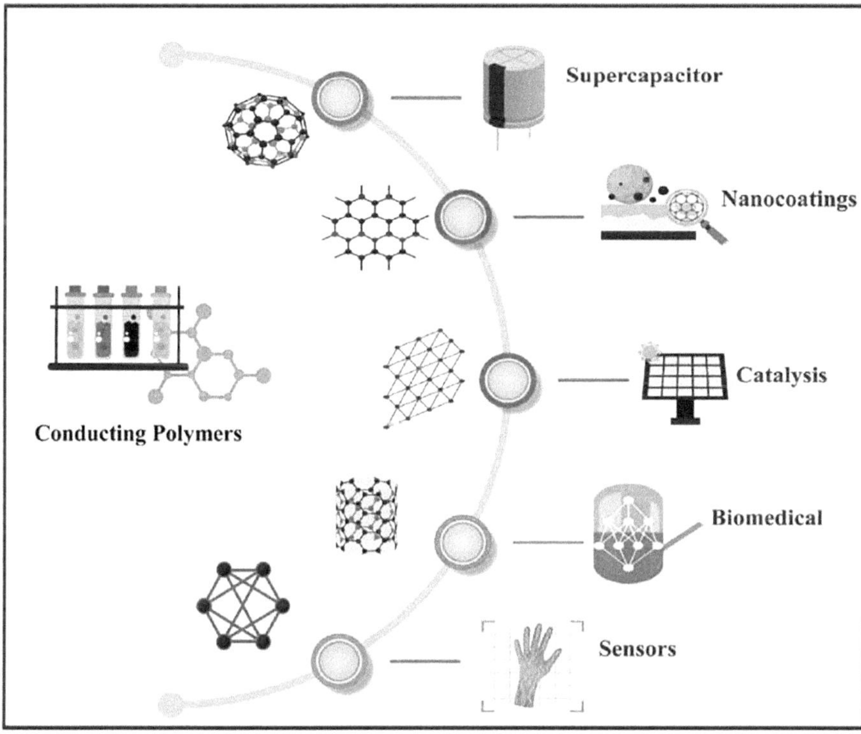

FIGURE 6.1 Conducting polymer applications and their composites.

reduced. On the other hand, almost all strong inhibitors have the potential to harm the environment when released in significant quantities. Since the release of inhibitors depends on leaching in nearly all pigments, coatings must be highly pigmented to guarantee an adequate presence of inhibitors over the years. The inhibitors are frequently released into the surrounding atmosphere, even when they are not needed, presenting a long-term environmental problem. By elevating the underlying metal's potential to a passive area, conducting polymers guard against corrosion by anodizing the material (Spinks et al., 2002). The polymer displays barrier properties in its electronically insulating state, while its capacity to store charge at the metal/coating contact is a key characteristic of its conducting state. It is possible to use this charge to oxidize the base metal and create a passive layer. The most investigated conducting polymers are polyaniline and polypyrrole because of their environmental stability, biocompatibility, and physical and electrical characteristics. According to reports, polypyrrole works well to prevent corrosion (Ruhi et al., 2015; L. Xu et al., 2005). Corrosion-resistant coatings made of polypyrrole can be created chemically or electrochemically and placed on a metal surface. It has been reported that aqueous media were used for the electrochemical implantation of polypyrrole films on metal or alloy substrates. The thickness of these coatings may be accurately regulated, and they exhibit good adhesion to metal surfaces (Hien et al., 2005; Ruhi et al., 2015; L. Xu et al., 2005). According to Schaftinghen et al., the corrosion protection effectiveness of polypyrrole coatings is significantly influenced by the surface pre-treatment of metal or alloy surfaces before electrochemical deposition (Ruhi et al., 2015). The durability of coatings in abrasive environmental conditions and their thermal stability, however, are still in doubt. Long-term interaction with diffusive anions, such as chlorides, causes the coatings' useful qualities to disappear (S. Sathiyanarayanan et al., 2007). Conducting polymer-based composite systems are intriguing alternatives that have generated significant industrial interest as a means of overcoming this. With small, evenly dispersed inorganic filler particles, the resulting composites are conducting polymers. Because conducting polymers and filler particles interact synergistically, polymer composites have better physical, chemical, and mechanical properties than polymer alone (Hosseini et al., 2006, 2007; Lenz et al., 2003). As a result, these composites can be investigated as a covering material with exceptional qualities. Metal or alloy substrates' corrosion resistance is improved by polyaniline/SiO_2 composite coatings [42, 43]. SiO_2 filler has a considerable impact on the thermal and mechanical characteristics of coatings when added to polymer matrices (Ciric-Marjanovic et al., 2009). Silica has a huge surface area and a smooth, non-porous surface, according to studies (S. H. Kim et al., 2003). This characteristic encourages interaction between the polymer matrix and the filler SiO_2. Despite this, the chemically produced polymer composites are challenging to utilize as coatings due to their insolubility in common organic solvents. Because it adheres to metal surfaces well and has strong scratch resistance, conventional polymer coatings like epoxy are believed to have better qualities than normal paints (Rani & Basu, 2012). Epoxy coating, on the other hand, falls apart due to cathodic disbondment after prolonged exposure to extreme environmental conditions. It is therefore insufficient for shielding the underlying metal. To achieve the desired outcomes, existing epoxy paints can be combined with polymer composites with greater corrosion-resistance. Epoxy paints with polyaniline and polypyrrole as anticorrosive agents have been the subject of very few studies. Polyaniline and polypyrrole were employed as additives innovatively by Armelin et al. in an epoxy paint system (Armelin et al., 2008). The epoxy paint's capacity to defend against corrosion under accelerated conditions is enhanced by the additional polymers, which also serve as adhesion promoters. According to Hema et al., investigation on the corrosion-preventive behavior of polyaniline composites, PANI-SiO_2 composites can be employed as efficient coatings for the prevention of corrosion in marine environments (Bhandari et al., 2012).

One of the promising possibilities for coating applications is chitosan, which has strong film-forming and barrier properties and a fondness for metals. It is a typical by-product of seafood, a type of polysaccharide that occurs in the shells of marine crustaceans. The correct disposal of seafood waste is a recurring issue since it raises serious environmental concerns related to ground and drinking

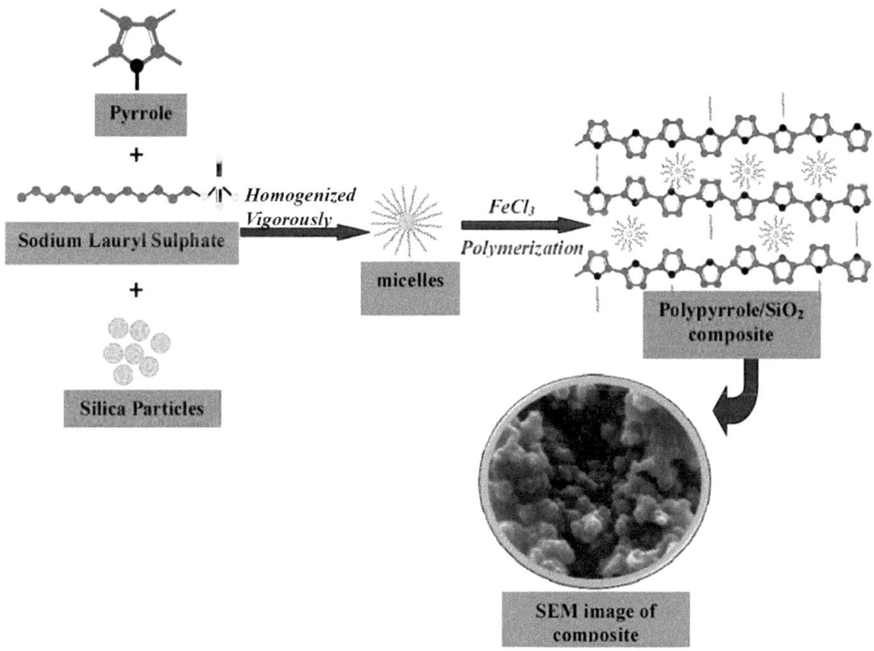

FIGURE 6.2 The process of synthesis of polypyrrole/SiO_2 composite.

water pollution. Because it is inexpensive, abundant in nature, biodegradable, and nontoxic, chitosan can be investigated as an eco-friendly alternative for coating purposes (Srinivasa & Tharanathan, 2007). Chitosan binds to electrically negative surfaces and forms compounds with metal ions. Chitosan's special chemical makeup can be used to create corrosion-resistant films for active metals. Research is being done to create chitosan-based coatings for metals that are resistant to corrosion. Chitosan is susceptible to absorbing moisture from the air, which might cause coating failure. Several studies have been conducted to increase the bond strength of chitosan and reduce its hydrophilicity (Bumgardner et al., 2012; Hahn et al., 2011). These steps, however, are insufficient for chitosan coatings to provide long-term corrosion protection. Chitosan and polypyrrole are said to work together synergistically to lower the size and increase the ability to dissolve polypyrrole (Li et al., 2011). Chitosan and polypyrrole have the interaction benefit of being corrosion-resistant films for materials like mild metal or alloy. Figure 6.2 illustrates the chemical oxidative polymerization of pyrrole into polypyrrole/SiO_2 composites (PCs) using ferric chloride as an oxidant, and Figure 6.3 depicts a schematic for the manufacture of a chitosan/polymer composite. This chapter's major goal is to investigate how various conductive polymer coatings affect materials in terms of their ability to resist corrosion.

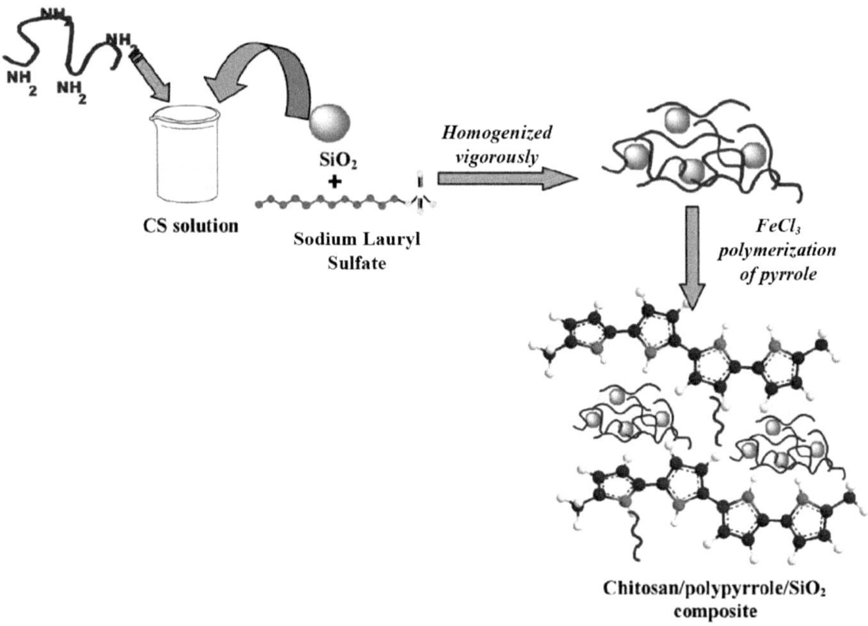

FIGURE 6.3 The process of synthesis of chitosan/polypyrrole/SiO_2 composite.

6.2 SYNTHESIS OF CONDUCTING POLYMERS

6.2.1 Polyacetylene

Polyacetylene exhibits multifunctional behavior, as do its derivatives. Some of its characteristics, including electrical conductivity, photoconductivity, liquid crystal qualities, and chiral identification, can be examined under detailed inspection. A linear polyene chain makes up the primary chain of polyacetylene. It is flexible and can have pendant groups added to it, meaning that the hydrogen molecules in the alternating carbon can be swapped out for another molecule to create mono- or disubstituted polyacetylene. There are numerous ways to make polyacetylene, including catalyst-assisted synthesis, precursor-assisted synthesis, catalytic polymerization of other polymers, and non-catalytic polymerization. When using the catalytic polymerization method, the synthesis involves catalysts like Ziegler-Natta catalysts or Luttinger catalysts. Acetylene is converted into polyacetylene and oligomers like vinyl acetylene and cyclooctatetraene. For the synthesis of polyacetylene, a variety of catalysts are available; however, Zeigler-Natta catalysts stand out for their excellent selectivity and degree of solubility in organic solvents. Combining $Ti(0\text{-}n\text{-}C_4H_9)_4$ and $(C_2H_5)_3A1$ results in extremely crystalline free-standing polyacetylene films on the side of the reaction flask that the catalyst is coated in (Ito et al., 1974). To prepare polyacetylene, Luttinger catalysts are also employed. These catalysts combine a hybrid reducing agent with a complex of a group [VIII] metal, such as nickel chloride. Without any evidence of oligomers, these catalysts yield high molecular weight

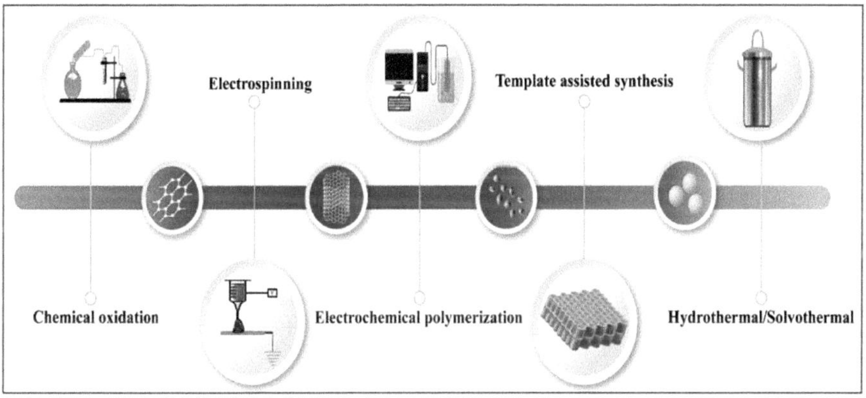

FIGURE 6.4 Different synthesis techniques for conducting polymers.

polyacetylene. As opposed to a Ziegler-Natta catalyst, a Luttinger catalyst catalyzes chemical reactions using hydrophilic solvents like water-ethanol, tetrahydrofuran (THF), or acetonitrile. However, compared to a Zeigler-Natta catalyst, it has less catalytic activity, and the products produced by these catalysts have essentially identical physical and chemical characteristics.

For synthesis, several electrochemical methods have been used, including galvanostatic charge-discharge, galvanostatic cyclic voltammetry, and potentiostatic approaches. Figure 6.4 illustrates several techniques for synthesizing conducting polymers. The fundamental benefit of this method is that a polymer film may be directly deposited on the metal, and the thickness of the film can be easily adjusted by adjusting the electrochemical parameters (Chen & Shy, 1985; Namsheer & Rout, 2021).

6.2.2 Polyaniline

One of the simplest ways to make polyaniline is through chemical oxidation; in this process, a monomer precursor of the appropriate polymer is combined with an oxidizing agent under the influence of an appropriate acid under ambient circumstances to yield products. The reaction medium's transformation to a green color shows that polyaniline has formed. The same process is used for creating the composite. In most cases, oxidizing agents, such as potassium bichromate, ceric nitrate, ceric sulfate, ammonium persulfate, and ammonium peroxy disulfate, are utilized. The conductivity effectively adjusts the physical parameters according to the pH of the acid dopant.

An aniline monomer becomes soluble in a solvent that is organic, like toluene, an oxidant solution, and an aqueous solution containing a dopant acid to create polyaniline using interfacial polymerization. When an oxidant solution is introduced to the monomer solution, polymerization occurs in the phase transition between these two immiscible liquids. The process of creating polyaniline likewise uses a microemulsion, but the surfactant utilized is different. This is because polymerization occurs at the interface of two immiscible liquids (El-Basaty et al., 2020; Österholm et al., 1993; Zeng et al., 2015).

The electropolymerization method is the same for polyacetylene and occurs without the use of an oxidant (AlMashrea et al., 2020; Ashokkumar et al., 2020). Depending on the electrical field, electrospinning is also employed to create fibrous polymer morphologies with nano- or micrometer-sized diameters (Namsheer & Rout, 2021).

6.2.3 Polypyrrole

Due to its high stability, improved conductivity, and relative ease of homopolymer and composite formation, polypyrrole is special due to its rising economic interest. A pyrrole monomer was chemically oxidized to create polypyrrole, which is a dark powdery substance that was first created. Highly conducting polypyrrole is produced frequently via electrochemical synthesis, which follows a similar process and methodology to other conducting polymers. Due to the smaller anode size, the product yield is constrained in this method. The primary advantage of this method over other approaches is the ability to regulate the thickness and shape by adjusting the electrochemical parameters (Namsheer & Rout, 2021).

6.2.4 Poly(p-phenylene)s

Benzoid aromatic nuclei are directly connected by C-C bonds to form the macromolecule known as poly(p-phenylene). Poly(p-phenylene)s are frequently created by directly oxidizing benzene molecules. In this process, benzene nuclei are dehydrocoupled using an oxidizing catalyst, resulting in the production of carbon-carbon bonds. Reagents with a binary or single system are used to carry out the polymerization reaction. A Lewis acid and an oxidant system are both components of the binary system, and in the case of a system containing only one reagent ($FeCl_3$), the system functions simultaneously as a Lewis acid and an oxidative system on its own. A prime instance of a binary reagent arrangement is the combination of the $AuCl_3$ and $CuCl_2$ system, in which $AuCl_3$ functions as a Lewis acid and $CuCl_2$ as an oxidant (Laforgue & Robitaille, 2008). A Wurtz-Fittig reaction, a metal coupling process, was used to produce the first poly(p-phenylene) to be chemically synthesized. The Ulman reaction is also employed to create poly(p-phenylene), and the results are more uniformly structured and have lower molecular weights. However, this technique helps create substituted phenyls like methyl and nitro groups (Threlfall, 1995). The precursor approach, developed by Marvels, Grubbs, and ICI, is well-known for producing poly(p-phenylene)s. The main disadvantage of the Marvels approach is that the final product will have a low molecular weight and less stereochemical control. Reductive polymerization and electrochemical synthesis have also been employed (Namsheer & Rout, 2021).

6.2.5 Poly(p-phenylene vinylene)

A variety of techniques are available for making poly(p-phenylene vinylene). There has been a lot of research done on the Wittig coupling reaction. Poly(p-phenylene vinylene) is produced in this reaction process by combining an aromatic bisphosphonium salt with a bisaldehyde. The synthesis of poly(p-phenylene vinylene) by the Pd-catalyzed coupling of alkyl-substituted aryldiboron acids with dibromo aromatic compounds was also studied. There have also been reports on other synthesis techniques, including chemical vapor deposition, ring-opening polymerization, benzoine condensation, electropolymerization, and metathesis polymerization (Namsheer & Rout, 2021).

6.2.6 Polythiophenes

The environmental stability, thermal stability, and good optical properties of polythiophene and its derivatives are widely explored in comparison to other conducting polymers. Doping engineering or chemical modifications can change the electrical and optical characteristics of polythiophenes. Depending on the dopant and side chain used, polythiophenes' band gap ranges from 3 to 1 eV (Kaloni et al., 2017; Vogel & Marvel, 1961). A significant polythiophene derivative known as poly(3,4-ethylenedioxythiophene, or PEDOT), this compound has received much electrical and electro-optical research. The primary issue related to the PEDOT derivative is that it is not water-soluble. This was remedied by adding a polyelectrolyte-like substance called polysulfonates (PSS) to the PEDOT matrix. Another class of polythiophene compounds is poly(3-hexylthiophene), or P3HT, whose primary uses are in the optoelectronics and semiconductor fields. P3HT is well-liked due to its extensive availability, affordability, well-known morphology, and simplicity of processing. Semicrystalline polymer P3HT has a backbone consisting of separate rings and straight-side chains. The freedom to explore conformational space is made possible by this structural arrangement. The Yamamoto and Lin-Dudek techniques are used to chemically produce polythiophene.

Other cutting-edge methods, including direct sol-gel, oxidative synthesis, organometallic coupling reaction, electropolymerization, template-assisted synthesis, and hydrothermal and solvothermal methods, have been successfully researched (Kiari et al., 2020; Zhang et al., 1995). The composite was made employing a variety of methods, including electropolymerization, green synthesis, and other innovative methods, for the production of polythiophene derivatives such as PEDOT, PEDOT:PSS, and P3TH (Kiari et al., 2020; Y. Kim et al., 2020). The structures of different conducting polymers are shown in Figure 6.5.

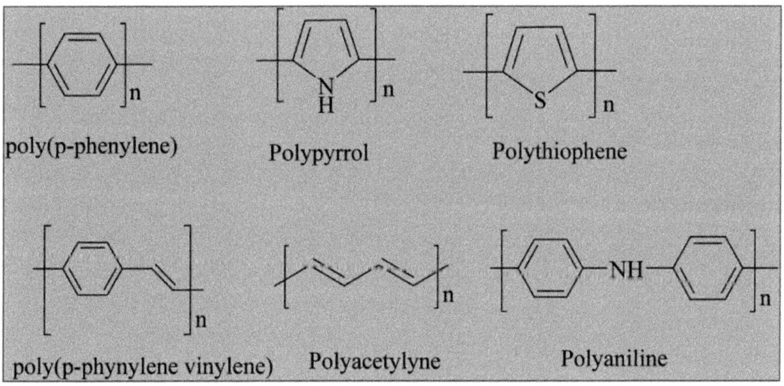

FIGURE 6.5 The structures of various conducting polymers.

6.3 THERMODYNAMICS OF CORROSION

Metals tend to transform back into their lower energy states, or metal oxides. Corrosion of metals is caused by this thermodynamic tendency. The degree of corrosion of a metal is influenced by both the pH of the aqueous solution and the electrode potential of the metal. Pourbaix diagrams provide an excellent summary of the combined impact of the aqueous medium's potential and pH on the corrosion-related byproducts.

(1) Utilization of inhibitors can be done to increase the pH of the electrolyte, and adding inhibitors will make it more alkaline.
(2) Anodic protection can be done by applying an external voltage that causes a change in the metal's potential, which is positive.
(3) Cathode protection can be achieved by providing a potential in a negative direction, for example, by employing zinc coating on a metal or alloy substrate.

To improve subsequent paintability while preserving the ability to provide galvanic protection and, as a result, provide corrosion resistance, zinc-based coatings may include different zinc-based alloys in addition to essentially pure zinc, as in the traditional electrodeposition or hot-dip galvanizing. Galfan (Zn with around 5% Al) and galvanneal (Zn with roughly 10% Fe) are two examples of these alloys. However, in humid air, the rate of corrosion of coated zinc rises quickly. To increase its corrosion resistance, various surface treatments, including organic painting, phosphate conversion coating, and chromate conversion coating, have been developed. For instance, chromate conversion coating is frequently employed because of its excellent corrosion resistance, outstanding paint adherence, affordability, and ease of application. Immersing the zinc-coated specimens in a chromic acid solution to create a passive layer of mixed trivalent and hexavalent chromium oxides and zinc oxides is the standard procedure for chromate conversion coating. Unfortunately, the poisonous and cancer-causing hexavalent chromium species are found in the chromate solution and layer (Deshpande et al., 2014; Mobin & Shabnam, 2010).

6.4 CONDUCTING POLYMER COATING

DeBerry was the first to disclose the use of conducting polymer films for metal anticorrosion, and he also claimed that stainless metal or alloy coated with polyaniline could maintain its passivation for a considerable duration in the presence of sulfuric acid (DeBerry, 1985). Since then, conducting polymer coating research for metal protection against corrosion has drawn a lot of attention because of its unique qualities, including simple synthesis, high conductivity, environmental friendliness, low cost, non-toxicity, and distinctive redox chemistries (Ohtsuka, 2012; Tallman et al., 2002). Though there are still some questions regarding the protective mechanism of conducting polymer coatings, it was generally accepted that they can shield metal by creating physical barriers, inhibition, and passive oxide layers (Herrasti et al., 2005).

The barrier effect stops hostile substances like water, dioxygen, and ions from penetrating the metal or alloy surface.

Because conducting polymer coatings are permeable and have a poor barrier function as a result, conventional polymers like epoxy and polyurethane are frequently utilized as topcoats. The active dissolving process of metal is suppressed by the high adsorption of electrically conducting polymers on the metallic substrate. The conducting polymer's robust oxidative properties aid in the metal's potential shift and protracted maintenance of its passive state, resulting in the creation of a barrier of protection on the metal-to-metal interface. Some of the conducting polymer's doped ions will be hydrolyzed and react with the metal ions on the defective site to produce an entirely novel passive oxide on the damaged site when the coating and passive oxide are locally damaged (Nguyen Thi Le et al., 2001). According to certain research, conducting polymers' protective capacity would decline as the amount of charge they could hold decreased (Kowalski et al., 2007, 2009). As a result, the polymer coating would need to be regularly changed for protection to be effective. The most often investigated polymers at the moment are polyaniline (PAni), polypyrrole (PPy), polythiophene (Pth), and their derivatives (H. Xu & Zhang, 2019).

6.4.1 Polyaniline (PAni)-based coatings

Because they are less expensive, simpler to make, and have higher environmental stability than other conducting polymers, polyaniline, and its derivatives are among the most widely investigated polymer coatings for corrosion prevention.

6.4.1.1 The polyaniline emeraldine salt form coatings

There are two different polyaniline forms: polyaniline emeraldine base form (PAni-EB) and polyaniline emeraldine salts (PAni-ES). The current studies have validated PAni-EB's barrier effect and revealed that it is nonconducting (McAndrew, 1997). However, some studies suggested that PAni-EB had an anodic protective effect, while others found inconclusive results. As a result, it is crucial to keep researching the causes and effects of PAni-EB-based coatings on metal or alloy. In comparison to PAni-EB, conducting PAni-ES as a coating over metal or alloy has received more research attention and has been applied in real-world projects (Fahlman et al., 1997; Wessling & Posdorfer, 1999). The primary acids employed in this process involve acids, such as phosphoric acid, sulfuric acid, and 5-sulphosalicylic acid. The conducting polyaniline is typically polymerized utilizing aniline in organic or inorganic acid solutions. PAni was unable to provide enough protection from corrosion over a chloride-containing medium, whereas the layer that sulfuric solution deposition on stainless metal or alloy produced had outstanding pitting and corrosion resistance. The different substrates, as well as the nature and amount of the deposition solutions, could be the cause of the contradictory results. Some PAni derivative coatings have superior corrosion protection capabilities than those of PAni coatings. A remarkable anodic protection was shown by PAni and its substitute, poly (o-anisidine), which were synthesized on stainless metal or alloy in monomers comprising 0.5 M sulfamic acid solution under low scan rates and showed effective anticorrosive behavior even after prolonged exposure to 3.5% NaCl solution. Chromium and nickel were added to the passive film under the poly (o-phenylenediamine) (PoPD) onto 304 stainless metal or alloy before it was coated in sulfuric acid solution, which provided good protection for the metal against pitting in a harsh chloride environment. The corrosion rate of mild metal or alloy coated with the synthetic terpolymer poly (2PA-co-AN-co-XY) polymer was much lower in corrosive settings such as 0.1 M HCl solution, 5% NaCl solution, simulated seawater, distilled water, and open-air (A. A. Hermas, 2008; Ozyilmaz et al., 2010; Tanveer & Mobin, 2012).

6.4.1.2 Bilayered conducting polymer coating

Metal or alloy is coated with an additional layer of conductive polymer to improve its anticorrosive qualities. However, the blend of PAni and Ni only functioned via the barrier mechanism (Ananda Kumar et al., 2008). Some bilayered coatings, such as PAni/Zn coating, may demonstrate the beneficial impact of the barrier process of the PAni and the sacrificial shielding of the zinc. When monomer coating alone is insufficient for avoiding corrosion in a hostile media, the combination coating may occasionally result in reduced corrosion (Martyak & McAndrew, 2007). The composition of the substrate, the order of deposition, the layer thickness, the presence of corrosive media, the pH of the medium, and the exposure time are additional aspects that should be thoroughly taken into account in future investigations.

6.4.1.3 Epoxy-PAni coatings

Because polyaniline tends to form brittle films with weak adhesion qualities and a poor barrier effect for porous surfaces, it is not a suitable coating binder. Some researchers combined PAni with other conventional organic coatings to maximize their synergies. Epoxy coating, the most common organic coating, offers great qualities like strong adhesion, high mechanical resistance, and significant chemical resistance in hostile situations. Epoxy coatings that combine PAni have therefore shown more corrosion resistance on metal or alloy than epoxy and PAni films alone. Toxic inorganic pigments can be replaced by epoxy combined with low concentrations of PAni-EB or PAni-ES (doped or undoped anions), which both exhibit outstanding anticorrosive behaviors on metal or alloy (Armelin et al., 2009). The corrosive shielding ability of the coating will be improved when suitable inorganic pigments, such as metal oxides and graphite, are used as pigments in PAni coatings. This is because particles with good conductivity can act as carriers of PAni to ensure the rapid movement of electrons associated with redox processes, as well as encourage the growth of a barrier by particles settling during the drying process (Kalendová et al., 2008). In addition to being blended with epoxy resin, PAni has been used as a pigment doping or undoping various anions in different organic films that include acrylic, vinyl acrylic, polyvinyl chloride, and alkyd coatings. In most cases, these blends showed excellent corrosive protection.

6.4.2 Polypyrrole (PPy)-based coatings

A lot of research has been done on polypyrrole for protecting metals against corrosion because of its excellent conductivity, stability, simplicity in synthesis, and eco-friendly characteristics. The low permeability and great stability of the polypyrrole coatings on stainless metal or alloy resulted in significant anodic protection behavior and improved metal or alloy's capacity to self-passivate with a protective oxide coating. Single PPy coatings have some restrictions when it comes to preventing corrosion. Porosity and ion exchanges of PPy films may be unfavorable in cases of localized corrosion brought on by tiny and aggressive anions, as protective behavior may be lost when the charges held within the polymer layer are spent during redox reactions (Riaz et al., 2007). As a result, the majority of studies concentrated on employing PPy as a component of composites and bilayer films. Numerous studies have been conducted on PPy-based bilayer coatings, which display superior protective behavior, including a stronger barrier effect, greater pitting protection, and longer durability (Kowalski et al., 2007, 2009; Tan & Blackwood, 2003; Tüken et al., 2004a). The performance of the corrosion protection system may be enhanced by depositing PPy on the surface of the zinc film and the SiO2 layer (Grari et al., 2015; Martins et al., 2004). The anticorrosion characteristics and other physical characteristics of coatings of polymers can be enhanced using another effective way. Due to the presence of zinc particles and improved barrier performance brought on by zinc centers' corrosion products filling up the pores of the polypyrrole film, the zinc particle-upgraded polypyrrole coating may offer efficient cathodic resistance to mild metal or alloy. On mild metal or alloy or stainless metal or alloy, integrating a thin graphite film between two PPy layers produced outstanding anticorrosion performance because the graphite layer reduced water permeability and mobility in corrosive environments (H. Xu & Zhang, 2019).

6.4.3 Polythiophene (Pth)-based coatings

Previous research mainly concentrated on synthesis techniques, even though polythiophene is an essential conducting polymer with outstanding electrical characteristics and stability (Ferraris et al., 1998). Studies on coatings that protect metals are uncommon since it is challenging to directly synthesize polythiophene on metal or alloy surfaces. Pth-based coatings were the subject of a recent investigation. By creating a passivating coating for the redox chemistry property, the electropolymerized polythiophene over mild metal or alloy in acetonitrile solution might prevent mild metal or alloy from corroding. In CAN-$LiClO_4$, the polythiophene could not be manufactured on mild metal or alloy, but it could be done on top of a thin PPy layer. In a 3.5% NaCl solution, the PPy/Pth coating exhibited minimal porosity and outstanding barrier qualities. In 3.5% NaCl solution, the Pth coating over nickel-coated mild metal or alloy might significantly boost barrier efficiency with time. Polythiophene nanoparticles (0.6 wt%) added to aqueous epoxy coatings on mild metal or alloy can provide high protection, and after 360 hours of immersion in a 3.5 wt% NaCl solution, the impedance values remained higher at 1106 cm^2. The mild metal or alloy that is coated with polythiophene and ash or MCM-41 particles that include iron (III) nitrate as an inhibitor shows outstanding anticorrosive efficacy in chloride environments. Several conducting polymers, including polycarbazole and polyvinyl alcohol, have also been investigated as corrosion-resistant films for metals or alloys, but they have not been widely studied and employed for various reasons.

The most well-known conducting polymers are PAni, PPy, and Pth (Kousik et al., 2001; Ocampo et al., 2005; Palraj et al., 2012; Tüken et al., 2004b, 2005).

6.5 NANOPOLYMER COMPOSITE COATINGS

There is a strong demand for coatings with characteristics including long-term resistance to corrosion, high adhesion, good durability, and high surface mechanical wear. It has been demonstrated that adding nanoparticles to conducting polymer coatings can improve a variety of properties, including adherence, thermal resistance, mechanical strength, hydrophobicity, and magnetostatic shielding, in addition to filling pores to increase the barrier effect. Nanoparticles of many different types, including TiO_2, Zn, ZnO, Si, SiO_2, Al_2O_3, CeO_2, nanoclays, and nanotubes, have been added to polymer coatings and have demonstrated superior protective behavior than conventional polymer coatings (H. Xu & Zhang, 2019).

6.5.1 Nanocomposite PAni coatings

On carbon metal or alloys, water-based epoxy coatings incorporating 0.01 wt% nanopolyaniline, 0.07 wt% of $ScCO_2$-produced self-doping nanoPAni, weight percent nanopolyaniline will increase anticorrosive effectiveness. It is thought that nanopolyaniline particles could improve the adherence of composite coatings and boost the barrier characteristics, resulting in the creation of a thin, homogenous oxide layer between the metal and coating interface (Patil & Radhakrishnan, 2006). When the corrosion process was started on the metal substrate, epoxy coatings containing 1.0 wt% nanoPAni-DBSA in the carbon metal or alloy substrate would give off dopant anions, and the resulting dopant anions would combine with the iron cations, becoming a secondary barrier layer, which would passivate the substrate. Typically, coatings use zinc metal as an addition for sacrificial protection. Nano-ZnO or nano-Zn combined with PAni could exhibit their synergy, according to research done on the anticorrosive efficacy of nano-Zn metal and PAni composites (Mostafaei & Nasirpouri, 2014; Olad et al., 2012; Patil & Radhakrishnan, 2006). The poly (vinyl acetate) blended with nano-ZnO-PAni hybrid material demonstrated a striking improvement in the protection against corrosion of metal or alloy in saline water, which can be attributed to a combination of the improvement in barrier properties caused by nanoparticulate additives and the redox behavior of PAni, leading to the formation of a protective oxide layer close to the substrate. Epoxy coatings incorporating PAni-nano-ZnO were shown to greatly enhance corrosion protection and barrier performance, according to Mostafaei and Nasirpouri (Mostafaei & Nasirpouri, 2014). The polyvinyl chloride combined with nano-ZnO-PAnihas shown better anticorrosion behavior. In contrast to PAni/Zn composite films, PAni/nano-Zn film composites exhibited better electrical conductivities and superior anticorrosive characteristics. Due to the improvement of barrier characteristics and electrochemical anticorrosion behavior, the polyaniline coating comprising both nano-Zn and epoxy resin as additives exhibits the best anticorrosion performance (Olad et al., 2012). The gas barrier, thermal stability, mechanical strength, fire-retardant, and anticorrosive properties of metal coatings can all be improved by combining nanoclay particles with polymer (Sun et al., 2004, 2007). The conductivities of PAni/hydrophilic nano-montmorillonite composite films were lower than those of pure polyaniline films and higher in the case of PAni/organophilic nano-montmorillonite composite films. Both of these hydrophilic and organophilic PAni/nano-montmorillonite composite films applied on iron samples possessed better anticorrosive properties than pure polyaniline coatings. The PAni/natural clinoptilolite nanocomposite having a 3 wt% clinoptilolite concentration provides the highest metal or alloy protection qualities in acidic conditions. The polyaniline-Cloisite 30B nanoclay can alter zinc-rich epoxy coatings, prevent the consumption of zinc powder, and lengthen the cathodic protection time by enhancing its early barrier effects. Water-based epoxy resin combined with polyaniline-Cloisite 30B nanoclay demonstrated greater long-term stability, high adhesion, and improved durability against corrosion. The epoxy/PAni/OMMT (alkyl ammonium ions in organo-modified montmorillonite) significantly outperformed epoxy/PAni/MMT (Na+ in natural clay) in terms of protecting metal or alloy panels against corrosion. Outstanding mechanical properties, strong electronic conductivity, nanoscale size, and a large surface area are all features of nanotubes (CNTs). In comparison to pure PAni coatings on metal or alloy, several researchers created CNT/PAni composites that displayed better anticorrosion behavior (A. E. A. Hermas et al., 2013; Kalendová et al., 2015; Madhan Kumar & Gasem, 2015; Mahmoudian et al., 2012). The increased surface area of the nanotubes reduces the PAni's porosity and creates a more corrosion-resistant passive film on the metallic material or alloy underneath the composite layer, which is how the coating with nanotubes has a higher anticorrosion behavior. Based on electrochemical impedance spectroscopy (EIS), the resistivity of the film comprising PAni nanotubes is greater compared to that of the film containing PAni nanofibers over mild metal or alloy in a 3.5% NaCl aqueous solution. In the H_2SO_4 solution, the composites made of PAni-multiwalled carbon nanotubes (PAni-mCNTs) have shown superior morphological structure, electrical characteristics, thermal stability, and anticorrosion behavior. A functional PAni/f-mCNT composite demonstrated electrochemical corrosion behavior over mild metal or alloy when exposed to a corrosive environment, increased mechanical characteristics, excellent hydrophobic nature, and uniform dispersion of CNTs in the PAni matrix. The CNTs can supplement Zn dust in metal or alloy coatings while also enhancing the coating's anticorrosion effectiveness. Additionally, investigated as possible additions in conducting polymer coatings are CeO_2 and TiO_2 nanoparticles. To protect mild metals or alloys, Ramezanzadeh et al. combined polyaniline (PAni) nanofibers with graphene oxide nanosheets that had CeO_2 grafted onto them. The test results showed that this combination improved the protection and active corrosion inhibition properties of the GO nanosheets. The anticorrosion effectiveness of an epoxy film that includes imidazole-tailored CeO_2 and unmodified CeO_2 nanoparticles on mild metal or alloy in NaCl 3.5 wt% solution. It is observed that the tailored coating possessed a greater resistance to corrosion and lower water absorption compared with unmodified coating, and the best percentage of embedded CeO_2 nanoparticles in the epoxy coating was 0.5 wt% (Hosseini & Aboutalebi, 2018). Polyaniline doped with phosphoric acid (PAni-PA) can passivate metals or alloys, and ceria nanoparticles can dramatically slow down electrolyte transport and enhance barrier protection. The corrosion prevention capabilities in the NaCl solution may be enhanced by combining PAni-PA and ceria nanoparticles with epoxy coatings. The ceria nanoparticles and PAni added to an alkyd coating significantly reduced blistering and delamination rates when compared to an alkyd blank coating. To create PAni in the presence of TiO_2-B_6 (PAni-TiO_2-B_6), the surface of TiO_2 nanoparticles was modified with vitamin B_6 (TiO_2-B_6). The test findings showed that compared to pure epoxy resin and epoxy/PAni nanocomposite, an epoxy-based coating comprising PAni-TiO_2-B_6 nanocomposites exhibited improved anticorrosion capabilities, higher hardness, and lower dielectric loss (H. Xu & Zhang, 2019).

6.5.2 Nanocomposite PPy coatings

Since the addition of nanoparticles to polypyrrole improved the polymers' solubility, durability, and adherence to the substrate as well as their inhibitory and passivation properties, it has been widely employed to protect metal substrates against corrosion. According to Ionită et al., the conductivity of PPy doped with carbon nanotubes was improved, as its ability to shield carbon metal or alloy from corrosion in a solution containing 3.5% NaCl. The area of synthesized PPy-based Ni-doped TiO_2 nanocomposites is enlarged, which improves its anticorrosion capacity in NaCl solution. The size of PPy/Ni-doped-TiO_2 nanocomposites is lower than PPy/TiO_2 nanocomposites. Mesoporous silica nanocontainer powders have been added to the polypyrrole matrix, which enhanced corrosion resistance for mild metal or alloy under a salt corrosion environment condition by forming a protective compound on the interface. The epoxy resin with polypyrrole/SiO_2 composite may passivate the defect up to 99.99% of the corrosion protection efficiency. Metal or alloy components coated with such coatings suppressed the spread of corrosion and produced very few blisters. The organic-inorganic nanocomposite (PPy/GNS/RE3+/DBSA) that includes pyrrole, graphene nanosheets (GNS), rare earth elements, and dodecyl benzene sulfonic acid (DBSA), and the nanocomposite coatings demonstrated excellent corrosion protection for carbon metal or alloy due to their barrier and passivation behavior. The montmorillonite-polypyrrole-based epoxy nanostructured composites showed a high aspect ratio, ease of dispersion, high electrical conductivity, high impedance, and improved barrier properties. Epoxy/Mt-PPy was found to provide a superior corrosion-protective covering for carbon metal or alloys (Ioni & Prun, 2011; H. Xu & Zhang, 2019).

6.6 CORROSION PROTECTION MECHANISMS OF CONDUCTING POLYMERS

The mechanisms underlying conducting polymers' corrosion prevention have not been fully understood till now, but the following four potential hypotheses have been proposed:

(1) **Anodic protection mechanism:** According to this mechanism, corrosion prevention (CP) coatings may cause the development of shielding layers of metal oxides on the metal surface, preventing corrosion.

(2) **Controlled inhibitor release mechanism:** According to this mechanism, the oxidized and consequently doped CP-based coating applied on a base metal substrate may release the anion dopant upon reduction, which is motivated by a defect in the coating that prevents it from binding to the base metal. If the acid dopant is soluble in water, the anions in the context of doped PANI can likewise be released by a straightforward elimination process.

(3) When a metal interacts with an electronically doped semiconductor or an electrically conducting polymer, it is thought that an electric field is created that restricts the flow of electrons from the metal to an oxidizing species, preventing or slowing the pace of corrosion.

(4) **Conducting polymer coatings:** Provide a dense, adherent, low porosity sheet that preserves a basic environment on a metal surface, limits the access of oxidants, and stops the oxidation of the metal surface. The barrier effectiveness and the rate at which O_2 and H_2O are transported into the polymer improve with increasing CP layer porosity. The O_2 reduction reaction site shifts from the M|CP-coating boundary to the CP coating|solution interface as a result of the substrate becoming more compact and adhering to the surface. Reduced reduction products (OH_) across the M|CP-coating interface as a result of the oxygen-reduction site's shift on the polymer surface shield the coating from disbonding and delamination.

While localized pinholes or small-sized flaws are created, O_2 reduction is engaged in the specific reoxidation of the CP and the continuous active activity of the CP coating. Deactivating CPs should not be used to increase the barrier effect. The open-circuit potential of the metal |CP-coating| solution is considered in a passive condition as far as the CP is concerned in its conductive form. For the coating's ability to provide long-term protection, the location and kinetics of the oxygen reduction are crucial. It is commonly acknowledged that improving the barrier effect involves dehydrating CP electrodeposited coatings on metals from aqueous media (Deshpande et al., 2014).

6.7 SUMMARY AND PERSPECTIVES

The most common and successful method of preventing corrosion of metal or alloy systems at the moment is the coating, while several barriers still require more research. The following findings can be taken from the available research. Most coatings have a physical barrier, electrochemical protection, or a combination of the two as part of their defense system. The physical barriers and outstanding electrochemical behavior of the monomer-conducting polymer coatings prevent metal or alloys from corrosion; however, the physical barriers, adhesion, mechanical strength, and durability behavior are not particularly effective. Due to their synergistic impact, most bilayered coatings perform better against corrosion than monomer-conducting polymer coatings. The anticorrosion capabilities of polymer coatings may be improved by doping them with certain anions, such as DS and PW12 ions. Conducting polymers that have nanoparticles added to them typically have improved electrochemical anticorrosion as well as adhesion, thermal stability, mechanical strength, and hydrophobic qualities, in addition to filling the holes to increase the barrier effect. The nanopolymer blends exhibit outstanding anticorrosion behavior and extended endurance when compared to other coatings. Future studies should carefully address a few factors for increased effectiveness and wider applications. To create more effective coatings, the mechanism of reaction and microstructure of nanoclay and nanotube-composite films should be further investigated. Given that weathering, surface treatment, environmental pH value, fracture resistance, and building craftwork have the most impact on the coating protection duration, the next research should consider all of these variables and come to a firm conclusion. The focus of future research should be on low-cost, environmentally friendly, practical construction, particularly on-site construction, and coatings with long service lives because these factors will determine whether a coating is appropriate to be utilized, in addition to how well it can protect metal or alloy. Standards and regulations for construction technology, as well as inspection, which are essential and required for applications, are in great demand.

REFERENCES

AlMashrea, B. A., Abla, F., Chehimi, M. M., Workie, B., Han, C., & Mohamed, A. A. (2020). Polyaniline coated gold-aryl nanoparticles: Electrochemical synthesis and efficiency in methylene blue dye removal. *Synthetic Metals*, *269*, 116528. https://doi.org/10.1016/J.SYNTHMET.2020.116528

Ananda Kumar, S., Shree Meenakshi, K., Sankaranarayanan, T. S. N., & Srikanth, S. (2008). Corrosion resistant behaviour of PANI-metal bilayer coatings. *Progress in Organic Coatings*, *62*(3), 285–292. https://doi.org/10.1016/J.PORGCOAT.2008.01.005

Armelin, E., Alemán, C., & Iribarren, J. I. (2009). Anticorrosion performances of epoxy coatings modified with polyaniline: A comparison between the emeraldine base and salt forms. *Progress in Organic Coatings*, *65*(1), 88–93. https://doi.org/10.1016/J.PORGCOAT.2008.10.001

Armelin, E., Pla, R., Liesa, F., Ramis, X., Iribarren, J. I., & Alemán, C. (2008). Corrosion protection with polyaniline and polypyrrole as anticorrosive additives for epoxy paint. *Corrosion Science*, *50*(3), 721–728. https://doi.org/10.1016/J.CORSCI.2007.10.006

Ashokkumar, S. P., Vijeth, H., Yesappa, L., Niranjana, M., Vandana, M., & Devendrappa, H. (2020). Electrochemically synthesized polyaniline/copper oxide nano composites: To study optical band gap and electrochemical performance for energy storage devices. *Inorganic Chemistry Communications*, *115*, 107865. https://doi.org/10.1016/J.INOCHE.2020.107865

Balint, R., Cassidy, N. J., & Cartmell, S. H. (2014). Conductive polymers: Towards a smart biomaterial for tissue engineering. *Acta Biomaterialia*, *10*(6), 2341–2353. https://doi.org/10.1016/J.ACTBIO.2014.02.015

Bhandari, H., Kumar, S. A., Dhawan, S. K., Bhandari, H., Kumar, S. A., & Dhawan, S. K. (2012). Conducting polymer nanocomposites for anticorrosive and antistatic applications. *Nanocomposites—New Trends and Developments*. https://doi.org/10.5772/50470

Brock, S. L. (2004). Nanostructures and nanomaterials: Synthesis, properties and applications by Guozhang Cao (University of Washington). Imperial College Press (distributed by World Scientific): London. 2004. xiv + 434 pp. $78.00. ISBN 1-86094-415-9. *Journal of the American Chemical Society*, *126*(44), 14679–14679. https://doi.org/10.1021/JA0409457

Bumgardner, J. D., Wiser, R., Gerard, P. D., Bergin, P., Chestnutt, B., Marini, M., Ramsey, V., Elder, S. H., & Gilbert, J. A. (2012). Chitosan: Potential use as a bioactive coating for orthopaedic and craniofacial/dental implants. *Journal of Biomaterials Science, Polymer Edition*, *14*(5), 423–438. https://doi.org/10.1163/156856203766652048

Chen, S. A., & Shy, H. J. (1985). Electrochemical polymerization of acetylene on a surface of platinum. *Journal of Polymer Science: Polymer Chemistry Edition*, *23*(9), 2441–2446. https://doi.org/10.1002/POL.1985.170230909

Ciric-Marjanovic, G., Dragičević, L., Milojević, M., Mojović, M., Mentus, S., Dojčinović, B., Marjanović, B., & Stejskal, J. (2009). Synthesis and characterization of self-assembled polyaniline nanotubes/silica nanocomposites. *Journal of Physical Chemistry B*, *113*(20), 7116–7127. https://doi.org/10.1021/JP900096B/ASSET/IMAGES/MEDIUM/JP-2009-00096B_0017.GIF

DeBerry, D. W. (1985). Modification of the electrochemical and corrosion behavior of stainless steels with an electroactive coating. *Journal of the Electrochemical Society*, *132*(5), 1022–1026. https://doi.org/10.1149/1.2114008/XML

Deshpande, P. P., Jadhav, N. G., Gelling, V. J., & Sazou, D. (2014). Conducting polymers for corrosion protection: A review. *Journal of Coatings Technology and Research*, *11*(4), 473–494. https://doi.org/10.1007/S11998-014-9586-7/FIGURES/4

El-Basaty, A. B., Moustafa, E., Fouda, A. N., & El-Moneim, A. A. (2020). 3D hierarchical graphene/CNT with interfacial polymerized polyaniline nano-fibers. *Spectrochimica Acta Part A: Molecular and Biomolecular Spectroscopy*, *226*, 117629. https://doi.org/10.1016/J.SAA.2019.117629

Fahlman, M., Jasty, S., & Epstein, A. J. (1997). Corrosion protection of iron/steel by emeraldine base polyaniline: An X-ray photoelectron spectroscopy study. *Synthetic Metals*, *85*(1–3), 1323–1326. https://doi.org/10.1016/S0379-6779(97)80256-1

Ferraris, J. P., Eissa, M. M., Brotherston, I. D., Loveday, D. C., & Moxey, A. A. (1998). Preparation and electrochemical evaluation of poly (3-phenylthiophene) derivatives: Potential materials for electrochemical capacitors. *Journal of Electroanalytical Chemistry*, *459*(1), 57–69. https://doi.org/10.1016/S0022-0728(98)00318-0

Grari, O., Taouil, A. E., Dhouibi, L., Buron, C. C., & Lallemand, F. (2015). Multilayered-polypyrrole-SiO2 composite coatings for functionalization of stainless steel: Characterization and corrosion protection behavior. *Progress in Organic Coatings*, *88*, 48–53. https://doi.org/10.1016/J.PORGCOAT.2015.06.019

Hahn, B. D., Park, D. S., Choi, J. J., Ryu, J., Yoon, W. H., Choi, J. H., Kim, H. E., & Kim, S. G. (2011). Aerosol deposition of hydroxyapatite-chitosan composite coatings on biodegradable magnesium alloy. *Surface and Coatings Technology*, *205*(8–9), 3112–3118. https://doi.org/10.1016/J.SURFCOAT.2010.11.029

Hermas, A. A. (2008). XPS analysis of the passive film formed on austenitic stainless steel coated with conductive polymer. *Corrosion Science*, *50*(9), 2498–2505. https://doi.org/10.1016/J.CORSCI.2008.06.019

Hermas, A. E. A., Salam, M. A., & Al-Juaid, S. S. (2013). In situ electrochemical preparation of multi-walled carbon nanotubes/polyaniline composite on the stainless steel. *Progress in Organic Coatings*, *76*(12), 1810–1813. https://doi.org/10.1016/J.PORGCOAT.2013.05.021

Herrasti, P., Recio, F. J., Ocón, P., & Fatás, E. (2005). Effect of the polymer layers and bilayers on the corrosion behaviour of mild steel: Comparison with polymers containing Zn microparticles. *Progress in Organic Coatings*, *54*(4), 285–291. https://doi.org/10.1016/J.PORGCOAT.2005.07.001

Hien, N. T. L., Garcia, B., Pailleret, A., &Deslouis, C. (2005). Role of doping ions in the corrosion protection of iron by polypyrrole films. *Electrochimica Acta*, *50*(7–8), 1747–1755. https://doi.org/10.1016/J.ELECTACTA.2004.10.072

Hosseini, M. G., & Aboutalebi, K. (2018). Improving the anticorrosive performance of epoxy coatings by embedding various percentages of unmodified and imidazole modified CeO2 nanoparticles. *Progress in Organic Coatings*, *122*, 56–63. https://doi.org/10.1016/J.PORGCOAT.2018.05.006

Hosseini, M. G., Sabouri, M., & Shahrabi, T. (2006). Comparison between polyaniline-phosphate and polypyrrole-phosphate composite coatings for mild steel corrosion protection. *Materials and Corrosion*, *57*(5), 407–410. https://doi.org/10.1002/MACO.200503940

Hosseini, M. G., Sabouri, M., & Shahrabi, T. (2007). Corrosion protection of mild steel by polypyrrole phosphate composite coating. *Progress in Organic Coatings*, *60*(3), 178–185. https://doi.org/10.1016/J.PORGCOAT.2007.07.029

Ioni, M., & Prun, A. (2011). Polypyrrole/carbon nanotube composites: Molecular modeling and experimental investigation as anti-corrosive coating. *Progress in Organic Coatings*, *72*(4), 647–652. https://doi.org/10.1016/J.PORGCOAT.2011.07.007

Ito, T., Shirakawa, H., & Ikeda, S. (1974). Simultaneous polymerization and formation of polyacetylene film on the surface of concentrated soluble Ziegler-type catalyst solution. *Journal of Polymer Science: Polymer Chemistry Edition*, *12*(1), 11–20. https://doi.org/10.1002/POL.1974.170120102

Kalendová, A., Sapurina, I., Stejskal, J., & Veselý, D. (2008). Anticorrosion properties of polyaniline-coated pigments in organic coatings. *Corrosion Science*, *50*(12), 3549–3560. https://doi.org/10.1016/J.CORSCI.2008.08.044

Kalendová, A., Veselý, D., Kohl, M., & Stejskal, J. (2015). Anticorrosion efficiency of zinc-filled epoxy coatings containing conducting polymers and pigments. *Progress in Organic Coatings*, *78*, 1–20. https://doi.org/10.1016/J.PORGCOAT.2014.10.009

Kaloni, T. P., Giesbrecht, P. K., Schreckenbach, G., & Freund, M. S. (2017). Polythiophene: From fundamental perspectives to applications. *Chemistry of Materials*, *29*(24), 10248–10283. https://doi.org/10.1021/ACS.CHEMMATER.7B03035/ASSET/IMAGES/MEDIUM/CM-2017-03035G_0047.GIF

Kiari, M., Berenguer, R., Montilla, F., & Morallón, E. (2020). Preparation and characterization of montmorillonite/PEDOT-PSS and diatomite/PEDOT-PSS hybrid materials. Study of electrochemical properties in acid medium. *Journal of Composites Science*, *4*(2), 51. https://doi.org/10.3390/JCS4020051

Kim, S. H., Ahn, S. H., & Hirai, T. (2003). Crystallization kinetics and nucleation activity of silica nanoparticle-filled poly(ethylene 2,6-naphthalate). *Polymer*, *44*(19), 5625–5634. https://doi.org/10.1016/S0032-3861(03)00623-2

Kim, Y., Kim, Y., & Kim, J. H. (2020). Highly conductive PEDOT:PSS thin films with two-dimensional lamellar stacked multi-layers. *Nanomaterials*, *10*(11), 2211. https://doi.org/10.3390/NANO10112211

Kousik, G., Pitchumani, S., & Renganathan, N. G. (2001). Electrochemical characterization of polythiophene-coated steel. *Progress in Organic Coatings*, *43*(4), 286–291. https://doi.org/10.1016/S0300-9440(01)00211-9

Kowalski, D., Ueda, M., & Ohtsuka, T. (2007). The effect of counter anions on corrosion resistance of steel covered by bi-layered polypyrrole film. *Corrosion Science*, *49*(8), 3442–3452. https://doi.org/10.1016/J.CORSCI.2007.03.007

Kowalski, D., Ueda, M., & Ohtsuka, T. (2009). Self-healing ability of conductive polypyrrole coating with artificial defect. *ECS Transactions*, *16*(52), 177–181. https://doi.org/10.1149/1.3229965/META

Laforgue, A., & Robitaille, L. (2008). Fabrication of poly-3-hexylthiophene/polyethylene oxide nanofibers using electrospinning. *Synthetic Metals*, *158*(14), 577–584. https://doi.org/10.1016/J.SYNTHMET.2008.04.004

Lenz, D. M., Delamar, M., & Ferreira, C. A. (2003). Application of polypyrrole/TiO2 composite films as corrosion protection of mild steel. *Journal of Electroanalytical Chemistry, 540*, 35–44. https://doi.org/10.1016/S0022-0728(02)01272-X

Li, Y., Li, G., Peng, H., & Chen, K. (2011). Facile synthesis of electroactive polypyrrole-chitosan composite nanospheres with controllable diameters. *Polymer International, 60*(4), 647–651. https://doi.org/10.1002/PI.2995

Madhan Kumar, A., & Gasem, Z. M. (2015). In situ electrochemical synthesis of polyaniline/f-MWCNT nanocomposite coatings on mild steel for corrosion protection in 3.5% NaCl solution. *Progress in Organic Coatings, 78*, 387–394. https://doi.org/10.1016/J.PORGCOAT.2014.07.009

Mahmoudian, M. R., Alias, Y., & Basirun, W. J. (2012). Effect of narrow diameter polyaniline nanotubes and nanofibers in polyvinyl butyral coating on corrosion protective performance of mild steel. *Progress in Organic Coatings, 75*(4), 301–308. https://www.academia.edu/2147929/Effect_of_narrow_diameter_polyaniline_nanotubes_and_nanofibers_in_polyvinyl_butyral_coating_on_corrosion_protective_performance_of_mild_steel

Martins, J. I., Reis, T. C., Bazzaoui, M., Bazzaoui, E. A., & Martins, L. (2004). Polypyrrole coatings as a treatment for zinc-coated steel surfaces against corrosion. *Corrosion Science, 46*(10), 2361–2381. https://doi.org/10.1016/J.CORSCI.2004.02.006

Martyak, N. M., & McAndrew, P. (2007). Corrosion performance of steel coated with copolyamides and polyaniline. *Corrosion Science, 49*(10), 3826–3837. https://doi.org/10.1016/J.CORSCI.2007.05.013

McAndrew, T. P. (1997). Corrosion prevention with electrically conductive polymers. *Trends in Polymer Science, 1*(5), 7–12. https://www.infona.pl//resource/bwmeta1.element.elsevier-21bc6237-e868-328a-bf0a-4a17cdbb4bd6

Mobin, M., & Shabnam, H. (2010). Corrosion behavior of mild steel and SS 304L in presence of dissolved copper. *Journal of Minerals & Materials Characterization & Engineering, 9*(12), 1113–1130.

Mostafaei, A., & Nasirpouri, F. (2014). Epoxy/polyaniline-ZnO nanorods hybrid nanocomposite coatings: Synthesis, characterization and corrosion protection performance of conducting paints. *Progress in Organic Coatings, 77*(1), 146–159. https://doi.org/10.1016/J.PORGCOAT.2013.08.015

Namsheer, K., & Rout, C. S. (2021). Conducting polymers: A comprehensive review on recent advances in synthesis, properties and applications. *RSC Advances, 11*(10), 5659–5697. https://doi.org/10.1039/D0RA07800J

Nguyen Thi Le, H., Garcia, B., Deslouis, C., & Le Xuan, Q. (2001). Corrosion protection and conducting polymers: Polypyrrole films on iron. *Electrochimica Acta, 46*(26–27), 4259–4272. https://doi.org/10.1016/S0013-4686(01)00699-5

Ocampo, C., Armelin, E., Liesa, F., Alemán, C., Ramis, X., & Iribarren, J. I. (2005). Application of a polythiophene derivative as anticorrosive additive for paints. *Progress in Organic Coatings, 53*(3), 217–224. https://doi.org/10.1016/J.PORGCOAT.2005.02.009

Ohtsuka, T. (2012). Corrosion protection of steels by conducting polymer coating. *International Journal of Corrosion, 2012*. https://doi.org/10.1155/2012/915090

Olad, A., Barati, M., & Behboudi, S. (2012). Preparation of PANI/epoxy/Zn nanocomposite using Zn nanoparticles and epoxy resin as additives and investigation of its corrosion protection behavior on iron. *Progress in Organic Coatings, 74*(1), 221–227. https://doi.org/10.1016/J.PORGCOAT.2011.12.012

Österholm, J. E., Cao, Y., Klavetter, F., & Smith, P. (1993). Emulsion polymerization of aniline. *Synthetic Metals, 55*(2–3), 1034–1039. https://doi.org/10.1016/0379-6779(93)90195-3

Ozyilmaz, A. T., Ozyilmaz, G., & Yigitoglu, O. (2010). Synthesis and characterization of poly(aniline) and poly(o-anisidine) films in sulphamic acid solution and their anticorrosion properties. *Progress in Organic Coatings*, *67*(1), 28–37. https://doi.org/10.1016/J.PORGCOAT.2009.09.010

Palraj, S., Selvaraj, M., Vidhya, M., & Rajagopal, G. (2012). Synthesis and characterization of epoxy-silicone-polythiophene interpenetrating polymer network for corrosion protection of steel. *Progress in Organic Coatings*, *75*(4), 356–363. https://doi.org/10.1016/J.PORGCOAT.2012.07.014

Patil, R. C., & Radhakrishnan, S. (2006). Conducting polymer based hybrid nano-composites for enhanced corrosion protective coatings. *Progress in Organic Coatings*, *57*(4), 332–336. https://doi.org/10.1016/J.PORGCOAT.2006.09.012

Rani, B. E. A., & Basu, B. B. J. (2012). Green inhibitors for corrosion protection of metals and alloys: An overview. *International Journal of Corrosion*, *2012*. https://doi.org/10.1155/2012/380217

Riaz, U., Ashraf, S. M., & Ahmad, S. (2007). High performance corrosion protective DGEBA/polypyrrole composite coatings. *Progress in Organic Coatings*, *59*(2), 138–145. https://doi.org/10.1016/J.PORGCOAT.2007.02.002

Rohwerder, M., & Michalik, A. (2007). Conducting polymers for corrosion protection: What makes the difference between failure and success? *Electrochimica Acta*, *53*(3), 1300–1313. https://doi.org/10.1016/J.ELECTACTA.2007.05.026

Ruhi, G., Bhandari, H., & Dhawan, S. K. (2015). Corrosion resistant polypyrrole/flyash composite coatings designed for mild steel substrate. *American Journal of Polymer Science*, *5*(1A), 18–27. https://doi.org/10.5923/S.AJPS.201501.03

Ruhi, G., Dhawan, S. K., Ruhi, G., & Dhawan, S. K. (2014). Conducting polymer nano composite epoxy coatings for anticorrosive applications. In *Modern Electrochemical Methods in Nano, Surface and Corrosion Science*. https://doi.org/10.5772/58388

Sathiyanarayanan, S., Azim, S. S., & Venkatachari, G. (2007). A new corrosion protection coating with polyaniline-TiO2 composite for steel. *Electrochimica Acta*, *52*(5), 2068–2074. https://doi.org/10.1016/J.ELECTACTA.2006.08.022

Spinks, G. M., Dominis, A. J., Wallace, G. G., & Tallman, D. E. (2002). Electroactive conducting polymers for corrosion control: Part 2. Ferrous metals. *Journal of Solid State Electrochemistry*, *6*(2), 85–100. https://doi.org/10.1007/S100080100211/METRICS

Srinivasa, P. C., & Tharanathan, R. N. (2007). Chitin/Chitosan—Safe, ecofriendly packaging materials with multiple potential uses. *Journal of Cellulose Chemistry and Technology*, *23*(1), 53–72. https://doi.org/10.1080/87559120600998163

Sun, Q., Deng, Y., & Wang, Z. L. (2004). Synthesis and characterization of polystyrene-encapsulated laponite composites via miniemulsion polymerization. *Macromolecular Materials and Engineering*, *289*(3), 288–295. https://doi.org/10.1002/MAME.200300219

Sun, Q., Schork, F. J., & Deng, Y. (2007). Water-based polymer/clay nanocomposite suspension for improving water and moisture barrier in coating. *Composites Science and Technology*, *67*(9), 1823–1829. https://doi.org/10.1016/J.COMPSCITECH.2006.10.022

Tallman, D. E., Spinks, G., Dominis, A., & Wallace, G. G. (2002). Electroactive conducting polymers for corrosion control: Part 1. General introduction and a review of non-ferrous metals. *Journal of Solid State Electrochemistry*, *6*(2), 73–84. https://doi.org/10.1007/S100080100212/METRICS

Tan, C. K., & Blackwood, D. J. (2003). Corrosion protection by multilayered conducting polymer coatings. *Corrosion Science*, *45*(3), 545–557. https://doi.org/10.1016/S0010-938X(02)00144-0

Tanveer, N., & Mobin, M. (2012). Anti-corrosive properties of poly (2-pyridylamine-co-aniline-co-2,3-xylidine) terpolymer coating on mild steel in different corrosive environments. *Progress in Organic Coatings*, *75*(3), 231–240. https://doi.org/10.1016/J.PORGCOAT.2012.05.002

Threlfall, T. L. (1995). Analysis of organic polymorphs. A review. *Analyst*, *120*(10), 2435–2460. https://doi.org/10.1039/AN9952002435

Tüken, T., Yazici, B., & Erbil, M. (2004a). A new multilayer coating for mild steel protection. *Progress in Organic Coatings*, *50*(2), 115–122. https://doi.org/10.1016/J.PORGCOAT.2004.01.002

Tüken, T., Yazici, B., & Erbil, M. (2004b). The use of polythiophene for mild steel protection. *Progress in Organic Coatings*, *51*(3), 205–212. https://doi.org/10.1016/J.PORGCOAT.2004.07.013

Tüken, T., Yazici, B., & Erbil, M. (2005). Electrochemical synthesis of polythiophene on nickel coated mild steel and corrosion performance. *Applied Surface Science*, *239*(3–4), 398–409. https://doi.org/10.1016/J.APSUSC.2004.06.006

Vogel, H., & Marvel, C. S. (1961). Polybenzimidazoles, new thermally stable polymers. *Journal of Polymer Science*, *50*(154), 511–539. https://doi.org/10.1002/POL.1961.1205015419

Wessling, B., & Posdorfer, J. (1999). Corrosion prevention with an organic metal (polyaniline): Corrosion test results. *Electrochimica Acta*, *44*(12), 2139–2147. https://doi.org/10.1016/S0013-4686(98)00322-3

Xu, H., & Zhang, Y. (2019). A review on conducting polymers and nanopolymer composite coatings for steel corrosion protection. *Coatings*, *9*(12), 807. https://doi.org/10.3390/COATINGS9120807

Xu, L., Chen, W., Mulchandani, A., & Yan, Y. (2005). Reversible conversion of conducting polymer films from superhydrophobic to superhydrophilic. *Angewandte Chemie International Edition*, *44*(37), 6009–6012. https://doi.org/10.1002/ANIE.200500868

Zeng, F., Qin, Z., Liang, B., Li, T., Liu, N., & Zhu, M. (2015). Polyaniline nanostructures tuning with oxidants in interfacial polymerization system. *Progress in Natural Science: Materials International*, *25*(5), 512–519. https://doi.org/10.1016/J.PNSC.2015.10.002

Zhang, X.-Q., Shen, X.-M., Yang, S.-Y., & Zhang, J.-Y. (1995). New conducting polythiophenes containing substituted polyfluorinated alkoxy chain. *Chinese Journal of Chemistry*, *13*(1), 79–84. https://doi.org/10.1002/CJOC.19950130112

7 Copolymers, multilayer conducting polymers, and doped conducting polymers as anti-corrosive materials

Rakesh Kumar Sahu and Veenu Chandraker

7.1 INTRODUCTION

Corrosion: Corrosion is a natural process that occurs when certain materials, particularly metals, react with their environment and undergo a deterioration or degradation. It is commonly seen as the rusting of iron or steel, but corrosion can affect a wide range of materials, including aluminum, copper, brass, and even non-metallic substances. Corrosion is primarily an electrochemical process, involving the transfer of electrons between the metal and the surrounding environment. It typically requires the presence of three key elements: a metal (or anode), an electrolyte (such as water or humidity), and oxygen (or another oxidizing agent). When these elements come together, a series of chemical reactions occur, leading to the breakdown of the metal and the formation of corrosion products [1].

Corrosion can occur in various forms, including uniform corrosion, localized corrosion, and galvanic corrosion. Uniform corrosion happens evenly over the entire surface of the metal, whereas localized corrosion occurs in specific areas, such as pitting corrosion and crevice corrosion. Galvanic corrosion happens when two different metals are in contact in the presence of an electrolyte, leading to the more active metal corroding [1, 2].

Corrosion can have significant economic and safety implications, as it can weaken structures, reduce the lifespan of equipment, and compromise the integrity of various objects. To prevent or mitigate corrosion, various protective measures can be employed, such as applying coatings or inhibitors, using corrosion-resistant materials, implementing proper design techniques, and controlling the environment (humidity, temperature, etc.). Regular inspection, maintenance, and corrosion management practices are also essential for addressing and minimizing the impact of corrosion [3, 4].

 DOI: 10.1201/9781003456025-7

To prevent or mitigate corrosion, several methods can be employed. These include the following:

1. **Protective Coatings**: Applying coatings like paint, enamel, or corrosion-resistant polymers can create a barrier between the metal and its environment, preventing direct contact and reducing corrosion.
2. **Cathodic Protection**: This method involves making the metal to be protected act as a cathode by connecting it to a sacrificial anode made of a more active metal. The sacrificial anode corrodes instead of the protected metal.
3. **Alloying**: Mixing metals together to form alloys can enhance corrosion resistance. Stainless steel, for example, contains chromium, which forms a passive oxide layer that protects the underlying metal from corrosion.
4. **Environmental Modifications**: Controlling the environment in which the metal is exposed can help reduce corrosion. This can include reducing moisture, controlling temperature, or adjusting the pH level of the surroundings.
5. **Regular Maintenance**: Inspecting and maintaining metal structures or objects can help identify and address corrosion issues early on, preventing further damage [1–4].

It's important to note that while these methods can significantly reduce corrosion, no method provides complete immunity. Continuous monitoring and appropriate measures should be taken to prevent and manage corrosion effectively.

Anti-Corrosive Materials: Anti-corrosive materials are substances or coatings that are specifically designed to protect surfaces from corrosion, which is the gradual deterioration of a material due to chemical reactions with its environment. Corrosion can occur due to factors such as moisture, oxygen, chemicals, and temperature [5].

There are several types of anti-corrosive materials commonly used to protect against corrosion:

1. **Metals and Alloys**: Some metals and alloys naturally possess resistance to corrosion. For example, stainless steel is an alloy containing chromium, which forms a passive oxide layer on its surface, protecting it from corrosion. Aluminum, titanium, and certain types of nickel alloys also exhibit good corrosion resistance.
2. **Protective Coatings**: Coatings are often applied to the surface of materials to act as a barrier between the material and its corrosive environment. These coatings can be organic or inorganic, and they form a protective layer that prevents direct contact between the material and corrosive substances. Examples include epoxy coatings, polyurethane coatings, and ceramic coatings.
3. **Galvanization**: Galvanization involves applying a protective layer of zinc onto the surface of a metal. The zinc layer acts as a sacrificial anode,

meaning it corrodes preferentially over the underlying metal, thereby protecting it from corrosion. Galvanization is commonly used to protect steel structures, such as fences, pipelines, and car bodies.

4. **Cathodic Protection**: Cathodic protection is a technique used to protect metal structures from corrosion by making them the cathode of an electrochemical cell. This is typically achieved by connecting the metal to be protected to a sacrificial anode, such as zinc or magnesium. The sacrificial anode corrodes instead of the protected metal, effectively preventing corrosion.
5. **Corrosion Inhibitors**: Corrosion inhibitors are chemicals that can be added to the environment surrounding a material to reduce or prevent corrosion. These inhibitors work by either forming a protective layer on the material's surface or altering the chemical environment to make it less corrosive. Common corrosion inhibitors include organic compounds like amines, phosphates, and silicates.
6. **Composite Materials**: Composite materials combine different substances to create a material with improved properties, including corrosion resistance. For example, fiberglass-reinforced plastics (FRP) are lightweight and corrosion-resistant, making them suitable for applications where metal corrosion is a concern [6, 7].

The selection of anti-corrosive materials depends on various factors, such as the type of material being protected, the corrosive environment, and the desired durability. It's important to choose the appropriate anti-corrosive material or combination of materials based on the specific application to effectively mitigate the effects of corrosion.

7.1.1 Copolymers, multilayer conducting polymers, and doped conducting polymers

Copolymers, multilayer conducting polymers, and doped conducting polymers can be used as anti-corrosive materials due to their unique properties and capabilities [8–11]. Here's an overview of each of these materials and their anti-corrosion applications.

Copolymers: Copolymers are polymers composed of two or more different monomer units. By combining different monomers, copolymers can exhibit a wide range of physical and chemical properties. In the context of anti-corrosion, copolymers can be designed to possess specific characteristics that make them resistant to corrosion. For example, copolymers containing fluorine atoms (fluoropolymers) are known for their excellent chemical resistance and can be used to protect surfaces from corrosive environments [8, 12, 13]. They create a barrier that prevents the interaction between the corrosive agents and the substrate.

Multilayer Conducting Polymers: Multilayer conducting polymers consist of alternating layers of conducting and insulating polymers. The conducting layers provide electrical conductivity, while the insulating layers act as barriers against corrosive species. These multilayer structures can be designed to have a high degree of corrosion resistance by tailoring the thickness and composition of each layer. The conducting layers can also provide active corrosion protection by acting as sacrificial anodes, where they undergo controlled oxidation to protect the underlying metal substrate [9, 14–16].

Doped Conducting Polymers: Doped conducting polymers are formed by introducing dopant molecules into the polymer matrix, which modify the electrical and chemical properties of the material. Doping can enhance the electrical conductivity of polymers, making them suitable for applications where electrochemical corrosion protection is required. Doped conducting polymers can act as corrosion inhibitors by forming a protective layer on the metal surface, which prevents corrosive species from reaching the substrate. They can also function as active corrosion protection materials by providing cathodic or anodic polarization to inhibit corrosion reactions [10, 11, 15].

These are well described in terms of the type of charge carriers within, that is, polarons and bipolarons. When these positively charged species reside on carbon atoms in the conjugated backbone, they represent carbocations. In the case of doped conjugated polymers, an electron is removed from the valence band by oxidation (p-doping) or is added to the conducting band by reduction (n-doping).

p-Doping: It is done by oxidation process. In this process, the conducting polymer is treated with a Lewis acid. P-doping increases the mobility of electrons in these delocalized orbitals, and the polymer becomes highly conductive.

$$\underset{\text{Polyacetylene}}{\left(C_2H_2\right)_n} + \underset{\text{Lewis acid}}{A} \rightleftharpoons \underset{p\text{–Doped polyacetylene}}{\left(C_2H_2\right)_n + A^-}$$

n-Doping: It is done by reduction process. In this process, the conducting polymer is treated with a Lewis base.

$$\underset{\text{Polyacetylene}}{\left(C_2H_2\right)_n} + \underset{\text{Lewis base}}{B} \rightleftharpoons \underset{p\text{–Doped polyacetylene}}{\left(C_2H_2\right)_n^{-}\ B^+}$$

7.2 SYNTHESIS METHOD OF COPOLYMERS, MULTILAYER CONDUCTING POLYMERS, AND DOPED CONDUCTING POLYMERS

Synthesis methods for copolymers, multilayer conducting polymers, and doped conducting polymers vary depending on the specific polymers and desired properties. Here's a general overview of the synthesis methods for each:

7.2.1 Synthesis method of copolymers

Copolymers are composed of two or more different monomers that are polymerized together. The development of controlled polymerization made possible the synthesis of polymers of varying structures, like block copolymers, macro-cycles polymers, cylindrical brushes, graft copolymers, gradient copolymers, hyper-branched polymers, and miktoarm stars. Each polymer has exclusive characteristics that are absent in related linear polymers with similar chemical composition. They can be synthesized using various methods, each offering different advantages and control over the resulting copolymer structure [17, 18].

There are several methods to synthesize copolymers:

1) **Free Radical Polymerization**: Free radical polymerization is a versatile method used to synthesize copolymers, which are polymers composed of two or more different monomers. In this process, monomers with different chemical structures are polymerized together using free radicals as initiators [19]. **Figure 7.1** shows free radical polymerization of vinyl acetate and methyl methacrylate.

   There are the steps involved in free radical copolymerization:

   a) **Monomer Selection**: Choose two or more monomers with different structures that are capable of undergoing free radical polymerization. For example, you might select monomer A and monomer B to form a copolymer.
   b) **Initiation**: The copolymerization reaction begins with the generation of free radicals, which are highly reactive species with unpaired electrons. Free radicals can be produced by various initiation methods, such as thermal decomposition of a suitable initiator or by using photo initiators that are activated by light.
   c) **Propagation**: The free radicals react with monomers A and B, resulting in the growth of polymer chains. For instance, a free radical can react with monomer A to form a growing chain with an active center, followed by the addition of monomer B to incorporate units of monomer B into the growing chain. This process continues, with alternating addition of monomers A and B, or can result in random incorporation of monomers, depending on the reactivity ratios of the monomers.
   d) **Termination**: Eventually, the polymer chains will terminate, either by recombination of two growing chains or by reaction with a terminating agent, which can be a molecule that readily reacts with the active centers of the growing chains. Termination reactions prevent further growth of the polymer chains.
   e) **Copolymer Composition and Structure**: The composition of the resulting copolymer is determined by the feed ratio of the monomers. The arrangement of the monomer units along the copolymer chain can be random (statistical copolymer), alternating, blocky (block copolymer), or have other specific patterns, depending on the reactivity ratios and conditions of the polymerization.

f) **Purification and Characterization**: After the polymerization, the copolymer needs to be purified to remove any residual monomers, unreacted initiators, or other impurities. Common purification methods include precipitation, extraction, or chromatography. The copolymer can then be characterized using techniques such as nuclear magnetic resonance (NMR), gel permeation chromatography (GPC), or infrared spectroscopy (IR) to determine its molecular weight, composition, and structure.

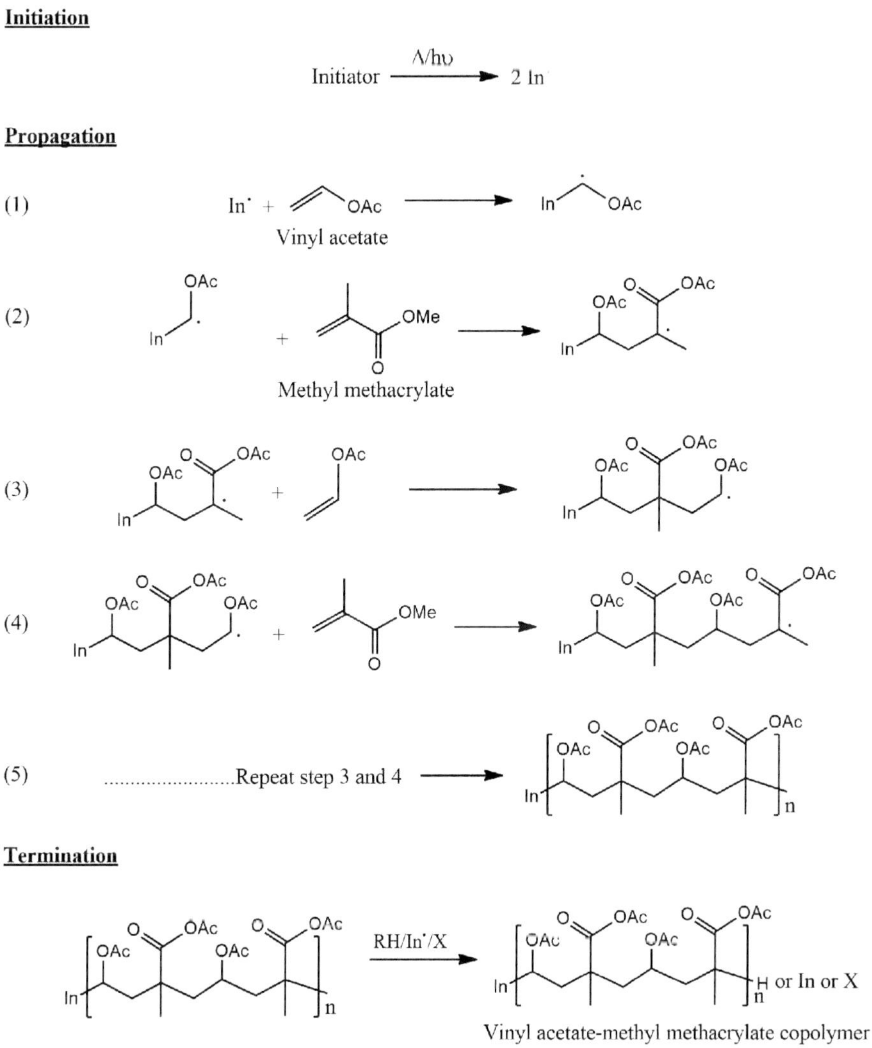

FIGURE 7.1 Free radical polymerization reaction of vinyl acetate and methyl methacrylate.

Free radical polymerization is a widely used method to produce copolymers due to its simplicity, versatility, and ability to incorporate various monomers. The properties of the resulting copolymers can be tailored by adjusting the monomer feed ratio, choice of initiators, reaction conditions, and copolymerization strategy, offering a wide range of applications in materials science, coatings, adhesives, and other fields.

2) **Condensation Polymerization**: Condensation polymerization is a type of polymerization reaction in which monomers react together to form a polymer while simultaneously producing a small molecule as a byproduct, such as water or alcohol. It is also known as step-growth polymerization because the polymer chain grows step by step, with each step involving the joining of two monomers.

In condensation polymerization, the monomers contain two or more functional groups that can react with each other. When these functional groups react, they form covalent bonds, resulting in the formation of a polymer chain. At the same time, a small molecule is eliminated as a byproduct. The reaction typically occurs under mild conditions, such as room temperature or slightly elevated temperatures. It can be carried out in the presence or absence of a catalyst, depending on the specific monomers involved. The process of condensation polymerization involves repeated cycles of monomer reaction and byproduct elimination. As a result, the polymer chain grows longer with each step, and the molecular weight of the polymer increases. The reaction continues until the monomers are completely consumed or until the desired molecular weight or polymer length is achieved [20].

Examples of polymers formed by condensation polymerization include polyesters, polyamides, polyurethanes, and certain types of silicone polymers. **Figures 7.2** and **7.3** show condensation polymerization.

It is worth noting that condensation polymerization is distinct from addition polymerization, where the polymer chain grows by the repeated addition of monomers without the production of small byproducts [21].

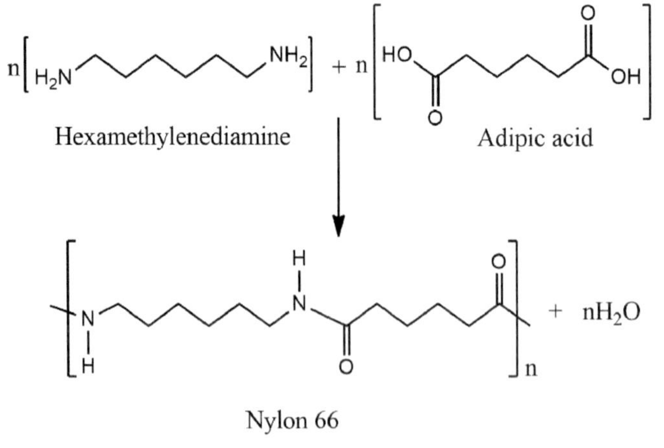

FIGURE 7.2 Condensation polymerization of hexamethylenediamine and adipic acid.

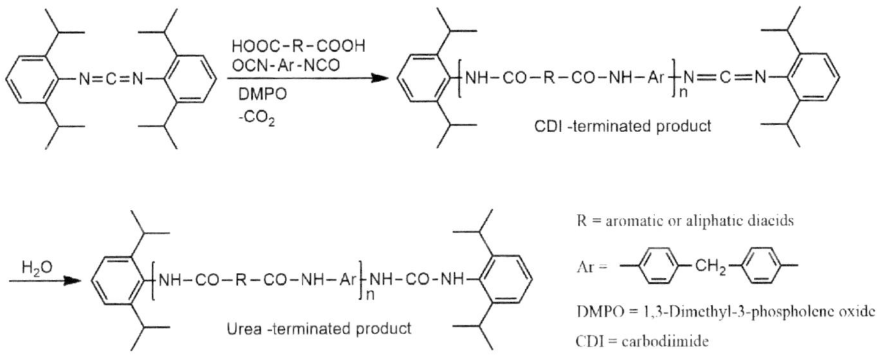

FIGURE 7.3 Condensation polymerization.

Condensation polymerization is a versatile method for synthesizing a wide range of polymers with different properties, and it finds applications in various industries such as plastics, fibers, adhesives, and coatings.

3) **Addition Polymerization**: Addition polymerization involves the sequential addition of monomers without the elimination of any by-products. Copolymers can be synthesized by adding different monomers to the reaction mixture, resulting in random or block copolymers. Common examples of copolymers synthesized through addition polymerization include styrene-butadiene rubber (SBR) and ethylene-vinyl acetate (EVA) copolymers [22].

   In the context of addition polymerization for copolymers, the process involves the simultaneous polymerization of two or more monomers with different chemical structures or functional groups. The resulting copolymer contains repeating units from each monomer, giving it a unique composition and properties that differ from those of homopolymers (polymers derived from a single monomer).

   There are two main types of copolymers formed through addition polymerization:

   a) **Random Copolymers**: In random copolymers, the monomer units are arranged in a random sequence along the polymer chain. The polymerization reaction involves the simultaneous addition of different monomers, resulting in the formation of copolymers with varying arrangements of monomer units. The composition of the copolymer can be controlled by adjusting the feed ratio of the monomers during polymerization [22, 23].
   b) **Block Copolymers**: Block copolymers are composed of long sequences or blocks of one monomer followed by another. These copolymers are formed by polymerizing one monomer to form a block and then adding another monomer to continue the polymerization and form a second block. Block copolymers exhibit distinct domains of each monomer along the polymer chain, leading to unique properties and phase separation [22, 24].

The specific polymerization mechanism and conditions for copolymers depend on the monomers used and the desired properties of the final copolymer. Different monomers may require different catalysts, initiators, or

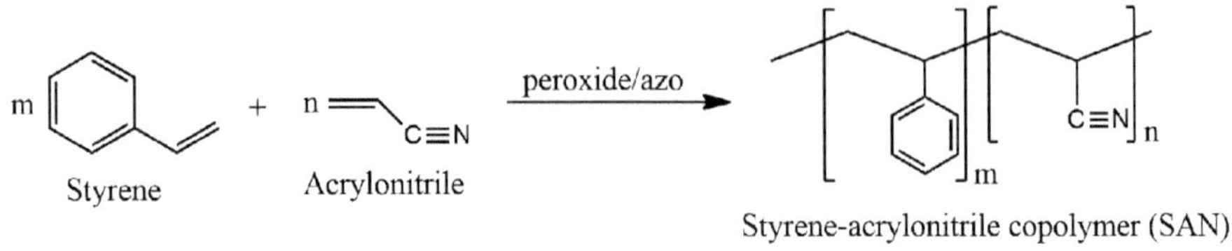

FIGURE 7.4 Addition polymerization of styrene and acrylonitrile.

reaction conditions to achieve efficient polymerization and control over the copolymer structure. **Figure 7.4** shows the addition polymerization of styrene and acrylonitrile.

Addition polymerization for copolymers is a versatile approach that allows the combination of different monomers to obtain materials with tailored properties and applications. It finds wide use in industries such as plastics, elastomers, coatings, adhesives, and more.

4) **Ring-Opening Polymerization**: This method is commonly used for the synthesis of copolymers derived from cyclic monomers, such as lactones or cyclic carbonates. In ring-opening polymerization, a reactive species, such as an initiator or a catalyst, is used to open the cyclic monomers and initiate the polymerization reaction. By using different monomers, copolymers with diverse properties can be obtained [25].

   Here's a general overview of the steps involved in the ring-opening polymerization of copolymers:

   a) **Selection of Monomers**: Choose two or more monomers that will undergo ring-opening polymerization to form a copolymer. These monomers should be cyclic and have reactive groups that can initiate polymerization.
   b) **Initiator**: An initiator is required to start the polymerization reaction. Initiators for ROP can include various types of catalysts, such as metal complexes, organic compounds, or even heat. The choice of initiator depends on the specific monomers and reaction conditions.
   c) **Polymerization**: The monomers are subjected to the polymerization reaction, which typically involves heating the reaction mixture to a specific temperature. The initiator activates the monomers, leading to the opening of their ring structures and the formation of polymer chains.
   d) **Copolymer Composition**: The copolymer composition is determined by the feed ratio of the different monomers. By controlling the molar ratio of the monomers, it is possible to produce copolymers with varying compositions and properties.
   e) **Termination**: The polymerization continues until the monomers are consumed or until the reaction is terminated by adding a suitable terminating agent. This step helps in controlling the molecular weight and dispersity of the copolymer.
   f) **Purification**: After the polymerization is complete, the copolymer is typically purified to remove any unreacted monomers, initiators, or other impurities. Purification methods may include solvent evaporation, precipitation, or chromatography, depending on the nature of the copolymer and its solubility.

An example of ROP for copolymers is the copolymerization of lactide and glycolide monomers to form a class of polymers known as poly(lactic-co-glycolic acid) or PLGA. **Figure 7.5** shows ring-opening polymerization reaction of L-lactide (LLA) and glycolide (GA).

Ring-opening polymerization can be controlled to achieve desired polymer properties by carefully selecting the monomer, initiator/catalyst, reaction conditions, and termination mechanisms. The choice of monomer and reaction conditions determines the type of polymer formed, while the selection of initiator or catalyst influences the rate of polymerization and the molecular weight of the resulting polymer. An example is shown in **Figure 7.6** using the ROP of trimethylene carbonate, initiated by a [Zn]-mediated initiator in the presence of an alcohol (R'OH) [25, 26].

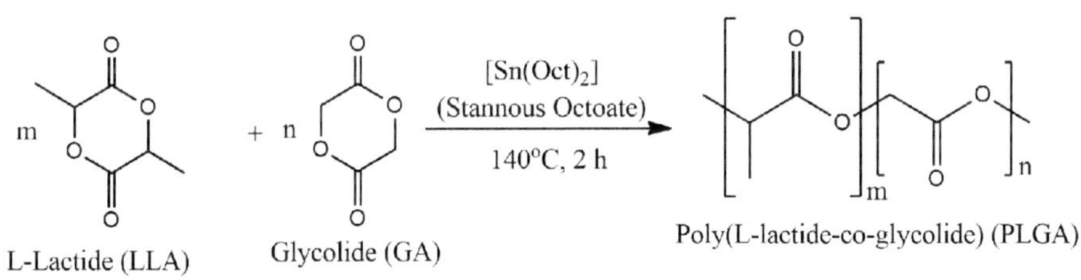

FIGURE 7.5 Ring-opening polymerization reaction of LLA and GA.

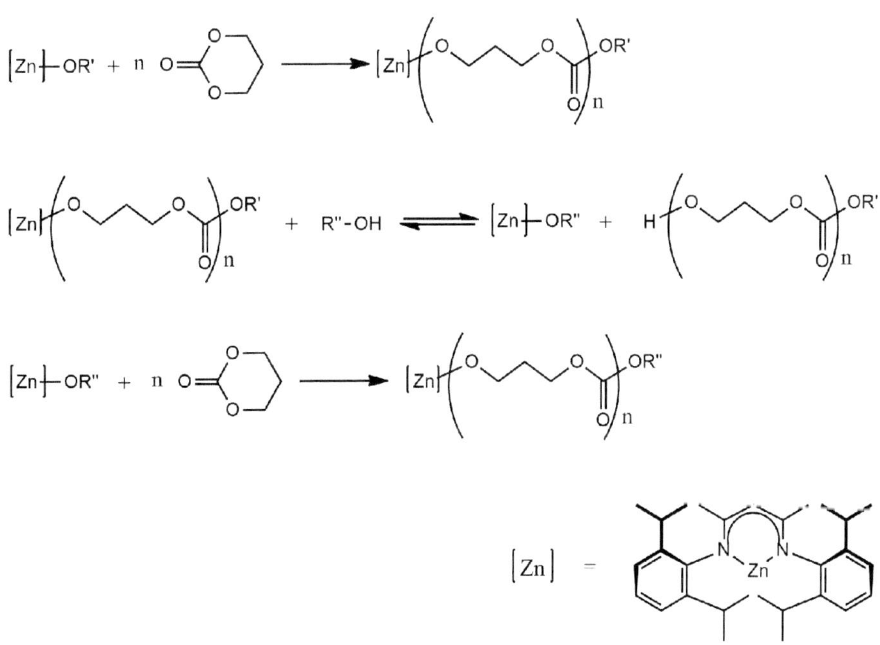

FIGURE 7.6 [Zn]-mediated living ring-opening polymerization of trimethylene carbonate.

Ring-opening polymerization has found wide application in various industries, including biomedical engineering, packaging, textiles, and coatings. The ability to control the molecular weight, structure, and properties of the resulting polymers makes ROP a versatile technique for the synthesis of functional materials [25, 27, 28].

5) **Living/Controlled Polymerization Techniques**: Living or controlled polymerization is a type of polymerization process in which the growth of polymer chains occurs in a controlled manner, allowing for precise control over the molecular weight and structure of the resulting polymers. Unlike traditional polymerization methods, which often result in polymers with broad molecular weight distributions and heterogeneous structures, living polymerization enables the synthesis of polymers with well-defined characteristics.

   In living polymerization, a reactive species called an initiator initiates the polymerization reaction. The initiator typically consists of a functional group that can initiate polymerization, such as a free radical or a Lewis acid or base. The key feature of living polymerization is that the reaction can be controlled to prevent chain termination and allow for the continuous growth of polymer chains [29, 30].

   There are several types of living polymerization techniques, including anionic polymerization, cationic polymerization, and controlled radical polymerization. Each technique has its own specific requirements and mechanisms, but they all share the common characteristic of controlled chain growth.

   a) Anionic polymerization involves the use of anionic initiators, such as alkyl lithium compounds or Grignard reagents, which can initiate the polymerization of monomers. The reaction proceeds by the addition of monomer units to the growing polymer chain through a nucleophilic attack on the active center. Anionic polymerization is typically used for the synthesis of polydiene rubbers and certain types of block copolymers.
   b) Cationic polymerization, on the other hand, relies on cationic initiators, such as Lewis acids or protic acids, to initiate the polymerization. The reaction proceeds through the formation of a carbocation intermediate, which reacts with monomer units to grow the polymer chain. Cationic polymerization is commonly used for the synthesis of polymers such as polyolefins, polyesters, and polyethers.
   c) Controlled radical polymerization techniques, such as atom transfer radical polymerization (ATRP), reversible addition-fragmentation chain transfer (RAFT) polymerization, and nitroxide-mediated polymerization (NMP), have gained significant attention in recent years. These methods utilize specific initiators and additives to control the radical polymerization process, allowing for the precise control of molecular weight, architecture, and end-group functionality of the resulting polymers.

   The three most promising CRP methods developed are stable-free radical polymerization (SFRP), which is commonly referred to as nitroxide mediated polymerization (NMP) but may also include organometallic species (**Figure 7.7 (1)** [31]); transition-metal-catalyzed atom transfer radical polymerization (ATRP) (**Figure 7.7 (2)** [31]); and reversible addition-fragmentation chain transfer (RAFT) polymerization, which involves degenerative transfer with alkyl iodides, methacrylate macromonomers, and dithioesters (**Figure 7.7 (3)** [31]) [31–34].

An example of the 2,2,6,6-tetramethylpiperidine-1-oxyl (MTEMPO)-mediated photocontrolled/living radical polymerization of Methacrylic acid (MAA) is shown in **Figure 7.8** [35].

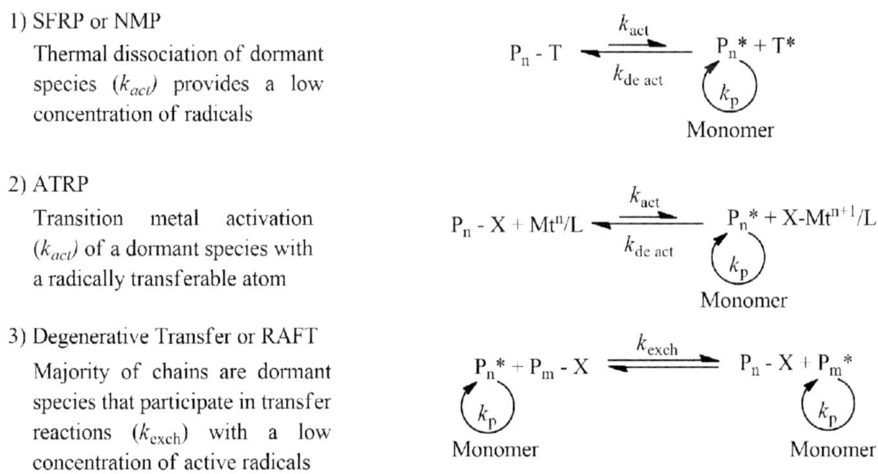

FIGURE 7.7 The three main CRP methods.

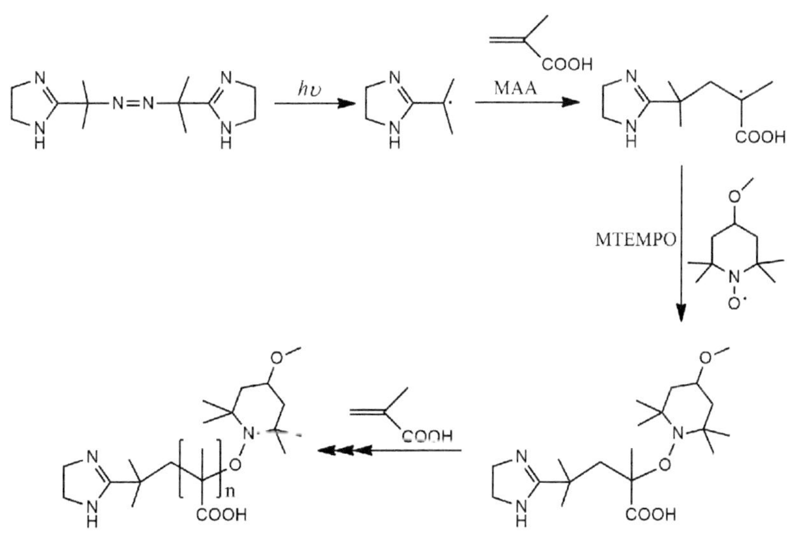

FIGURE 7.8 The mechanism of the MTEMPO-mediated photocontrolled/living radical polymerization of MAA.

Living polymerization techniques have found wide applications in various areas, including the synthesis of advanced materials, drug delivery systems, coatings, adhesives, and surface modifications. The ability to control the polymerization process and tailor the properties of polymers has opened up new avenues for the development of functional materials with specific properties and applications [36, 37].

7.2.2 Synthesis Method of Multilayer Conducting Polymers

The synthesis of multilayer conducting polymers can be achieved through various methods, including the layer-by-layer (LBL) assembly technique, electro-polymerization, and chemical vapor deposition (CVD). These are the synthesis methods [38, 39]:

1) **Layer-by-Layer (LBL) Assembly Technique**
   The LBL assembly technique involves the sequential deposition of conducting polymers and oppositely charged species onto a substrate [40]. The process typically involves the following steps:
   a) **Substrate Preparation**: Prepare a suitable substrate, such as glass, silicon, or a conducting material like indium tin oxide (ITO). Clean the substrate to ensure its surface is free from contaminants.
   b) **Polymer Deposition**: Deposit the first layer of a conducting polymer onto the substrate using techniques such as spin-coating, dip-coating, or vapor deposition.
   c) **Counterion Adsorption**: Immerse the substrate in a solution containing a counterion species that can electrostatically interact with the charged polymer. Repeat this step by depositing alternating layers of conducting polymer and counterion species until the desired number of layers is achieved.
   d) **Rinse and Dry**: Rinse the substrate to remove any unbound species and dry it gently to ensure the adhesion of the layers.
   e) **Characterization and Modification**: Characterize the multilayer conducting polymer structure using techniques such as scanning electron microscopy (SEM), atomic force microscopy (AFM), or spectroscopic methods. You can also modify the structure by incorporating dopants, functional groups, or other molecules to enhance its properties.

   The assembly technologies used to assemble such films form five distinct categories, namely: (i) immersive, (ii) spin, (iii) spray, (iv) electromagnetic, and (v) fluidic assembly, shown in **Figure 7.9**. [41].
2) **Electropolymerization**
   Electropolymerization is a technique used to deposit thin films of conducting polymers onto a conductive substrate through an electrochemical process. Multi-layer conducting polymerization refers to the sequential deposition of multiple layers of conducting polymers, which can offer enhanced functionality or tailored properties [42, 43].

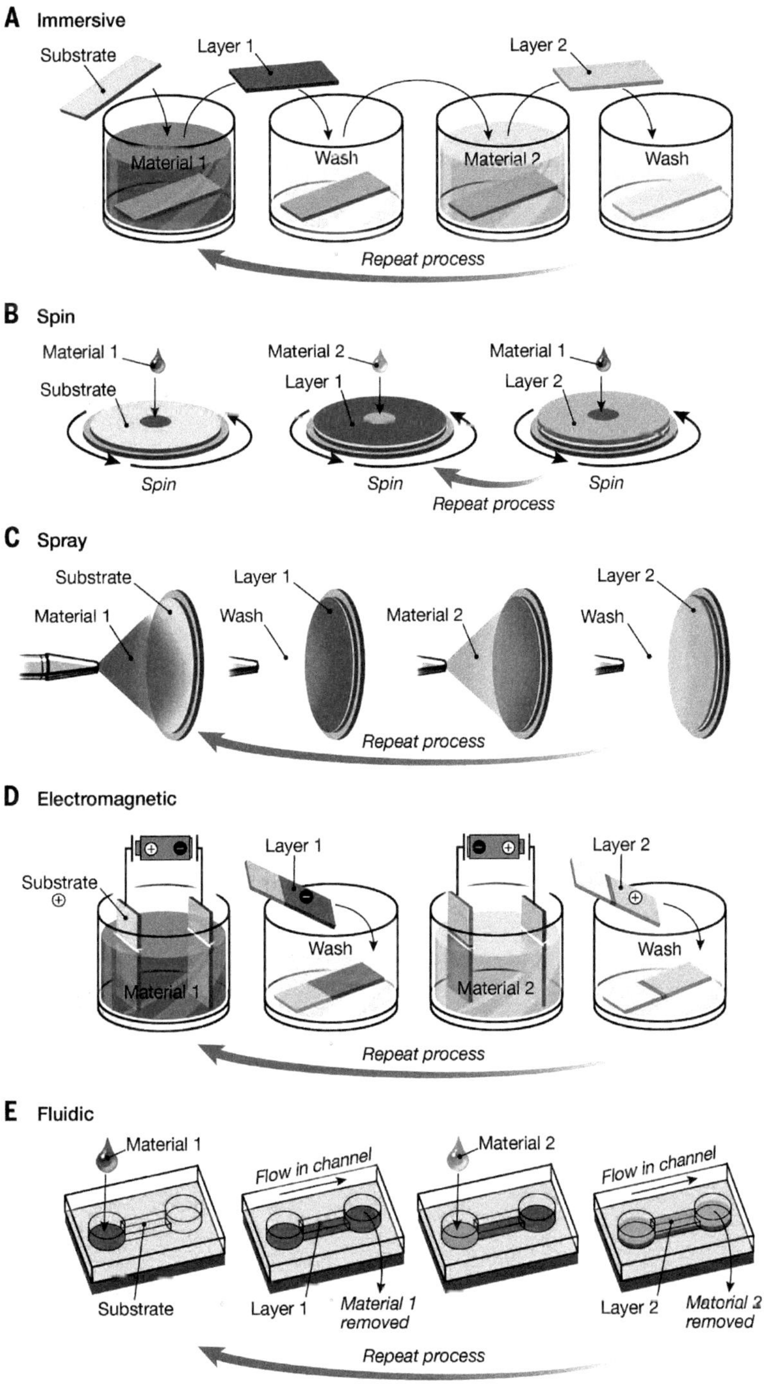

FIGURE 7.9 Layer-by-layer assembly technologies. (a) Immersive assembly, (b) Spin assembly, (c) Spray assembly, (d) Electromagnetic assembly, (e) Fluidic assembly.

The process of multi-layer electropolymerization typically involves the following steps:

a) **Substrate Preparation**: The substrate on which the conducting polymers will be deposited should be properly cleaned and prepared. It is important to ensure a clean and conductive surface for effective polymerization.
b) **Electrolyte Selection**: An appropriate electrolyte solution is chosen based on the specific conducting polymer being used. The electrolyte should be compatible with the monomer and provide the necessary ionic species for the polymerization reaction.
c) **Electrochemical Cell Setup**: The substrate is placed in an electrochemical cell along with a counter electrode and a reference electrode. The cell is filled with the electrolyte solution, creating a complete electrochemical circuit.
d) **Polymerization Process**: The electrochemical cell is subjected to a controlled potential or current, depending on the specific polymerization conditions. The potential or current is applied for a specific duration, allowing the monomer to undergo oxidation or reduction and form a conducting polymer layer on the substrate.
e) **Layer-by-Layer Deposition**: After the first layer of conducting polymer is deposited, the process can be repeated to form subsequent layers. This can be achieved by adjusting the electrochemical conditions, such as the potential or current, to control the growth of each layer. Alternatively, different monomers can be used for each layer to obtain desired properties.
f) **Characterization and Optimization**: Throughout the multi-layer electropolymerization process, it is essential to monitor and characterize the properties of the deposited layers. This can be done using techniques such as cyclic voltammetry (CV), impedance spectroscopy, or scanning electron microscopy (SEM). The process parameters can be optimized based on the desired properties of the conducting polymer film.

By repeating the layer-by-layer deposition process, it is possible to create complex structures and tailor the properties of the conducting polymer film. This technique finds applications in various fields, including electronics, sensors, energy storage, and bioelectronics [42, 43].

Wei Feng et al. synthesized the GO/PANI composite and studied the result of UV-vis spectra and FTIR indicate that PANI (polyaniline) in the composite is in the p-doped state. The p-doped state refers to the presence of positively charged carriers (holes) within the PANI structure. This p-doped state is beneficial for the performance of the supercapacitor. Graphene oxide (GO) and PANI are combined in the composite through various interactions, including electrostatic interactions, hydrogen bonding, and pi-pi stacking. These interactions contribute to the structural and functional properties of the composite material [44, 45].

Electrostatic interactions occur between the charged components of GO and PANI, leading to their attraction and binding. Hydrogen bonding involves the formation of hydrogen bonds between functional groups present in GO and PANI, providing additional stability to the composite structure. π-π stacking refers to the stacking of aromatic rings in PANI and GO, which further enhances the structural integrity and properties of the composite.

The mentioned interactions, namely electrostatic interactions, hydrogen bonding, and π-π stacking, are depicted in **Figure 7.10**, which presumably provides a visual representation or diagram of the proposed composite structure [44, 45].

3) **Chemical Vapor Deposition (CVD)**

Chemical vapor deposition (CVD) is a commonly used technique for depositing thin films of various materials, including conducting polymers [46]. Conducting polymers are a class of polymers that exhibit electrical conductivity due to their ability to delocalize electrons within their molecular structure.

To achieve multi-layer conducting polymerization using CVD, you would typically follow these steps:

a) **Substrate Preparation**: Start by preparing the substrate on which the multi-layer conducting polymer film will be deposited. The substrate should be clean and free of any contaminants that could affect film quality.
b) **Reactant Selection**: Choose appropriate monomers or precursors for the conducting polymer you want to deposit. Different conducting polymers have different requirements, so select the materials based on your desired properties and applications.

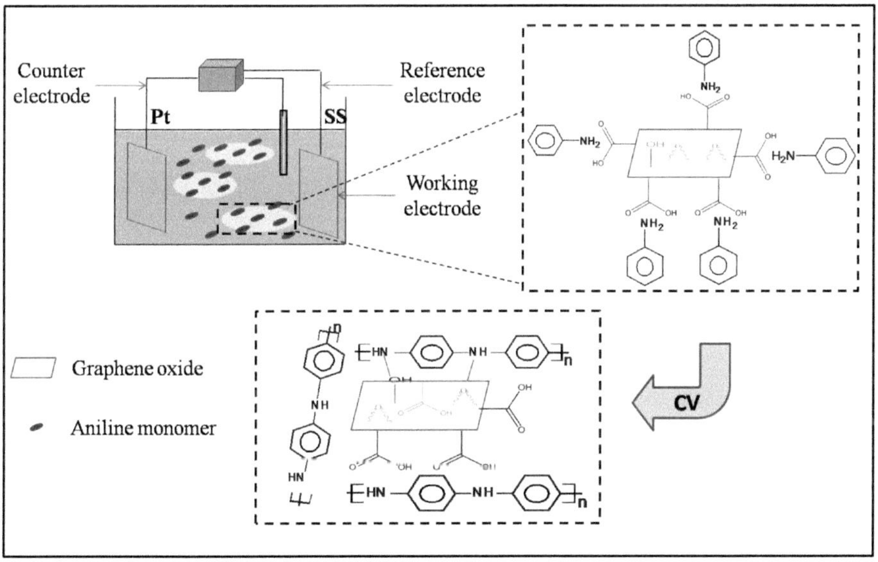

FIGURE 7.10 Schematic Illustration of the proposed possible formation process of GO/PANI composite.

c) **Reactor Setup**: Set up a CVD reactor system suitable for your specific requirements. The reactor typically consists of a reaction chamber, gas sources, temperature control, and a means to introduce the monomers or precursors into the reactor.
d) **Pre-treatment**: Some substrates may require a pre-treatment step to enhance adhesion between the substrate and the deposited film. This can include surface cleaning, functionalization, or the deposition of a thin adhesion layer.
e) **Deposition Process**: Begin the deposition process by introducing the precursor gases into the reactor chamber. The precursors should be chosen to enable the polymerization of the conducting polymer under the conditions of the CVD process.
f) **Polymerization**: Initiate the polymerization reaction by applying energy to the system. This can be achieved through various methods such as thermal energy, plasma, or light irradiation, depending on the specific conducting polymer and reaction conditions.
g) **Layer-by-Layer Deposition**: To achieve multi-layer deposition, you will need to repeat the deposition process multiple times. After each deposition cycle, you can modify the reaction conditions or switch to different precursor gases to create distinct layers with desired properties.
h) **Characterization**: Once the desired number of layers has been deposited, the resulting multi-layer conducting polymer film should be characterized to evaluate its properties. This can involve various techniques such as electrical conductivity measurements, spectroscopy, microscopy, and other relevant analyses.

The specific details of the CVD process for multi-layer conducting polymerization can vary depending on the conducting polymer being used, the desired film properties, and the available equipment [46, 47].

Conducting polymer synthesis by CVD can be challenging due to the requirement for a volatile monomer and dopant combination, as well as the need for suitable oxidants and initiators that can be vaporized. Moreover, the deposition conditions and choice of substrate must be carefully optimized to ensure successful polymerization and film quality [47].

While CVD is commonly employed for depositing thin films of various materials, including polymers, conducting polymerization using CVD is less prevalent. Solution-based methods remain the primary approach for synthesizing conducting polymers, but CVD could offer an alternative route with specific advantages in certain applications.

7.2.3 Doped conducting polymers

Doping is the process of introducing dopant molecules into conducting polymers to enhance their electrical conductivity [48, 49]. Doped conducting polymers can be synthesized by two main methods:

1) **In-Situ Doping:** It is a technique used in conducting polymerization to introduce dopant molecules into the polymer matrix during the polymerization

process itself. The monomers and dopants are mixed together, and polymerization occurs in the presence of the dopant. This approach allows for the simultaneous synthesis and doping of conducting polymers, resulting in materials with enhanced electrical conductivity.

Conducting polymers are a unique class of polymers that can exhibit electrical conductivity similar to metals and semiconductors. However, their inherent electrical conductivity is generally low, and they need to be doped to achieve the desired conducting properties. Doping involves the introduction of dopant molecules that donate or accept charge carriers, thereby increasing the electrical conductivity of the polymer.

In contrast, in-situ doping allows for the simultaneous incorporation of dopant molecules during the polymerization reaction. This approach offers several advantages, including improved control over the doping process, enhanced doping efficiency, and better dispersion of dopant molecules within the polymer matrix. In-situ doping can be achieved through various techniques, such as chemical oxidation, electrochemical methods, or the use of pre-doped monomers.

Chemical oxidation is a common approach for in-situ doping of conducting polymers. In this method, an oxidizing agent is added to the reaction mixture during the polymerization process. The oxidizing agent facilitates the formation of charged sites within the growing polymer chain, effectively doping the polymer. The choice of the oxidizing agent and its concentration can be tailored to achieve the desired doping level.

Electrochemical methods can also be employed for in-situ doping. In these approaches, the polymerization reaction takes place in an electrochemical cell where the monomers are electrochemically polymerized in the presence of a dopant species. The electrochemical process allows for precise control over the doping level by adjusting the applied potential and the concentration of the dopant species.

Another strategy for in-situ doping involves the use of pre-doped monomers. In this case, the monomers themselves are pre-doped with dopant molecules before the polymerization reaction. During the polymerization process, the pre-doped monomers undergo polymerization, resulting in the simultaneous synthesis and doping of the conducting polymer.

In-situ doping techniques offer advantages in terms of improved control over the doping process, enhanced doping efficiency, and better distribution of dopant molecules within the polymer matrix. These techniques have been widely used to fabricate conducting polymers with tailored electrical conductivity for various applications, including organic electronics, sensors, batteries, electrochromic devices, and coatings [50–52].

2) **Post-Synthesis Doping:** Post-synthesis doping is another method used to introduce dopant species into conducting polymers after their synthesis or formation. Unlike in-situ doping, post-synthesis doping involves treating preformed conducting polymers with dopant molecules to enhance their conductivity. This method offers flexibility in controlling the dopant concentration and allows for fine-tuning of the electrical properties of the conducting polymer.

The general procedure for post-synthesis doping of conducting polymers involves the following:

a) **Molecular doping from an orthogonal solvent**: In this mechanism, a dopant molecule is introduced into a conducting polymer from a separate solvent. The dopant molecules interact with the polymer chains, altering their electronic properties and enhancing the overall conductivity. **Figure 7.11(a)** shows a representation of the conducting polymer and the dopant molecule, along with arrows indicating the interaction between them.
b) **Electrochemical doping**: In this mechanism, the doping process occurs through an electrochemical reaction. The conducting polymer is placed in an electrolyte solution, and an external voltage or current is applied. This causes the transfer of ions between the electrolyte and the polymer, leading to changes in the polymer's charge carrier concentration. **Figure 7.11(b)** illustrates the setup with the conducting polymer, electrolyte solution, and the application of an electric potential.
c) **Ion-exchange doping**: Ion-exchange doping involves the substitution of ions within a conducting polymer with dopant ions. **Figure 7.11(c)** shows the conducting polymer structure with ions embedded in its lattice, and arrows indicating the exchange of ions between the dopant and the polymer.
d) **Reaction scheme for p-type molecular doping with ion-exchange**: **Figure 7.11(d)** would present a schematic representation of the chemical reactions involved in p-type molecular doping with ion-exchange. It includes the molecular structures of the dopant, conducting polymer, and relevant intermediates or transition states.
e) **Molecular structures of PBTTT, BMP, TFSI, and $FeCl_3$**: This part would consist of the molecular structures of the mentioned compounds. PBTTT is a type of conducting polymer, while BMP, TFSI, and $FeCl_3$ are commonly used dopant molecules in organic electronics. **Figure 7.11(e)** displays the chemical structures of these compounds.

Post-synthesis doping provides the advantage of controlling the doping level and distribution in conducting polymers without affecting the polymerization process. It allows for the introduction of different dopant species to achieve specific electrical properties and performance characteristics. Post-synthesis doping is commonly used in applications such as organic electronic devices, flexible electronics, energy storage devices, and electrochemical sensors [53–56].

The choice of doping agent depends on the desired properties and applications of the conducting polymer. Common dopants include acids, such as p-Toluenesulfonic acid (p-TSA), and oxidizing agents, such as ferric chloride ($FeCl_3$) or iodine (I_2).

The specific synthesis methods may vary depending on the polymers and applications involved [54, 56].

Researchers continue to explore and develop new techniques to synthesize copolymers, multilayer conducting polymers, and doped conducting polymers with improved properties and performance.

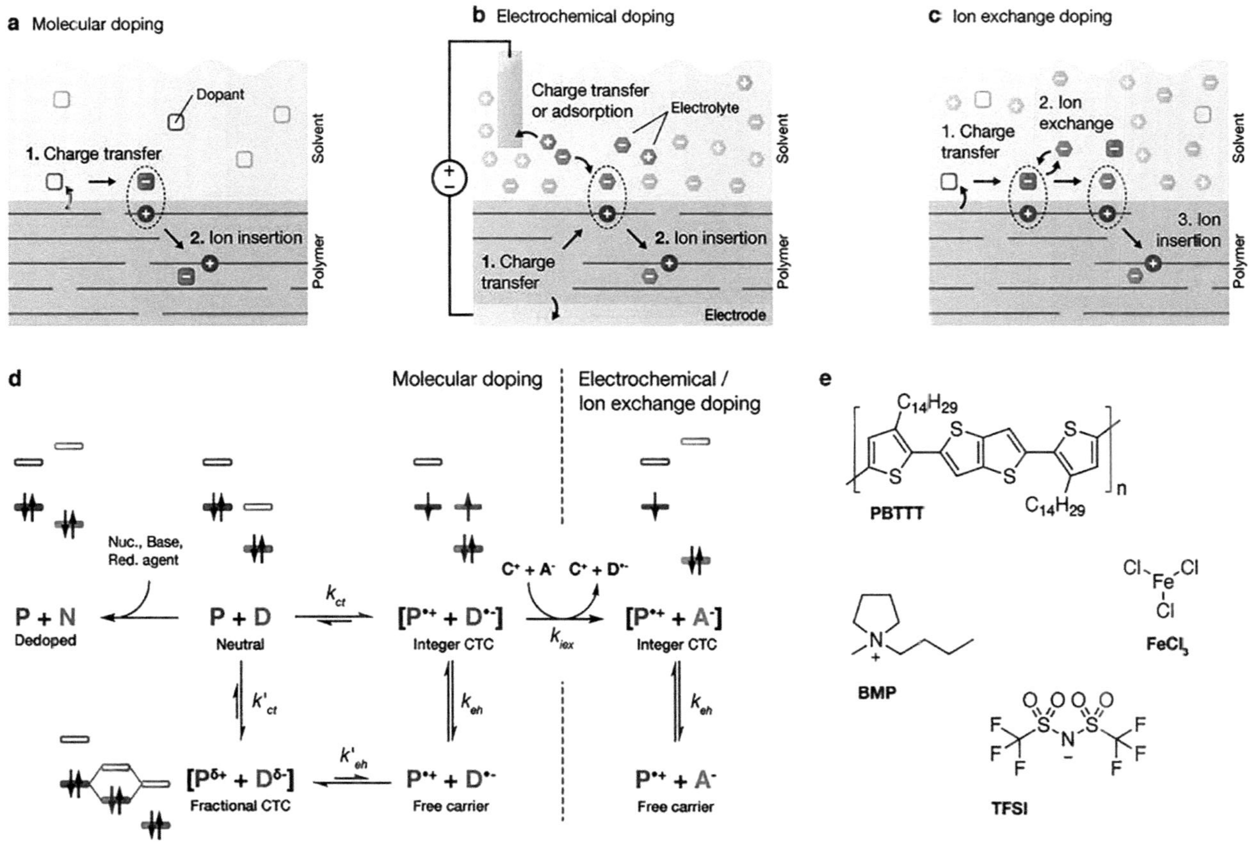

FIGURE 7.11 Doping mechanism. (a) Molecular doping from an orthogonal solvent. (b) Electrochemical doping. (c) Ion-exchange doping. (d) Reaction scheme for p-type molecular doping with ion-exchange. Representative electron configurations for each state are shown; reorganization effects are neglected for clarity. (e) Molecular structures of PBTTT, BMP, TFSI, and $FeCl_3$.

7.3 COPOLYMERS AS SMART ANTI-CORROSIVE MATERIALS

Copolymers have gained significant attention in the field of materials science due to their versatile properties and potential applications. When it comes to anti-corrosive materials, copolymers offer unique advantages and can be engineered to possess "smart" characteristics [57–59]. Here's an overview of copolymers as smart anti-corrosive materials:

Corrosion Protection Mechanisms: Copolymers can provide corrosion protection through various mechanisms. They can act as barriers, preventing the diffusion of corrosive agents such as water and oxygen to the underlying metal surface. Additionally, copolymers can release corrosion inhibitors, either passively or in response to environmental stimuli, to protect the metal from corrosion [60, 61].

Stimuli-Responsive Behavior: Smart copolymers can exhibit stimuli-responsive behavior, meaning they can change their properties in response to external stimuli such as pH, temperature, light, or mechanical stress. By incorporating responsive units into the copolymer structure, their anti-corrosive properties can be tailored to specific environmental conditions. For example, pH-responsive copolymers can release corrosion inhibitors in acidic environments, where corrosion is more likely to occur [57].

Self-Healing Capabilities: Copolymers can be designed to possess self-healing properties, allowing them to repair damage and maintain their protective function. When the copolymer coating is damaged, encapsulated healing agents within the material can be released to initiate healing processes, such as polymerization or cross-linking, to seal the defects and restore the barrier properties [62, 63].

Controlled Release of Corrosion Inhibitors: Copolymers can be engineered to incorporate corrosion inhibitors into their structure or as separate entities within the coating. By controlling the release rate of these inhibitors, copolymers can provide long-term protection against corrosion. Stimuli-responsive copolymers can release inhibitors selectively when exposed to specific environmental conditions or upon detecting corrosion signals, enhancing their effectiveness [57].

Compatibility with Coating Formulations: Copolymers offer versatility in terms of compatibility with various coating formulations. They can be synthesized to have specific molecular weights, composition ratios, and functional groups, enabling their incorporation into different coating systems such as paints, varnishes, or polymer films. This flexibility allows copolymers to be used as additives or as the primary coating material in anti-corrosive applications [64].

Tailored Mechanical and Chemical Properties: Copolymers can be designed with specific mechanical and chemical properties to meet the requirements of anti-corrosive applications. By adjusting the monomer composition and copolymerization conditions, the copolymer's flexibility, adhesion, toughness, and chemical resistance can be optimized for corrosion protection in different environments [65].

Copolymers provide a promising avenue for developing smart anti-corrosive materials with enhanced performance and tailored functionalities. Further research and development in this field can lead to the design of advanced coatings that offer long-lasting corrosion protection and adaptability to changing environmental conditions.

Poly(phenylene methylene) (PPM) has a unique combination of special properties, such as high thermal stability, hydrophobicity, resistance to oxidation, and good barrier properties. Therefore, poly(phenylene methylene) has recently been discovered as an optimistic new candidate in the field of corrosion protection. In addition, fluorescence of PPM is helpful in detecting layer malfunctions upon corrosion [65–70].

The addition of plasticizers to PPM or the use of PPM copolymers with alkoxy side chains along the polymer backbone (**Figure 7.12**) can overcome the cracking issue typically observed in PPM coatings, allowing for their application in real systems and as reference samples in comparative studies. This modification enables the production of continuous films that exhibit anti-corrosion properties. The corrosion resistance of coatings incorporating PPM blended with the plasticizer benzyl butyl phthalate (BBP) or PPM copolymers containing octyloxy side groups, which serve as internal plasticizers, has been investigated using anodic polarization techniques and potentiostatic measurements. These studies have mainly focused on evaluating the barrier-like behavior over a wide potential window and under long anodic polarization, with preliminary reports indicating the potential for self-healing ability in these coatings [66, 67].

The corrosion protection properties of poly(phenylene methylene) (PPM)-based coatings with thicknesses of 50 μm and 30 μm were evaluated using the ACET (Accelerated Corrosion Test). The main focus of the study was to compare the performance of coatings formed by a blended PPM, which contains an external plasticizer (BBP), with a PPM copolymer that incorporates an internal plasticizer (octyloxy-PPM copolymer). The results of the investigation showed that both substrates exhibited stable $|Z|_{0.01Hz}$ values of approximately 10^8 Ω cm^2 when coated with 50 μm thick coatings. However, when the coating thickness was reduced to 30 μm for the blended PPM coatings, a significant decrease in corrosion protection capability was observed, as indicated by a decrease in $|Z|_{0.01Hz}$ to 10^6 Ω cm^2 after two aging cycles [71].

In contrast, when the PPM copolymer coating was optimized to a thickness of 30 μm, an unexpected behavior was observed during the accelerated test. The $|Z|_{0.01Hz}$ value actually increased from 10^7 Ω cm^2 to values higher than 10^8 Ω cm^2 over time. This trend in the experimental electrochemical parameters of the PPM copolymer can be explained by the healing of pre-existing pores in the polymer layer or even the formation of new pores. The healing of the coating surface was further confirmed by

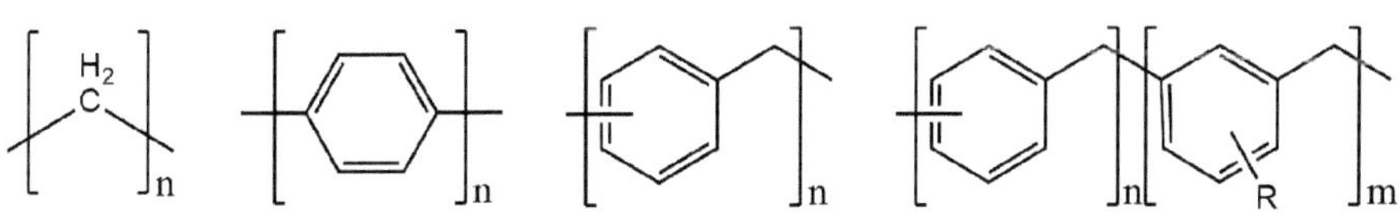

FIGURE 7.12 Chemical structure (from left) of polyethylene, poly(p-phenylene), poly(phenylene methylene), and poly(phenylene methylene) containing side chains (R = 4-octyloxy).

optical microscopy images taken under 254 nm light irradiation, which stimulated the fluorescence emission of PPM [71].

In summary, the investigation showed that the corrosion protection capability of the blended PPM coatings decreased significantly when the thickness was reduced to 30 μm. However, the PPM copolymer coatings exhibited an unexpected improvement in corrosion protection at the same thickness, likely due to the healing of pores within the coating layer [71].

Furthermore, in the ACET conducted on a substrate with artificial damage, it was found that self-healing (which is exclusive to PPM copolymer) occurs during both cathodic polarization and the relaxation period. This was evident from the decrease in current density and the effective electrical insulation of the working electrode, as indicated by extremely low and physically significant OCP values. These observations strongly suggest that exothermic heat fluxes generated during both processes are responsible for this phenomenon. The primary characteristic of the PPM copolymer that can account for the observed self-healing behavior is its thermoplastic nature. The localized corrosion of aluminum alloy in the presence of chloride ions triggered the localized flow of the polymer, facilitated by the heat released by the associated reactions.

7.4 MULTILAYER CONDUCTING POLYMERS AS SMART ANTI-CORROSIVE MATERIALS

The recent interest in conducting polymers has focused on their potential use as corrosion inhibitors and protective coatings. Among conducting polymers, the polyaniline (Pani) family, known for their stability and ease of synthesis, has been extensively studied. Some studies have demonstrated the effectiveness of ortho-methoxy substituted Panis as corrosion inhibitors for steel in acidic chloride solutions, while others have focused on Pani coatings [9, 72–74].

DeBerry reported in 1985 that Pani could reduce the corrosion rate of stainless steel in sulfuric acid by stabilizing the passive oxide film [72]. However, studies by Sekine et al. [73] and Troch-Nagels et al. [74] found that Pani coatings deposited from sulfuric or nitric acids were not effective in protecting steel from corrosion. Camalet et al. [75] discovered that Pani films formed in oxalic acid, in the presence of a passive Fe(II)-oxalate layer, provided good protection to steel in acidic chloride solutions. More recent studies by Bernard et al. [76] showed that Pani deposited from metanilic acid, incorporating metanilate into the polymer backbone, yielded superior copolymer coatings.

Chemically deposited emeraldine base, a form of Pani, was found to protect stainless steel in acidic chloride environments, as demonstrated by Ahmad and MacDiarmid. Santos et al. later showed similar protection for carbon steel in a 3% NaCl solution. Pani has also been used as a primer in combination with epoxy or polyurethane topcoats, providing corrosion protection superior to the topcoat alone. However, when compared to traditional primer/topcoat combinations, Pani/topcoat systems have shown relatively poor performance [77, 78].

In comparisons between the conducting emeraldine salt and non-conducting emeraldine base forms of Pani, the non-conducting form generally provides better protection in NaCl solutions, surpassing polyvinyl chloride coatings. However, in HCl environments, the conducting form appears to offer better protection, as the undoped non-conducting form exhibits poor adhesion. Nonetheless, the protection mechanism in both cases is attributed to the conjugation responsible for the unique electronic properties of conducting polymers, as non-conjugated polymers do not provide the same level of protection [78].

Polymers based on polypyrrole (Ppy) have also been explored for their corrosion protection capabilities. One drawback of Ppy is that it typically needs to be deposited from non-aqueous solutions. The effectiveness of Pani and Ppy coatings in corrosion resistance strongly depends on the acid used for their deposition. Pani deposited from nitric acid was found to be inferior to Ppy by Troch-Nagels et al. [74], whereas Camalet et al. [75] found that Pani deposited from oxalic acid exhibited superior corrosion protection compared to Ppy.

Despite the successes claimed thus far with conducting polymer coatings, they have not yet reached the point of replacing traditional coating systems. Challenges remain, such as the dispersion of Pani and the limited availability of monomers with conjugated p-bonds necessary for electrical conductivity. However, one potential solution is to synthesize combinations of polymers, such as copolymers, bilayers, or composites, to enhance their properties. Sekine et al. [73] demonstrated that double-electropolymer films based on 2-vinylpyridine and acrolein provide superior protection to steels compared to single polymers.

Three different combinations of polyaniline (Pani) and polypyrrole (Ppy) coatings were applied to carbon steel and 304L stainless steel. The morphology of the coatings was found to be independent of the underlying metal. However, the adhesion strength and corrosion protection abilities of the multipolymer coatings varied depending on the substrate [9].

The results of the Rockwell scratch test, summarized in Table 7.1 [9], demonstrate the critical forces for delamination (Lc) of the polymer coatings. These Lc values were consistently below 2N for all the coatings. This indicates that the adhesion of the coatings was significantly lower than that of commercial paint films, which typically have Lc values of 5N [79]. Consequently, it can be inferred that the polymer coatings were not efficient physical barriers.

Moreover, when comparing the relative adhesion of the three different types of coatings on a given substrate, the variation was only around 0.2N. Additionally, the adhesive strength of the copolymer coatings to stainless steel was approximately equal to that of a single Pani film. However, it is important to note that the Pani film provides substantially less protection against corrosion [78].

For carbon steel, the performance of the multilayered coatings was not significantly better than that of single Pani coatings. Therefore, the more complex deposition procedures involved in the multilayered coatings did not justify their use for carbon steel.

In contrast, the multilayered coatings exhibited significantly improved performance on stainless steel compared to single Pani coatings, especially in terms of protection against pitting corrosion. This improvement may be attributed to the multipolymer coatings acting as better chemical barriers, effectively trapping chloride ions and preventing them from attacking the stainless steel substrate.

TABLE 7.1
Critical Forces for Delamination (*Lc*) of the Polymer Coating, as Measured by the Rockwell Scratch Test (Reproduced with permission reference number [9], © 2003, ELSEVIER)

Substrate	Coating	*Lc* (N)
Carbon steel	Ppy/Pani	1.2 (0.02)
	Pani/Ppy	1.1 (0.02)
	Mixed	1.5 (0.10)
304L stainless steel	Ppy/Pani	2.1 (0.08)
	Pani/Ppy	1.8 (0.02)
	Mixed	1.9 (0.01)
Carbon steel	Pani only	2.0 (0.01)
	Paint	5.0

The degree of protection and the morphology of the coatings strongly depended on the order in which the two conducting polymers were coated. The best performance was observed when a Ppy underlayer was deposited, followed by a Pani top layer (Ppy/Pani). This particular coating completely eliminated pitting corrosion and reduced the corrosion rate by a factor of approximately 2000. The Ppy/Pani coating had a densely packed morphology, making it suitable as both physical and chemical diffusion barriers. It was also believed to be the most conductive, providing an effective electronic barrier [9].

Additionally, the Ppy/Pani coating exhibited the strongest adhesion to the stainless steel substrate, although its adhesion strength was slightly less than that of a typical paint film (about 0.2N). This suggests that the ability of a conducting polymer film to act as electronic and chemical diffusion barriers is more crucial in providing corrosion protection than its ability to serve as a physical barrier [9].

7.5 DOPED CONDUCTING POLYMERS AS SMART ANTI-CORROSIVE MATERIALS

Conducting polymers and their composites are now commonly employed as corrosion inhibitors on metal surfaces. The ability of these composites to restrict the flow of corrosion-causing chemicals on metal surfaces accounts for their corrosion-protecting properties. The development of a passive film by the polyaniline coating on the steel surface protects it against corrosion, according to research. The corrosion-inhibiting properties of a polyaniline epoxy mixed coating on steel have been investigated. On mild steel surfaces, polyaniline/polypyrrole (Ppy) and polyaniline-polypyrrole phosphotungstate composites were utilized as corrosion inhibitors. Composite films outperform bare polyaniline and polypyrrole in terms of corrosion resistance [80, 81].

Mechanism of Corrosion Protection: For the corrosion protection, two mechanisms, the barrier effect and the polymer coating, work as a barrier against

> the penetration of oxidants and aggressive anions, protecting the substrate metals. This effect is similar to paint coating, which inhibits the substances from penetrating the substrate steel. On the anodic protection, the conducting polymer with the strongly oxidative property works as an oxidant to the substrate steel, the potential of which is shifted to that in the passive state. In solution at neutral pH, the corrosion potential (or open circuit potential in corrosion) of bare steel is located in the active potential region, and the corrosion rate of the steel is usually relatively high. Owing to the coating of conducting polymer, the maximum current in the active-passive transition region was limited by the barrier effect, and then the potential can be easily shifted to a higher potential in the passive state by the strongly oxidative property of the conducting polymer [82, 83] (**Figure 7.13**).

In the passive state, the corrosion rate of steel becomes much lower. It is assumed that both the barrier effect and the oxidative property induce the anodic protection. Finally, the potential of the substrate steel may be in agreement with a redox potential of the PPy layer in the following reaction, and thus, depends on the degree of oxidation state of the PPy layer [82].

$$PPy^{n+}\cdot\left(\frac{n}{x}\right)A^{x-}\cdot+\cdot me \rightleftarrows PPy^{(n-m)+}\cdot\left(\frac{n-m}{x}\right)A^{x-}\cdot+\cdot\left(\frac{m}{x}\right)A^{x-}\text{aq.}$$

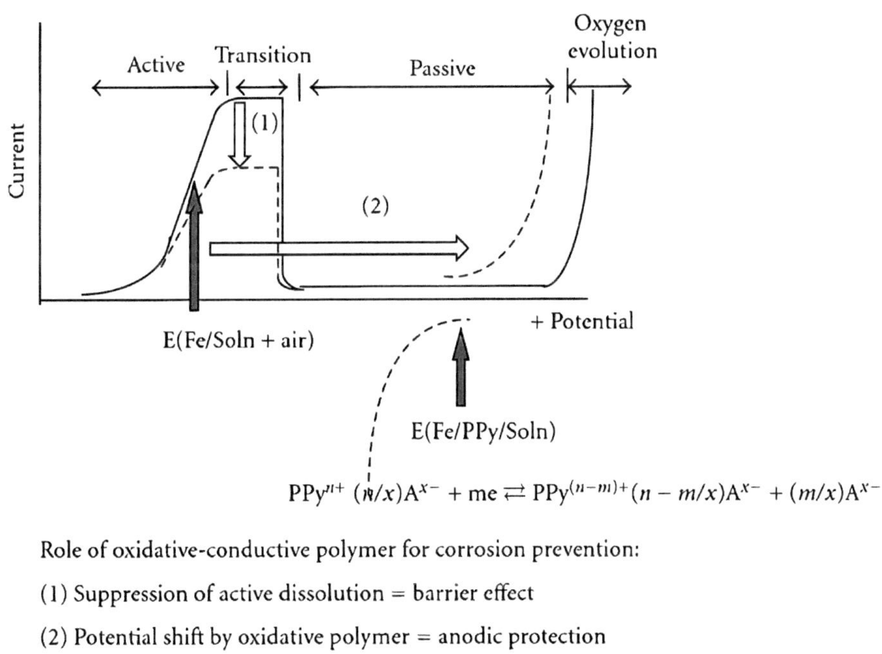

FIGURE 7.13 Potential-current relation of steels covered by oxidative conducting PPy. A barrier effect of PPy suppresses active dissolution of the steel, and an oxidative property of PPy shifts the potential into passive state.

The conductivity of the PPy layer affects the oxidative power, which brings about the passive state. If the coating layer has little conductivity, the role of the coating as the oxidant is limited in the neighborhood of the passive oxide. If the layer has enough high conductivity, however, the oxidant power of the whole layer is available and the power increases with the increase of the layer thickness. The oxidation degree and the conductivity are assumed to decline with longer exposure to the environment. If oxidants in the environment reoxidize the degraded PPy layer, the oxidation degree and conductivity can be recovered. When the oxidant in the environment, typically oxygen gas in air, can recover the PPy layer, the duration to maintain the oxidative power of the PPy layer can be prolonged, and the passive state of the steel underneath the PPy layer can be kept for a longer period [82, 83]. The recovery process is illustrated in **Figure 7.14**.

Ion Exchange in the Conducting Polymer and Its Effect on Corrosion Protection: In the anodic protection, the largest problem is breakdown of the passive oxide due to the attack of aggressive anions such as chloride and bromide ions in solution, and the breakdown is followed by large damage from localized corrosion of pitting and crevice corrosion. In contrast with cathodic protection, there is a large risk of localized corrosion associated with anodic protection. When we control the doping ions in the PPy layer, we possibly prevent penetration of the aggressive anions into the PPy layer. When the steels covered with the conducting PPy are immersed in the sodium chloride solution, the anions doped in the PPy can be exchanged with the chloride anions in the aqueous solution. The chloride anions penetrate the PPy to the substrate steels and then induce the breakdown of the surface passive oxide film, followed by the pitting corrosion [84].

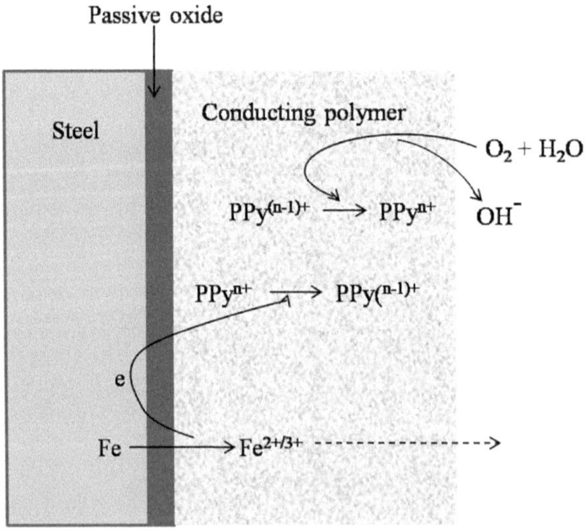

FIGURE 7.14 Degradation of oxidative property of PPy and recovery, which was done from reduction of oxygen on the PPy surface.

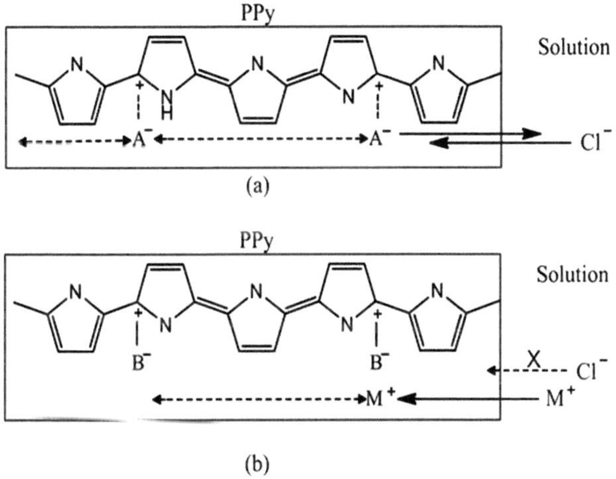

FIGURE 7.15 Ionic perm-selectivity of PPy film. (a) PPy film with anionic perm-selectivity, in which small-sized anions are doped in PPy, and (b) PPy film with cationic perm-selectivity, in which large-sized anions are doped in PPy film.

The mobility of the dopant anions in the PPy is affected by their mass and volume. When we adopted organic acid ions as the dopant ions in the PPy, they possessed enough mass and volume to be immobile in the PPy. In general, organic acid anions with a large mass are assumed to have small mobility and diffusion in the PPy layer. Accompanied by the oxidation and reduction of the PPy, small anions are doped into and dedoped out of the PPy, respectively, to maintain neutrality, as described in reaction (1) and shown in **Figure 7.15(a)** [82, 85].

When the mobility and diffusion of the doped anions are restricted to a small value, conversely, the cations are dedoped out of and doped into the PPy during the oxidation and reduction, respectively. The dedoping process of cations in the PPy during the oxidation and the doping during the reduction are described in the following reaction:

$$\mathrm{PPy}^{n+}\left(\frac{n}{x}\right)\mathrm{B}^{x-}+\left(\frac{m}{y}\right)\mathrm{M}_{aq}^{y+}+me \rightleftarrows \mathrm{PPy}^{(n-m)+}\left(\frac{n}{x}\right)\mathrm{B}^{X-}+\left(\frac{m}{y}\right)\mathrm{M}^{y+}$$

When one considers the conducting PPy as a charged membrane, the immobile anions with large mass are assumed to have fixed sites with negative charge in the PPy. In the channel between the negatively charged sites, the cations can be mobile, and the movement of the anions is greatly inhibited; that is, the membrane exhibits cationic perm-selectivity. As illustrated in **Figure 7.15(b)**, under the situation where the dopant anions are large enough, the anions in the solution are excluded from the PPy, and the substrate steel is protected against the pitting corrosion by chloride attack [82, 86].

Design for Corrosion Protection by the PPy. The anodic protection greatly depends on the passivity and passive oxide on the steel. For the protection, the passivity and passive oxide must be kept stable. Further, the prevention of penetration of aggressive anions plays an important role in the protection (**Figure 7.16**). Anodically prepared a PPy film on steel from an oxalate solution containing Py monomer and PPy layer protected the steel in sodium chloride solution for a long period [84–86]. They assumed that the ferric oxalate layer, which was formed underneath the PPy film by the polymerization, worked as a passivation film against corrosion. They also presented that the overcoat layer of PPy doped with dodecylsulfate, $C_{12}H_{25}OSO_3^-$(DoS), anions was effective for the corrosion protection and that a bilayer coating of PPy-oxalate/PPy-DoS could maintain the passivation state for longer than 500 h, in which no corrosion products were observed. DoS ion is a surfactant and forms micelles in aqueous solution at concentrations higher than critical concentration. Py monomers, which are probably incorporated in the micelle of DoS in aqueous solution, start to be polymerized when the micelles are collapsed on the electrode to which anodic potential is applied. DoS ions have relatively large masses and work as an immobile dopant in the PPy. The PPy doped with DoS, thus, is considered as a membrane with negatively charged fixed sites and, thus, with cationic perm-selectivity. The outer layer of PPy-DoS can, therefore, exclude the insertion of aggressive anions such as chloride ions. In **Figure 7.16**, the mass change is plotted with anodic oxidation and cathodic reduction of a gold electrode covered with the PPy layers.

The mass change was measured by electrochemical quartz crystal microbalance (EQCM) with gold coating. The gold coating was covered by PPy doped with phosphomolybdate ions, $PMo_{12}O_{40}^{3-}$(PMo), and a bilayered PPy of PPy-PMo/PPy-DoS. The mass change of the PPy-PMo film in **Figure 7.17** indicates the uptake of mass during the oxidation and, inversely, the removal during the reduction.

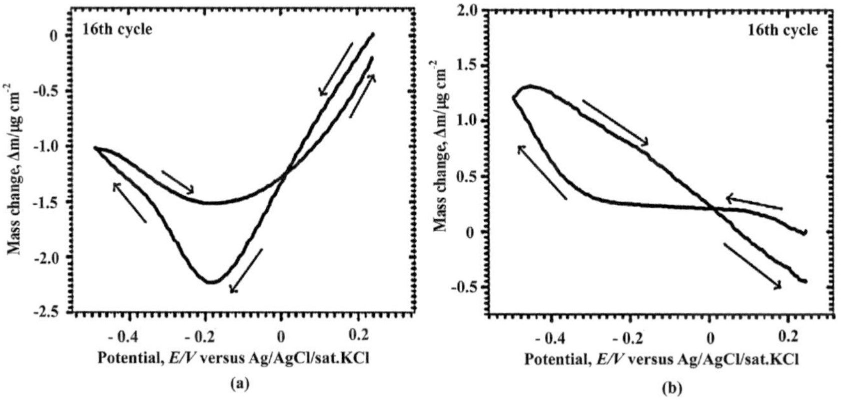

FIGURE 7.16 Mass change of the PPy-PMo layer and bilayer of PPy-PMo/PPy-DoS during the potentiodynamic reduction and oxidation in 3.5% NaCl solution. The data was a result of the 16th redox cycle.

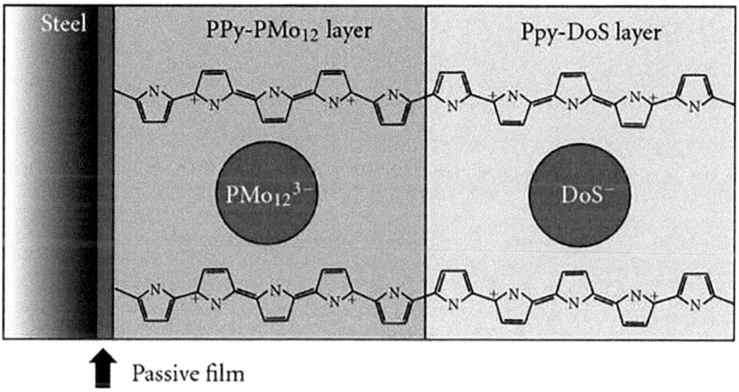

FIGURE 7.17 The mass change of the PPy-PMo film.

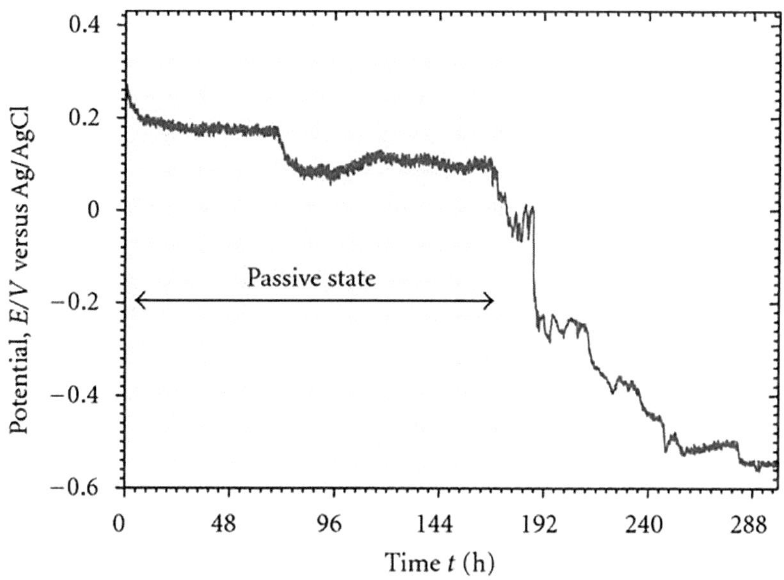

FIGURE 7.18 Model of bilayered PPy film and transient of open circuit potential of steel covered by the bilayered PPy in 3.5% NaCl solution.

The behavior of the mass change during the oxidation reflects the removal of electrons from the PPy and simultaneous insertion of anions from the electrolyte to the PPy and vice-versa during the reduction. When one introduces the outer layer of PPy-DoS, the mass change is inversely different from the result given above, as shown in **Figure 7.18**.

During oxidation, the mass increases, and during reduction it decreases. In PPy-DoS layer, in which negatively charged ions are fixed, the cations are mobile; during

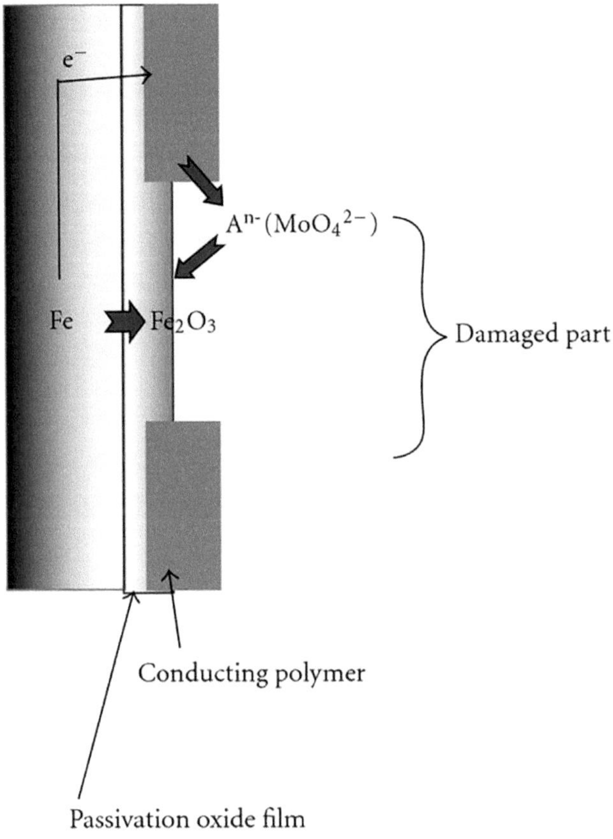

FIGURE 7.19 Schematic model of self-healing property of PPy-PMo12/PPy-DoS bilayered PPy. Molybdate anions, dissolving from PPy film, reform a passive oxide at the damaged part.

oxidation, the simultaneous removal of both electrons and cations from PPy and during the reduction vice versa. It can be understood that the PPy doped with DoS functions as a cationic perm-selective membrane. Kowalski et al. designed the corrosion protection PPy layer of steels [87]. For the inner layer, the PPy was doped with PMo. PMo works as a passivator, which stabilizes the passive state of steels and facilitates the formation of passive oxide. For the outer layer, the PPy doped with DoS was prepared. The outer layer can inhibit the anions from penetrating the PPy layer. The results by Kowalski et al. [88] are shown in **Figure 7.19**, where the open circuit potential of the steel covered with the bilayered PPy is plotted during the immersion in 3.5% sodium chloride solution.

The steel covered with the bilayered PPy, about 5 μm thick, consisting of PPy-PMo/PPy-DoS, exhibited the passivation for 190 h during which no corrosion products were observed. If the steel was covered with a single PPy-DoS layer of the same thickness, the passivation is kept for 10 h. It is assumed that PMo ion-doped in the inner PPy stabilizes the passive oxide and helps the maintenance of the passive state of the substrate steel. The design that combines the inner layer stabilizing the passive oxide with the outer layer inhibiting anions from penetrating through PPy to the steel may be suitable for the corrosion protection of steel.

7.6 ADSORPTION MECHANISM

Adsorption is a process in which molecules or ions from a substance adhere to the surface of another substance. In the case of anti-corrosive materials, the adsorption mechanism plays a crucial role in protecting the surface of metals from corrosion. There are different mechanisms involved in the adsorption of anti-corrosive materials, depending on the specific material and the corrosion inhibition mechanism it employs. There are a number of adsorption mechanisms like physical, chemical, coordinate, charge transfer, and physical barrier adsorption mechanisms. It is important to note that these adsorption mechanisms are not mutually exclusive, and a combination of multiple mechanisms can be involved in the anti-corrosion process. The specific mechanism depends on the nature of the anti-corrosive material, the composition of the metal surface, and the corrosive environment in which the inhibition is required.

The different steel types are alloys susceptible to corrosion, the majority element in their composition being iron, which is the most vulnerable to oxidation processes at the surface contact with aggressive media. Thus, in order to restrict the damage of the metallic surfaces and at the same time to increase the corrosion resistance, anti-corrosive coatings can be applied by the adsorption of some polymers/copolymers, leading to the occurrence of thin films that attach to the metallic surfaces through the chemical/physical bonds. Polymers and copolymers are an efficient choice for the functionalization of metal surfaces because they have numerous active centers distributed along their macromolecular chains, through which they can adsorb on the substrate, forming coherently organized films with anti-corrosive properties.

Many polymers are ecological, biodegradable, biocompatible compounds, without toxic and carcinogenic effects, being applied in numerous industrial and medical fields. Moreover, macromolecular networks based on polyvinyl butyral have special characteristics such as "shape memory" [89–91], thus having the ability to self-regenerate, recovering their initial shapes after the application of an electrical, magnetic, or thermal stimulus, leading to the preservation of the metallic surface properties covered with polymer film, after exposure in aggressive environments for a long time. The application of polymers as corrosion inhibitors in aqueous media is sometimes limited, as these are water-insoluble compounds but soluble in various organic solvents. For this reason, their testing as anti-corrosive agents in aqueous media requires the use of appropriate film deposition methods, which remove the inconveniences caused by insolubility.

Thus, the assembly of protective layers based on polymers can be performed by several methods, as follows: (1) electro-polymerization on metallic electrodes using cyclic voltammetry [92]; (2) preparation of an aqueous dispersion medium by emulsion polymerization, which allows the investigation of the inhibitory and adsorption properties of the respective polymer [93]; (3) functionalization of metallic surfaces by polymer adsorption from a specific organic solvent [91].

In this study, it was determined the protection performance of copolymer based on poly(vinyl butyral-co-vinyl alcohol-co-vinyl acetate), named PVBA, adsorbed on the carbon steel and 304L stainless steel surfaces from methanol solution using the simple dipping method of metal samples, a well-determined time. This investigation represents the anti-corrosive performance of the copolymer film adsorbed on copper surface.

The copolymer molecular structure is shown in **Figure 7.20**. It is observed that the PVBA presents three types of structural units afferent to the polyvinyl butyral (with the greatest polymerization degree), to the polyvinyl alcohol, and to the polyvinyl acetate (in a very low percentage). Butyral groups with a hydrophobic character and the -OH groups from the polyvinyl alcohol macromolecule, presenting hydrophilic properties, favor the binding of PVBA to certain metal substrates such as iron, aluminum, tin, and lead through noncovalent interactions (hydrogen bonds) or metallic coordination [94, 95], creating a host-guest interaction between the metal matrix and copolymer, respectively.

For both the steels (carbon and stainless), there are competitive adsorption processes of methanol molecules and PVBA macromolecules. Most likely, the methanol molecules are primarily adsorbed on the steel surface because the macromolecular chain steric arrangement can prevent their diffusion at the metal/electrolyte interface. Thus, the hydrogen bond can form between adsorbed methanolic hydroxyl groups and those of the polyvinyl alcohol fraction from the PVBA macromolecule.

At the same time, the coordinative bonds are formed between the free electron pairs of the oxygen atoms belonging to the butyral and acetate groups and the vacant

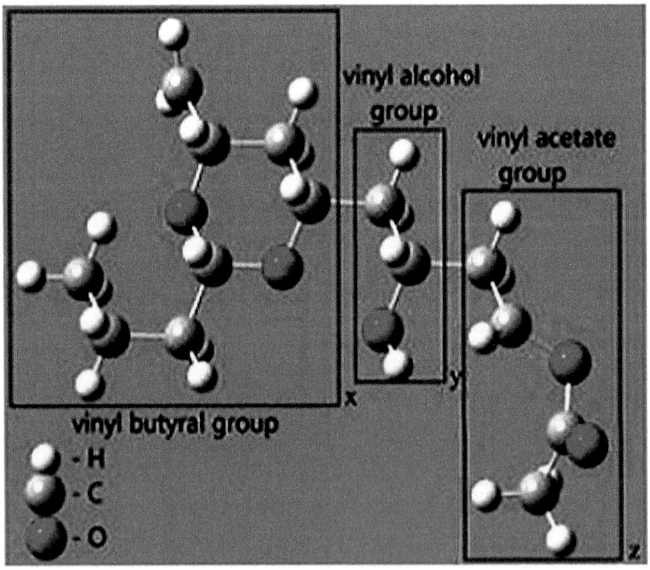

FIGURE 7.20 Molecular structure of poly(vinyl butyral-co-vinyl alcohol-co-vinyl acetate); x:y:z = 39:29:1.

"*d* "orbital of the iron atoms, which ensures a moderate chemical adsorption of the PVBA macromolecules. Thus, a synergistic mechanism can be assumed between physical adsorption through hydroxyl groups and chemical adsorption primarily through the butyral groups, the acetate groups participating to a lesser extent.

Due to the hydrophilic groups, when the samples are immersed in NaCl solution, the increase of the swelling capacity of the upper-surface layer takes place, followed by the polymeric film thickness increase and its extension development on the metallic surface. This can favor a superficial detachment of the coating on the metal surface, more pronounced in the case of carbon steel than that of stainless steel, which also explains the lower protection performance for carbon steel than that for stainless steel.

After consecutive electrochemical measurements performed on bare steel electrodes and PVBA-coated samples, it was found that the copolymer provides a protection performance of approximately 66% for carbon steel and 74% for stainless steel.

7.7 CONCLUSION

Conducting polymers are not just conducting. Their use as adsorbents of organic dyes is a typical example of an application that does not require electrical conductivity. However, in the next step, their conductivity allows for the construction of electrodes by their in situ deposition on non-conducting substrates and subsequent use in electrochemistry. In this experiment, an electric potential was applied to the conducting polymer adsorbent. The electrochemical switching between the redox polymer forms may change the adsorption properties due to a change in molecular structure, degree of protonation, and/or hydrophilicity. For example, the electro-remediation of aqueous effluents containing Congo red on polyaniline/titanium dioxide composite points in this direction, similar to the electro-oxidation of this dye on the polyimide membrane coated with polyaniline. Such systems might be regarded as intelligent adsorbents controlled by the applied potential. The array of adsorbents may be broadened by testing the polymers of ring-substituted anilines, viz., phenylenediamines and various copolymers, with aniline.

Most of the known protective mechanisms were suggested on the basis of evaluation results regarding the anti-corrosion properties of PAN-based coatings. Different mechanisms seem to proceed in parallel depending on the CP, metal substrate, and corrosive environment. The unique properties of CPs make them very suitable candidates for the development of multifunctional coatings, as the ultimate goal of automotive, aerospace, and other manufacturers is the design of self-healing coatings that could be recovered after mechanical and chemical damages. CP-based coatings are considered multifunctional coatings in the sense that they may offer more than one function to the metal, such as self-healing properties, corrosion indicators, physical barriers, or antistatic and antifouling properties as composites. Within this frame, the genesis of a new generation of CP-based coatings that can offer more than anti-corrosion protection to the metal is in progress, as seen by numerous publications exploring CP composites and nanocomposites. CP composites/nanocomposites combine CP properties with characteristics of other inorganic or organic materials, resulting in innovative smart coatings of a better quality and a lower cost without undesired effects caused when CP primers are utilized alone. However, further efforts are required toward a better understanding of the potential and limitations of CP-based smart coatings in anti-corrosion technology.

7.8 FUTURE OUTLOOK

Conducting polymers (CPs) for protection against the corrosion of metals and alloys have been utilized in a variety of formulations to ensure good adherence to the metal substrate and long-term functionality, considering environmental factors. Evaluation of the protective efficiency of different CP-based coatings in various configurations shows very promising results, and owing to restrictions on the use of heavy metals and chromating processes as environmentally non-friendly and toxic for human health, an important research field has developed during the past decades. Numerous studies have been published dealing, at the beginning, mostly with the protection mechanisms and giving emphasis lately to new strategies, which aim at improving protection efficiency, considering also technical and economic factors. Recent innovations in CP-based coatings are based on the technology of "smart" or "intelligent" coatings, terms used lately to describe multifunctional coatings, which can provide more than one function to the metal substrate. In this sense, CP-based coatings belong to smart coatings and constitute the cornerstone for the development of a new technology. In fact, most of these coating technologies are currently under development in academia and industry. In particular, the scaling up of innovative CP-based coatings for possible industrial applications is still in its infancy. A CP film cannot alone offer complete protection to a metal. The simplest requirement for this configuration is the presence of a topcoat (i.e., another polymer film) to eliminate diffusion of corrosion species through the CP film toward the metal substrate and/or prevent ion exchange processes. Without a topcoat, within a few weeks, CP films fail to protect metals and alloys under constant immersion conditions in corrosive media, especially in the presence of chlorides. There are two basic methods to deposit CP films on oxidizable metal surfaces: electrochemical and chemical, by using different CP formulations as dispersions/solutions or blends, for instance, with epoxy and acrylic resins. The selection of a suitable method depends on the advantages/disadvantages encountered by the specific multi-interfacial system of interest, namely, the metal surface/CP coating/corrosive environment. Blending of CPs with epoxy or acrylic polymers takes advantage of the combination of the good mechanical properties and ability of conventional polymers with the reversible redox properties of CPs. Future trends in the technology of CP-based coatings might be expected as follows:

- Preparation of advanced CP-based composites/nanocomposites with desired protective properties.
- Functionalized CP-based coatings of unique chemistry and nanostructure will be further explored owing to innovations in the synthesis of conducting polymers, curing methods, and combined utilization of inhibiting agents with CPs
- Development of new surface pre-treatments, to improve strong adherence for long-term protection with environmentally friendly methods, will be a popular research topic of the next decade.

- More attention will be placed on optimizing "smart" CP-based coatings in terms of functionality (self-healing properties, corrosion indicators, self-cleaning, anti-icing, antifouling, antireflection, etc.), considering the nature of specific metals and alloys to be protected, along with economic and environmental factors.
- Automatic diagnosis of the metal state in several structures will be critical to their "smart" maintenance by CP-based coatings, which present a great potential for built-in coating condition monitoring.

REFERENCES

[1] Palanisamy G. *Corrosion inhibitors* (1–24). Intech Open, 2019.

[2] Tan Y. Monitoring localized corrosion processes and estimating localized corrosion rates using a wire-beam electrode. *Corrosion*. 1998;54(5):403–413.

[3] Farh HM, Seghier ME, Zayed T. A comprehensive review of corrosion protection and control techniques for metallic pipelines. *Engineering Failure Analysis*. 2022:106885.

[4] Zarras P, Stenger-Smith JD. Corrosion processes and strategies for prevention: An introduction. In *Handbook of Smart Coatings for Materials Protection* (pp. 3–28). Woodhead Publishing, 2014.

[5] Nurkulov F, Ziyamukhamedova U, Rakhmatov E, Nafasov J. Slowing down the corrosion of metal structures using polymeric materials. In *E3S Web of Conferences 2021* (Vol. 264, p. 02055). EDP Sciences.

[6] Mouritz AP. *Introduction to Aerospace Materials* (pp. 232–250). Woodhead Publishing, 2012.

[7] Kausar A. *Conducting Polymer-Based Nanocomposites: Fundamentals and Applications*. Elsevier, 2021.

[8] Gao C, Zhao X, Fatehi P, Dong X, Liu K, Chen S, Wang S, Kong F. Lignin copolymers as corrosion inhibitor for carbon steel. *Industrial Crops and Products*. 2021;168: 113585.

[9] Tan CK, Blackwood DJ. Corrosion protection by multilayered conducting polymer coatings. *Corrosion Science*. 2003;45(3):545–557.

[10] Al-Gamal AG, Farag AA, Elnaggar EM, Kabel KI. Comparative impact of doping nano-conducting polymer with carbon and carbon oxide composites in alkyd binder as anti-corrosive coatings. *Composite Interfaces*. 2018;25(11):959–980.

[11] Deshpande PP, Jadhav NG, Gelling VJ, Sazou D. Conducting polymers for corrosion protection: A review. *Journal of Coatings Technology and Research*. 2014;11:473–494.

[12] Han X, Wu J, Zhang X, Shi J, Wei J, Yang Y, Wu B, Feng Y. Special issue on advanced corrosion-resistance materials and emerging applications. The progress on antifouling organic coating: From biocide to biomimetic surface. *Journal of Materials Science & Technology*. 2021;61:46–62.

[13] Husain E, Abdel Nazeer A, Alsarraf J, Al-Awadi K, Murad M, Al-Naqi A, Shekeban A. Corrosion behavior of AISI 316 stainless steel coated with modified fluoropolymer in marine condition. *Journal of Coatings Technology and Research*. 2018;15:945–955.

[14] Advincula RC. Conducting polymers with super hydrophobic effects as anticorrosion coating. In *Intelligent Coatings for Corrosion Control* (pp. 409–430). Butterworth-Heinemann, 2015.

[15] Riaz U, Nwaoha C, Ashraf SM, Laboratory MR. Recent advances in corrosion protective composite coatings based on conducting polymers and natural resource derived polymers. *Progress in Organic Coatings*. 2014;77(4):743–756.

[16] Wolz A, Zils S, Michel M, Roth C. Structured multilayered electrodes of proton/electron conducting polymer for polymer electrolyte membrane fuel cells assembled by spray coating. *Journal of Power Sources*. 2010;195(24):8162–8167.
[17] Gregory A, Stenzel MH. Complex polymer architectures via RAFT polymerization: From fundamental process to extending the scope using click chemistry and nature's building blocks. *Progress in Polymer Science*. 2012;37(1):38–105.
[18] Mayo FR, Walling C. Copolymerization. *Chemical Reviews*. 1950;46(2):191–287.
[19] Krstina J, Moad G, Rizzardo E, Winzor CL, Berge CT, Fryd M. Narrow polydispersity block copolymers by free-radical polymerization in the presence of macromonomers. *Macromolecules*. 1995;28(15):5381–5385.
[20] Ajioka M, Suizu H, Higuchi C, Kashima T. Aliphatic polyesters and their copolymers synthesized through direct condensation polymerization. *Polymer Degradation and Stability*. 1998;59(1–3):137–143.
[21] Kausar A. Advances in condensation polymer containing zero-dimensional nanocarbon reinforcement—fullerene, carbon nano-onion, and nanodiamond. *Polymer-Plastics Technology and Materials*. 2021;60(7):695–713.
[22] Sharma V, Kundu PP. Addition polymers from natural oils—A review. *Progress in Polymer Science*. 2006;31(11):983–1008.
[23] Tsitsilianis C, Gotzamanis G, Iatridi Z. Design of "smart" segmented polymers by incorporating random copolymers as building blocks. *European Polymer Journal*. 2011;47(4):497–510.
[24] Keddie DJ. A guide to the synthesis of block copolymers using reversible-addition fragmentation chain transfer (RAFT) polymerization. *Chemical Society Reviews*. 2014;43(2):496–505.
[25] Nuyken O, Pask SD. Ring-opening polymerization—an introductory review. *Polymers*. 2013;5(2):361–403.
[26] Li J, Rothstein SN, Little SR, Edenborn HM, Meyer TY. The effect of monomer order on the hydrolysis of biodegradable poly (lactic-co-glycolic acid) repeating sequence copolymers. *Journal of the American Chemical Society*. 2012;134(39):16352–16359.
[27] Gregory GL, López-Vidal EM, Buchard A. Polymers from sugars: Cyclic monomer synthesis, ring-opening polymerisation, material properties and applications. *Chemical Communications*. 2017;53(14):2198–2217.
[28] Balla E, Daniilidis V, Karlioti G, Kalamas T, Stefanidou M, Bikiaris ND, Vlachopoulos A, Koumentakou I, Bikiaris DN. Poly (lactic Acid): A versatile biobased polymer for the future with multifunctional properties—From monomer synthesis, polymerization techniques and molecular weight increase to PLA applications. *Polymers*. 2021;13(11):1822.
[29] Matyjaszewski K. Introduction to living polymeriz. Living and/or controlled polymerization. *Journal of Physical Organic Chemistry*. 1995;8(4):197–207.
[30] Yagci Y, Tasdelen MA. Mechanistic transformations involving living and controlled/living polymerization methods. *Progress in Polymer Science*. 2006;31(12):1133–1170.
[31] Matyjaszewski K, Spanswick J. Controlled/living radical polymerization. *Materials Today*. 2005;8(3):26–33.
[32] Hawker CJ, Bosman AW, Harth E. New polymer synthesis by nitroxide mediated living radical polymerizations. *Chemical Reviews*. 2001;101(12):3661–3688.
[33] Destarac M, Charmot D, Franck X, Zard SZ. Dithiocarbamates as universal reversible addition-fragmentation chain transfer agents. *Macromolecular Rapid Communications*. 2000;21(15):1035–1039.
[34] Moad CL, Moad G, Rizzardo E, Thang SH. Chain transfer activity of ω-unsaturated methyl methacrylate oligomers. *Macromolecules*. 1996;29(24):7717–7726.

[35] Yoshida E. Nitroxide-mediated photo-controlled/living radical polymerization of methacrylic acid. *Open Journal of Polymer Chemistry*. 2013;3(01):16–22.
[36] Yagci Y, Jockusch S, Turro NJ. Photoinitiated polymerization: Advances, challenges, and opportunities. *Macromolecules*. 2010;43(15):6245–6260.
[37] Zhang H. Controlled/"living" radical precipitation polymerization: A versatile polymerization technique for advanced functional polymers. *European Polymer Journal*. 2013;49(3):579–600.
[38] Lee T, Min SH, Gu M, Jung YK, Lee W, Lee JU, Seong DG, Kim BS. Layer-by-layer assembly for graphene-based multilayer nanocomposites: Synthesis and applications. *Chemistry of Materials*. 2015;27(11):3785–3796.
[39] Idumah CI, Ezeani EO, Nwuzor IC. A review: Advancements in conductive polymers nanocomposites. *Polymer-Plastics Technology and Materials*. 2021;60(7):756–783.
[40] Li Y, Wang X, Sun J. Layer-by-layer assembly for rapid fabrication of thick polymeric films. *Chemical Society Reviews*. 2012;41(18):5998–6009.
[41] Richardson JJ, Björnmalm M, Caruso F. Technology driven layer-by-layer assembly of nanofilms. *Science*. 2015;348(6233):aaa2491.
[42] Allison L, Hoxie S, Andrew TL. Towards seamlessly-integrated textile electronics: Methods to coat fabrics and fibers with conducting polymers for electronic applications. *Chemical Communications*. 2017;53(53):7182–7193.
[43] Baba A, Tian S, Stefani F, Xia C, Wang Z, Advincula RC, Johannsmann D, Knoll W. Electropolymerization and doping/dedoping properties of polyaniline thin films as studied by electrochemical-surface plasmon spectroscopy and by the quartz crystal microbalance. *Journal of Electroanalytical Chemistry*. 2004;562(1):95–103.
[44] Zhang Q, Li Y, Feng Y, Feng W. Electropolymerization of graphene oxide/polyaniline composite for high-performance supercapacitor. *Electrochimica Acta*. 2013;90:95–100.
[45] Wang H, Hao Q, Yang X, Lu L, Wang X. Effect of graphene oxide on the properties of its composite with polyaniline. *ACS Applied Materials & Interfaces*. 2010;2(3):821–828.
[46] Heydari Gharahcheshmeh M, Gleason KK. Device fabrication based on oxidative chemical vapor deposition (oCVD) synthesis of conducting polymers and related conjugated organic materials. *Advanced Materials Interfaces*. 2019;6(1):1801564.
[47] Hecht DS, Hu L, Irvin G. Emerging transparent electrodes based on thin films of carbon nanotubes, graphene, and metallic nanostructures. *Advanced Materials*. 2011;23(13):1482–1513.
[48] Ma Z, Shi W, Yan K, Pan L, Yu G. Doping engineering of conductive polymer hydrogels and their application in advanced sensor technologies. *Chemical Science*. 2019;10(25):6232–6244.
[49] Tu Z, Ma Z, Li J, Liang J, Li S, Shi Y, Pan L. Prospective on doping engineering of conductive polymers for enhanced interfacial properties. *Applied Physics Letters*. 2021;119(15).
[50] Sadki S, Schottland P, Brodie N, Sabouraud G. The mechanisms of pyrrole electropolymerization. *Chemical Society Reviews*. 2000;29(5):283–293.
[51] Alhweij H, Emanuelsson EA, Shahid S, Wenk J. High performance in-situ tuned self-doped polyaniline (PANI) membranes for organic solvent (nano) filtration. *Polymer*. 2022;245:124682.
[52] Hu SW, Yang LW, Tian Y, Wei XL, Ding JW, Zhong JX, Chu PK. Simultaneous nanostructure and heterojunction engineering of graphitic carbon nitride via in situ Ag doping for enhanced photoelectrochemical activity. *Applied Catalysis B: Environmental*. 2015;163:611–622.

[53] Long YZ, Li MM, Gu C, Wan M, Duvail JL, Liu Z, Fan Z. Recent advances in synthesis, physical properties and applications of conducting polymer nanotubes and nanofibers. *Progress in Polymer Science*. 2011;36(10):1415–1442.
[54] Balint R, Cassidy NJ, Cartmell SH. Conductive polymers: Towards a smart biomaterial for tissue engineering. *Acta Biomaterialia*. 2014;10(6):2341–2353.
[55] Bhadra S, Khastgir D, Singha NK, Lee JH. Progress in preparation, processing and applications of polyaniline. *Progress in Polymer Science*. 2009;34(8):783–810.
[56] Jacobs IE, Lin Y, Huang Y, Ren X, Simatos D, Chen C, Tjhe D, Statz M, Lai L, Finn PA, Neal WG. High-efficiency ion-exchange doping of conducting polymers. *Advanced Materials*. 2022;34(22):2102988.
[57] Cao Y, He J, Wu J, Wang X, Lu W, Lin J, Xu Y, Chen G, Zeng B, Dai L. A smart anticorrosive epoxy coating based on environmental-stimuli-responsive copolymer assemblies for controlled release of corrosion inhibitors. *Macromolecular Materials and Engineering*. 2022;307(7):2100983.
[58] Albeedh NS. *Study of Anti-Corrosive Properties of Copolymer and Bi-Layer Coatings of Conducting Polymers on Stainless Steel Materials.*
[59] Pramanik NB, Mondal P, Mukherjee R, Singha NK. A new class of self-healable hydrophobic materials based on ABA triblock copolymer via RAFT polymerization and Diels-Alder "click chemistry." *Polymer*. 2017;119:195–205.
[60] Kumar SS, Bashir S, Ramesh K, Ramesh S. New perspectives on Graphene/Graphene oxide based polymer nanocomposites for corrosion applications: The relevance of the Graphene/Polymer barrier coatings. *Progress in Organic Coatings*. 2021;154:106215.
[61] Rahimi SK, Potrekar R, Dutta NK, Choudhury NR. Anticorrosive interfacial coatings for metallic substrates. *Surface Innovations*. 2013;1(2):112–137.
[62] Hentschel J, Kushner AM, Ziller J, Guan Z. Self-healing supramolecular block copolymers. *Angewandte Chemie International Edition*. 2012;51(42):10561–10565.
[63] Laysandra L, Chuang CH, Kobayashi S, Au-Duong AN, Cheng YH, Li YT, Mburu MM, Isono T, Satoh T, Chiu YC. Design of self-cross-linkable poly (n-butyl acrylate)-co-poly [N-(hydroxymethyl) acrylamide] amphiphilic copolymers toward elastic and self-healing properties. *ACS Applied Polymer Materials*. 2020;2(12):5432–5443.
[64] Govers SP, Alexander N, Al-Masri M, Omeis J, van der Ven LG, de With G, Esteves AC. Surface segregation of polydimethylsiloxane-polyether block copolymers in coatings driven by molecular architecture. *Progress in Organic Coatings*. 2021;150:105991.
[65] Tessmar JK, Göpferich AM. Customized PEG-derived copolymers for tissue-engineering applications. *Macromolecular Bioscience*. 2007;7(1):23–39.
[66] Steiger D, Tervoort T, Weder C, Smith P. Poly (p-phenylene alkylene) s—A forgotten class of polymers. *Macromolecular Rapid Communications*. 2000;21(8):405–422.
[67] D'Elia MF, Braendle A, Schweizer TB, Ortenzi MA, Trasatti SP, Niederberger M, Caseri W. Poly (phenylene methylene): A multifunctional material for thermally stable, hydrophobic, fluorescent, corrosion-protective coatings. *Coatings*. 2018;8(8):274.
[68] D'Elia MF, Magni M, Trasatti SP, Schweizer TB, Niederberger M, Caseri W. Poly (phenylene methylene)-based coatings for corrosion protection: Replacement of additives by use of copolymers. *Applied Sciences*. 2019;9(17):3551.
[69] Braendle A, Schwendimann P, Niederberger M, Caseri WR. Synthesis and fractionation of poly (phenylene methylene). *Journal of Polymer Science Part A: Polymer Chemistry*. 2018;56(3):309–318.
[70] Braendle A, Perevedentsev A, Cheetham NJ, Stavrinou PN, Schachner JA, Mösch-Zanetti NC, Niederberger M, Caseri WR. Homoconjugation in poly (phenylene methylene) s: A case study of non-π-conjugated polymers with unexpected fluorescent properties. *Journal of Polymer Science Part B: Polymer Physics*. 2017;55(9):707–720.

[71] D'Elia MF, Magni M, Romanò T, Trasatti SP, Niederberger M, Caseri WR. Smart anti-corrosion coatings based on poly (phenylene methylene): An assessment of the intrinsic self-healing behavior of the copolymer. *Polymers*. 2022;14(17):3457.
[72] DeBerry DW. Modification of the electrochemical and corrosion behavior of stainless steels with an electroactive coating. *Journal of the Electrochemical Society*. 1985;132(5):1022.
[73] Sekine I, Kohara K, Sugiyama T, Yuasa M. Syntheses of polymerized films on mild steels by electro-oxidation and electroreduction and their corrosion resistance. *Journal of the Electrochemical Society*. 1992;139(11):3090.
[74] Troch-Nagels G, Winand R, Weymeersch A, Renard L. Electron conducting organic coating of mild steel by electropolymerization. *Journal of Applied Electrochemistry*. 1992;22(8):756–764.
[75] Camalet JL, Lacroix JC, Aeiyach S, Chane-Ching K, Lacaze PC. Electrodeposition of protective polyaniline films on mild steel. *Journal of Electroanalytical Chemistry*. 1996;416(1–2):179–182.
[76] Bernard MC, Joiret S, Hugot-Le Goff A, Long PD. Protection of iron against corrosion using a polyaniline layer: II. Spectroscopic analysis of the layer grown in phosphoric/metanilic solution. *Journal of the Electrochemical Society*. 2001;148(8):B299.
[77] Ahmad N, MacDiarmid AG. Inhibition of corrosion of steels with the exploitation of conducting polymers. *Synthetic Metals*. 1996;78(2):103–110.
[78] Santos Jr. JR, Mattoso LH, Motheo AD. Investigation of corrosion protection of steel by polyaniline films. *Electrochimica Acta*. 1998;43(3–4):309–313.
[79] Vaughn GD, Frushour BG, Dale WC. Scratch indentation, a simple adhesion test method for thin films on polymeric supports. *Journal of Adhesion Science and Technology*. 1994;8(6):635–650.
[80] Abd El-Ghaffar MA, Abdelwahab NA, Fekry AM, Sanad MA, Sabaa MW, Soliman SM. Polyester-epoxy resin/conducting polymer/barium sulfate hybrid composite as a smart eco-friendly anti-corrosive powder coating. *Progress in Organic Coatings*. 2020;144:105664.
[81] Riaz U, Ashraf SM, Ahmad S. High performance corrosion protective DGEBA/polypyrrole composite coatings. *Progress in Organic Coatings*. 2007;59(2):138–145.
[82] Ohtsuka T. Corrosion protection of steels by conducting polymer coating. *International Journal of Corrosion*. 2012;2012.
[83] Zadeh MK, Yeganeh M, Shoushtari MT, Esmaeilkhanian A. Corrosion performance of polypyrrole-coated metals: A review of perspectives and recent advances. *Synthetic Metals*. 2021;274:116723.
[84] Le HN, Garcia B, Deslouis C, Le Xuan Q. Corrosion protection and conducting polymers: Polypyrrole films on iron. *Electrochimica Acta*. 2001;46(26–27):4259–4272.
[85] Hien NT, Garcia B, Pailleret A, Deslouis C. Role of doping ions in the corrosion protection of iron by polypyrrole films. *Electrochimica Acta*. 2005;50(7–8):1747–1755.
[86] Van Schaftinghen T, Deslouis C, Hubin A, Terryn H. Influence of the surface pre-treatment prior to the film synthesis, on the corrosion protection of iron with polypyrrole films. *Electrochimica Acta*. 2006;51(8–9):1695–1703.
[87] Kowalski D, Ueda M, Ohtsuka T. Self-healing ion-permselective conducting polymer coating. *Journal of Materials Chemistry*. 2010;20(36):7630–7633.
[88] Kowalski D, Ueda M, Ohtsuka T. The effect of ultrasonic irradiation during electropolymerization of polypyrrole on corrosion prevention of the coated steel. *Corrosion Science*. 2008;50(1):286–291.
[89] Bai Y, Chen Y, Wang Q, Wang T. Poly (vinyl butyral) based polymer networks with dual-responsive shape memory and self-healing properties. *Journal of Materials Chemistry A*. 2014;2(24):9169–9177.

[90] Kumar P, Khan N, Kumar D. Polyvinyl butyral (PVB), versatile template for designing nanocomposite/composite materials: A review. *Green Chemistry & Technology Letters*. 2016;2(4):185.

[91] Samide A, Merisanu C, Tutunaru B, Iacobescu GE. Poly (vinyl butyral-co-vinyl alcohol-co-vinyl acetate) coating performance on copper corrosion in saline environment. *Molecules*. 2020;25(3):439.

[92] Samide A, Brătulescu G, Merişanu C, Cioateră N. Anticorrosive coating based on poly (vinyl acetate) formed by electropolymerization on the copper surface. *Journal of Applied Polymer Science*. 2019;136(14):47320.

[93] Samide A, Tutunaru B, Merişanu C, Cioateră N. Thermal analysis: An effective characterization method of polyvinyl acetate films applied in corrosion inhibition field. *Journal of Thermal Analysis and Calorimetry*. 2020;142:1825–1834.

[94] Jiang ZC, Xiao YY, Kang Y, Pan M, Li BJ, Zhang S. Shape memory polymers based on supramolecular interactions. *ACS Applied Materials & Interfaces*. 2017;9(24):20276–20293.

[95] Carrot C, Bendaoud A, Pillon C. Chapter 3. Polyvinyl Butyral. In *Handbook of Thermoplastics* (p. 128). CRC Press, 2015.

8 Computational modeling of conducting polymers used as anti-corrosive materials

Omar Dagdag, Rajesh Haldhar, Abhinay Thakur, Elyor Berdimurodov, and Hansang Kim

8.1 INTRODUCTION

The destructive attack on metals by the environment is known as corrosion [1]. Currently, a lot of researchers have been using organic coatings to protect steel structures from corrosion [2]. Since organic coatings are readily accessible, easy to employ, and inexpensive, they are typically used to protect metallic surfaces from corrosion [3]. A potential method for preventing corrosion on metal substrates is to apply organic coatings to their surfaces [4]. Two heteroatoms (P, O, N, and S) and aromatic rings make up the majority of the components of epoxy resins [5, 6]. These have the ability to transport electrons [7] to the surface of the metal, filling open orbitals and creating covalent bonds.

Conductive polymer coatings are among the most efficient, affordable, and environmentally benign solutions for anti-corrosion applications [8].Although the capacity of conducting polymers to prevent corrosion of metals in aqueous circumstances is widely recognized, the processes involved are complex, as the mechanisms are not fully understood. Given that the dopant affects the conducting polymer's conductivity and permeability and is released when the polymer and metal are galvanically coupled, it is predicted that the dopant used with the conducting polymer will have an effect on the corrosion processes.

During World War II, computer simulations first became popular [9]. Since then, and particularly recently, computer power has increased dramatically. As a result, computer simulation is now a crucial part of the experimental method used to study a variety of research issues in numerous fields. Nowadays, predictions from computer simulations can generally accord with experimental work, thanks to the creation of accurate force-field parameters and the construction of sensible simulation methodologies and tactics. **Figure 8.1** depicts the connection between computer simulations and experimental research.

DFT and MD simulations are the most significant of the several computer simulation techniques that have been utilized to forecast the physical characteristics of

DOI: 10.1201/9781003456025-8

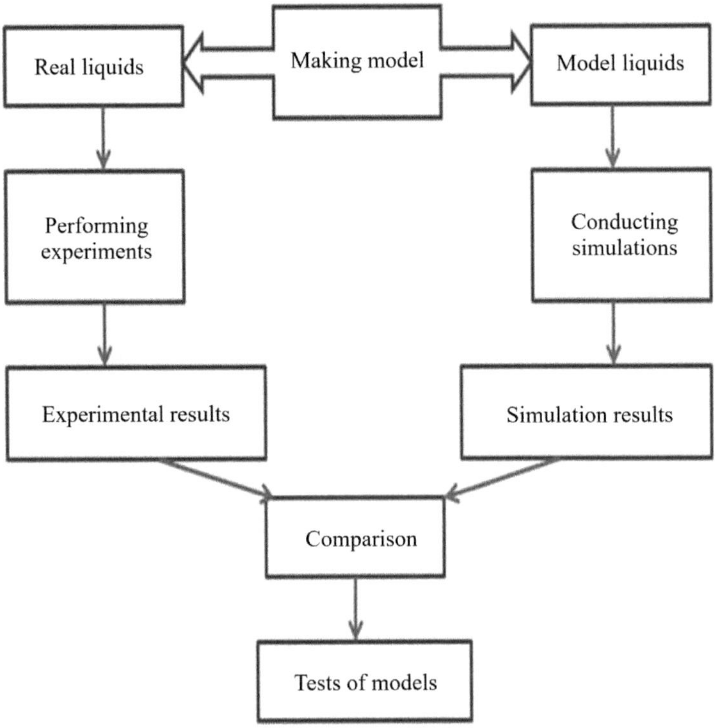

FIGURE 8.1 The connection between computer simulation and experiment.
Reprinted with permission from [9] © 2017 Elsevier Publications.

conducting polymers employed as anti-corrosive materials. The main objective of this chapter is to cover various simulation techniques, particularly the DFT and MD simulation techniques used to comprehend the physical characteristics of conducting polymers as anti-corrosive materials.

8.2 COMPUTATIONAL MODELING OF CONDUCTING POLYMERS AS ANTI-CORROSIVE MATERIALS

According to Dagdag et al. [10], an epoxy coating formulation based on DGEBA that was cured with polyaminoamide (PAA) was employed as an anti-corrosive material for Al alloy substrates. **Figure 8.2** illustrates the chemical composition of DGEBA, PAA, the process of the thermoset polymer's structure, and the curing process. The Biovia program was used to carry out the DFT computation [11, 12]. MC and MD using Materials Studio 8.0 [13–15].

HOMO and LUMO (**Figure 8.3**) are widely used as recommendations in molecular modeling, along with mechanism verification to ascertain if a reaction is going as planned and which molecules are in control [16]. While HOMO displays the electrons

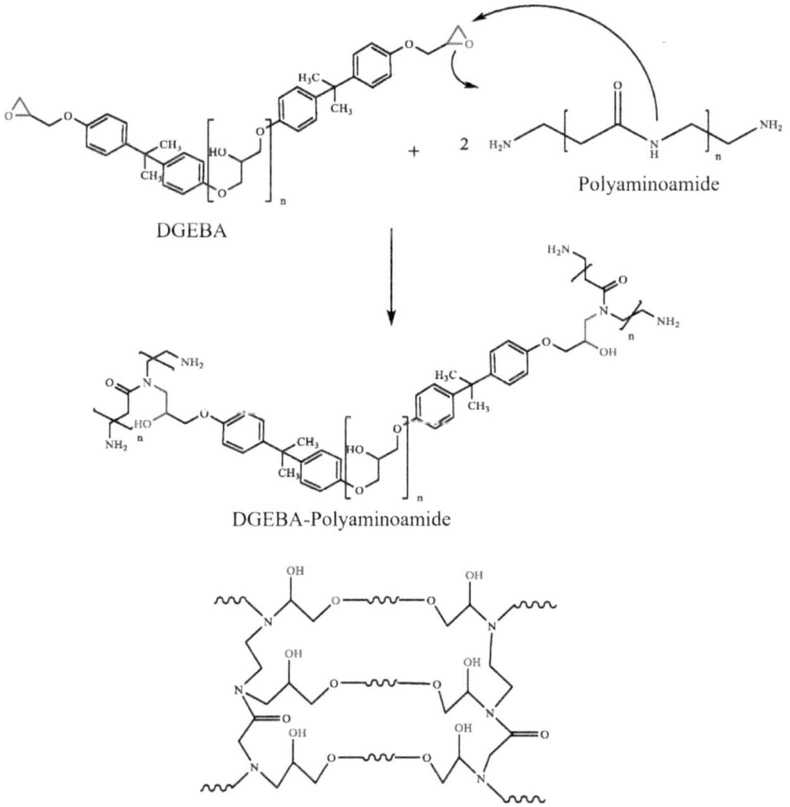

FIGURE 8.2 The curing reaction between PAA and DGEBA.

Reprinted with permission from [10] © 2022 Elsevier Publications.

being provided to the electron-accepting areas of the molecule, LUMO displays the net-donor parts of the molecule. The MEP, LUMO, along with the HOMO of the epoxy coating, are shown in **Figure 8.3**. The molecule bearing the heteroatoms (O in addition to N) has the HOMO and LUMO substantially distributed over one of the phenyl rings. For epoxy symbols, HOMO and LUMO are especially advantageous since they may interact with the surface of Al (111) via both electron-donation and electron-acceptance [17].

Analyzing its ionization potential and electron affinity reveals the same behavior, producing a similar capacity. Additionally, the reasonable softness value and relatively low relative hardness value promote epoxy's propensity for adhering to metal surfaces [13]. With the use of MAC, it is now possible to postulate the preferred spots (atoms) in coatings for metal adsorption. It is generally believed that the atom with the biggest negative charge may successfully host the functions of the middle iron surface as well as the coating molecules [13]. **Figure 8.3** displays the MAC for the overlay. The most disagreeable charges are found in the O and N molecules of the covering, suggesting that these concentrations have the most significant electron

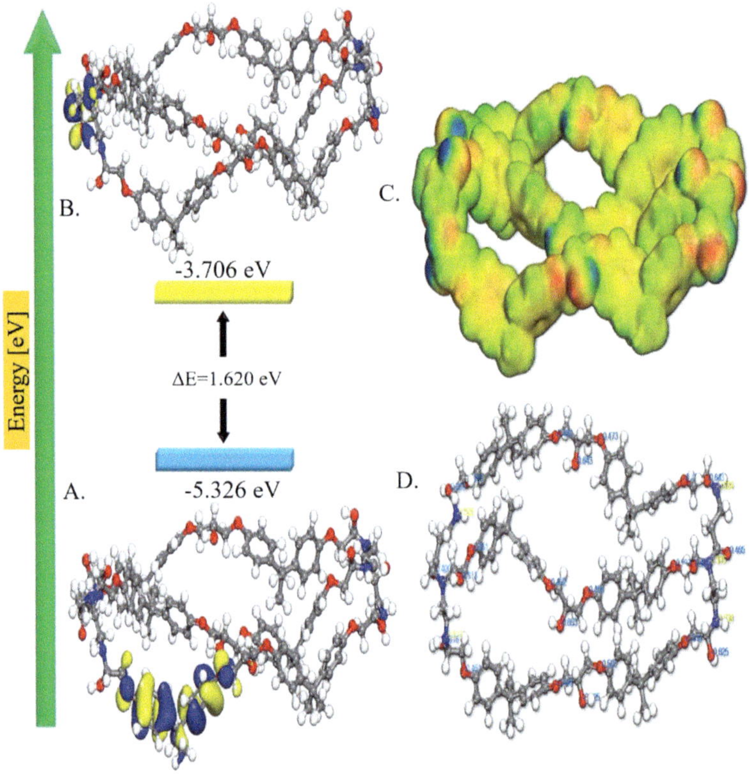

FIGURE 8.3 (a) HOMO, (b) LUMO surfaces, (c) ESP, and (d) MAC of the epoxy molecule. Reprinted with permission from [10] © 2022 Elsevier Publications.

thickness and are linked to surface iron. In **Figure 8.3**, electrostatic potential (ESP) is also depicted in an open manner after manifesting for the cover (red region).

Figure 8.4 demonstrates the epoxy component's minimal energy usage on steel in an erosion-prone environment (as previously chosen). Calculations using covering adsorption, as recommended by the MAC, reveal that O particles encourage diffusion. This solid adsorption forms a protective coating on the metal surface to prevent erosion and to protect the material. Based on the MC calculations, the O and N atoms are in charge of the epoxy polymer's adsorption.

In the same group [18], an EP coating based on DGEBA-DAA was created and applied to 15CDV6 steel in a 3% NaCl solution. MD simulations were used to show the DGEBA-DAA coating's anti-corrosive performance.

Figure 8.5a displays the side and top views of the DGEBA-DAA equilibrium configurations that are adsorbed on the metal surface. It is conceivable that DGEBA-DAA gets the parallel orientation necessary to ensure its efficient adsorption. **Figure 8.5b** shows the DGEBA-DAA adsorption density field on a metal substrate. Evidently, the investigated inhibitor will probably adhere to the metal surface to create a dense, hydrophobic barrier and prevent corrosion. In the current investigation,

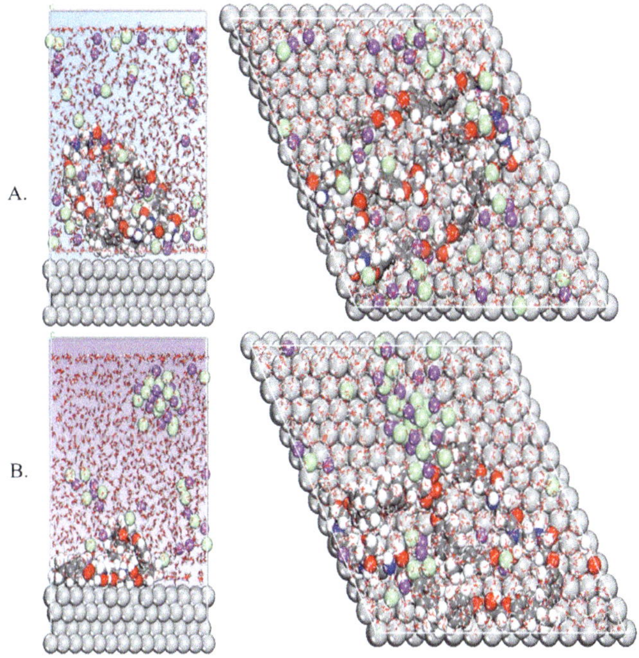

FIGURE 8.4 (a) MC and (b) MD were acquired from the Epoxy coating's adsorption configurations in the simulated corrosive medium on the surface of Al (111).

Reprinted with permission from [10] © 2022 Elsevier Publications.

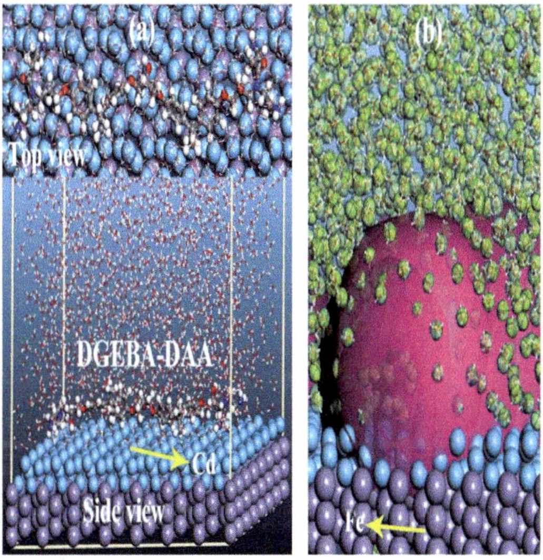

FIGURE 8.5 (a) Equilibrium configuration and (b) density fields of DGEBA-DAA molecule adsorbed on Cd(111)/Fe(111) surface.

Reprinted with permission from [18] © 2020 Elsevier Publications.

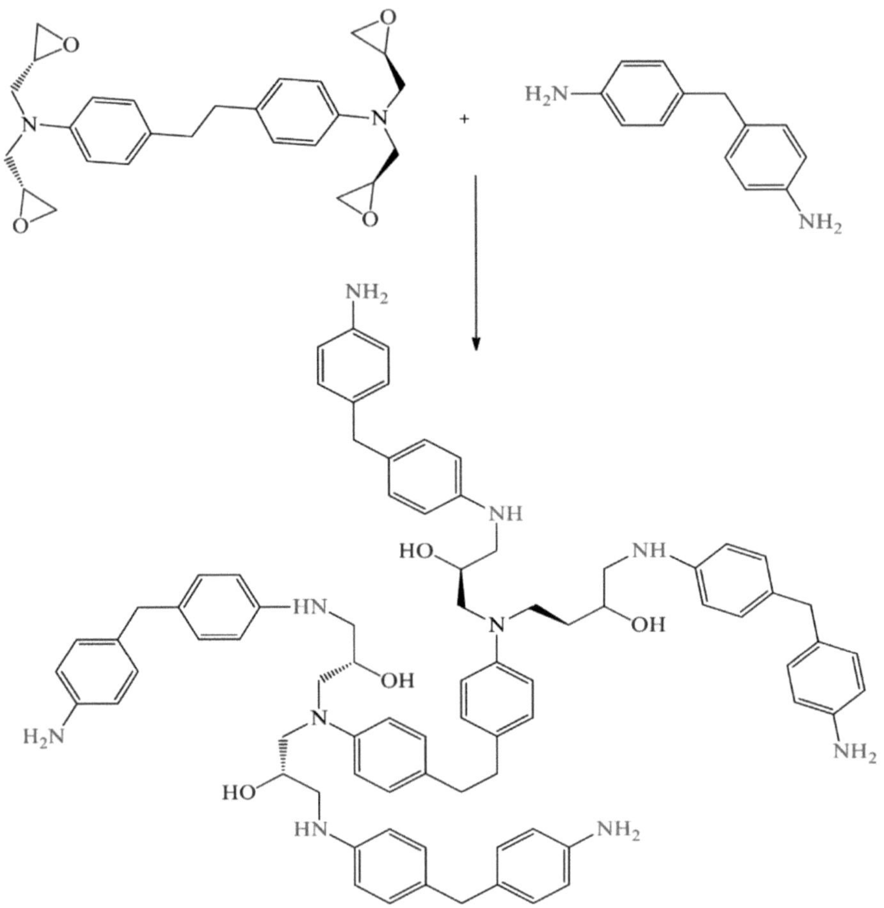

FIGURE 8.6 TGEDA-MDA's molecular structure acts as a protective layer.
Reprinted with permission from [19]© 2020 Elsevier Publications.

the E_{ads} value for DGEBA-DAA was determined to be –573.4 kJ/mol. This low number suggests that the adsorptive system is stable and that spontaneous adsorption is possible. This is consistent with its superior effectiveness in corrosion inhibition.

Other research conducted by this group [19] aimed to investigate a different anti-corrosive coating composition for CS corrosion in 3 wt% NaCl media. A methylene dianiline (MDA) is used to cure an epoxy resin (TGEDA) in the formulation (TGEDA-MDA). **Figure 8.6** shows the TGEDA-MDA molecular structure.

The optimized HOMO, LUMO, along with the ESP picture of the TGEDA-MDA, is shown in **Figure 8.7**. When the results were examined, it was discovered that HOMO is mostly present over one of the MDA moieties, suggesting that this area of the TGEDA-MDA molecule is connected to a high electron density and is likely to function as an electron donor during interactions with metals. On the other hand, LUMO is largely concentrated over the TGEDA-MDA molecule's core area,

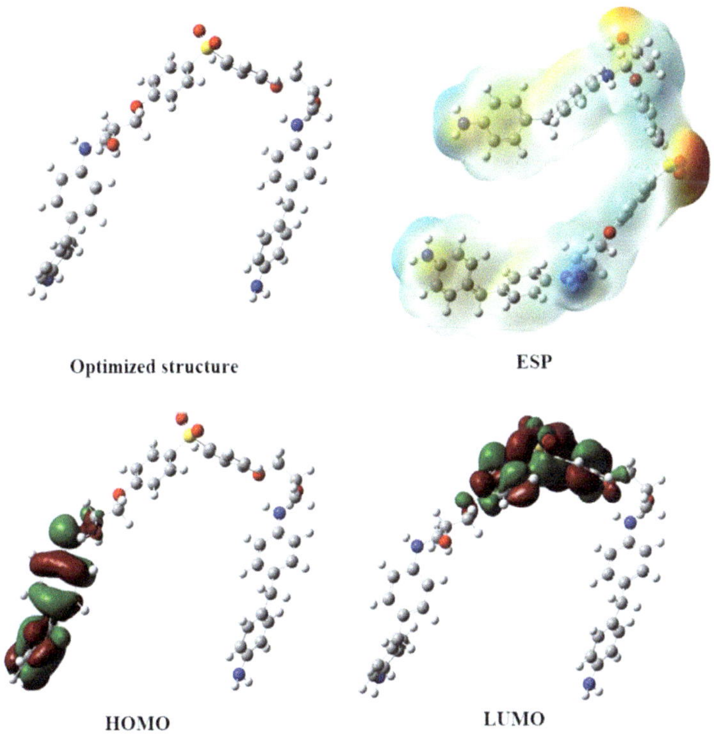

FIGURE 8.7 TGEDA-MDA's ESP, HOMO, along with LUMO surfaces.

Reprinted with permission from [19]© 2020 Elsevier Publications.

indicating that this area most likely acts as an electron acceptor during interactions with the metal. It is widely known that organic coatings, along with metallic surfaces, can interact as donor-acceptor. The relationship between a high E_{HOMO} value along with a low E_{LUMO} value is strong metal-coating interactions. The relationship between a high E_{HOMO} value along with a low E_{LUMO} value is strong metal-coating interactions. Due to its comparatively high E_{HOMO} and low E_{LUMO} values, TGEDA-MDA has a remarkable potential to react/adsorb on the metallic surface in this case. Similar to this, low levels of ΔE are associated with high reactivity along with coating effectiveness. TGEDA-MDA is more highly reactive and can operate as a powerful adsorbate on metallic surfaces since it has a lower value for ΔE. According to theory, higher levels of the dipole moment enhance the dipole-dipole interactions among the adsorbed inhibitor along with the metallic surface, resulting in better adsorption and a greater inhibitory effect [20].

A computational investigation based on MD simulation was carried out to get fundamental knowledge of the surface bonding nature and improved interfacial adhesion between the Fe(110) in NaCl medium, as well as Fe(110) (blank) and the TGEDA-MDA coating. On the Fe (110) in NaCl medium surface, the equilibrium adsorption configuration that was investigated using TGEDA-MDA is shown in **Figure 8.8** from

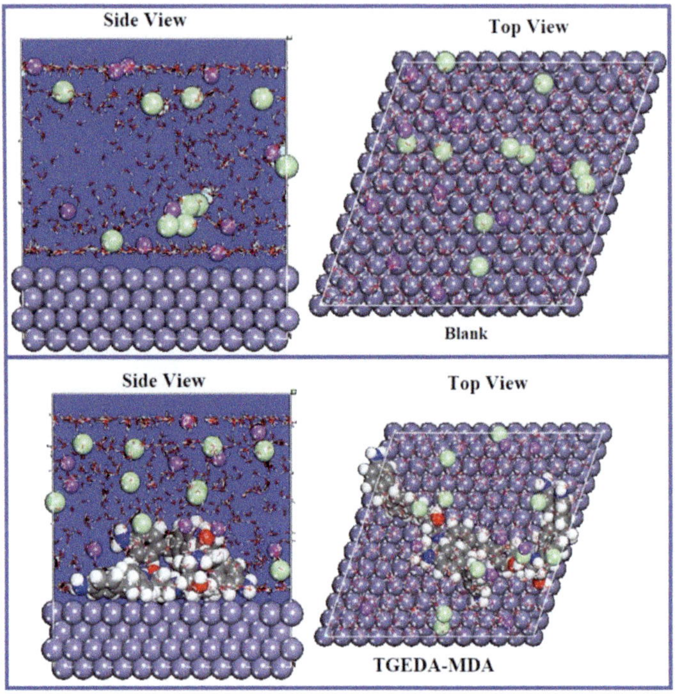

FIGURE 8.8 The most stable low-energy designs for the adsorption of Fe (110)/without and with TGEDA-MDA are shown in top and side views, respectively.

Reprinted with permission from [19]© 2020 Elsevier Publications.

the top and side perspectives. It appears that TGEDA-MDA has linked itself to Fe (110) in the form shown. The TGEDA-MDA coating is able to adhere to the film that has been applied over the steel sample, as shown by visual evidence.

As can be observed from the side view of the finished structure, the TGEDA-MDA structure's molecular skeleton is adherent to Fe (110) in a parallel orientation. This direction often increases the structure's ability to cover more surface area. Furthermore, a closer inspection of the side view of the TGEDA-MDA film reveals that the O, along with N heteroatoms, specifically the C-O and C-N atoms, are pointing in the direction of the surface.

MC simulations were performed to gain additional information regarding the TGEDA-MDA's propensity toward steel-covered materials. **Figure 8.9** depicts the top and side views of the equilibrium adsorption configuration of Fe(110) (blank) along with TGEDA-MDA on Fe (110) in the NaCl medium surface. The adsorption of TGEDA-MDA is greatly aided by O, along with N atoms as well as aromatic rings. The TGEDA-MDA structure has clearly attached to the surface of the Fe (110) atom in this picture, demonstrating its inclination to form a protective layer on the Fe (110)-covered surface. The TGEDA-MDA structure may protect the steel-covered sample from corrosive elements by being positioned above the Fe (110)-coated surface, as can be seen from the side view.

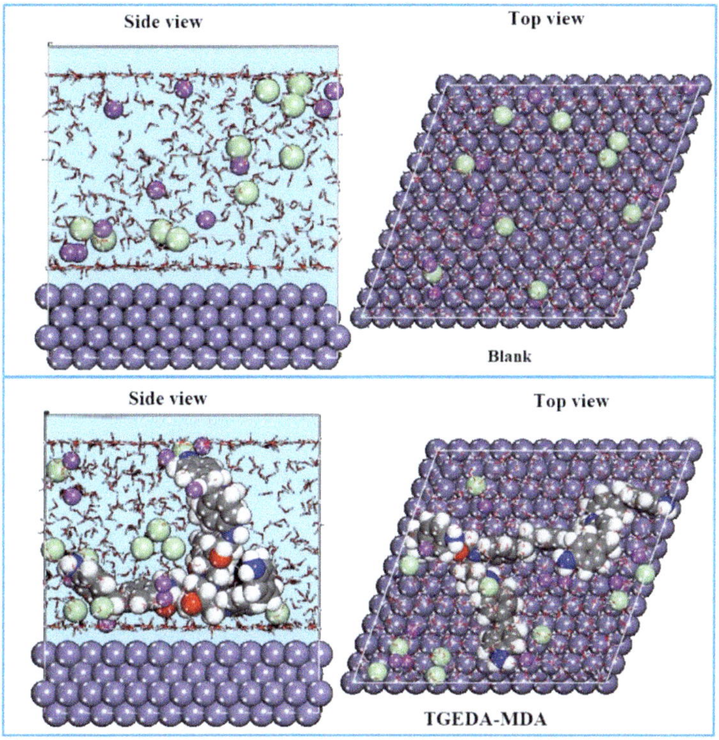

FIGURE 8.9 The most stable low-energy designs for the adsorption of Fe (110)/without along with TGEDA-MDA are shown in top and side views, respectively.

Reprinted with permission from [19]© 2020 Elsevier Publications.

For this study, our group has assessed the anti-corrosive capabilities of two epoxy resins (ERs), specifically TGEDA along with HGEDPAT, as well as their polymer composite reinforced with Zn [21]. The creation of ERs (ER1 and ER2) is seen in **Figure 8.10**.

Quantum chemical calculations employing DFT were carried out to confirm the electrochemical findings and demonstrate the adsorption behavior of ER1 along with ER2 on the copper surface. **Figure 8.11** shows the FMOs (HOMO as well as LUMO) electron distributions for the neutral ER1 along with ER2 molecules.

E_{HOMO} and E_{LUMO} are unquestionably related to ER1 along with ER2's capacities for electron donation and acceptance, respectively [22]. In the ER2 molecule, the HOMO is concentrated on one of the diphenylmethane moieties along with the adjacent N and O atoms, whereas the LUMO is primarily concentrated on a different diphenylmethane moiety (**Figure 8.11**). This result demonstrates that one of the diphenylmethane moieties, as well as the surrounding N and O atoms, functions as an electron donor in metal-ER2 interactions, whereas the other diphenylmethane moiety functions as an electron acceptor. The electron-rich oxygen and nitrogen heteroatoms in π-electrons conjugations are associated with high electron density. While the LUMO is largely located on one of the four 4-benzylbenzenamine

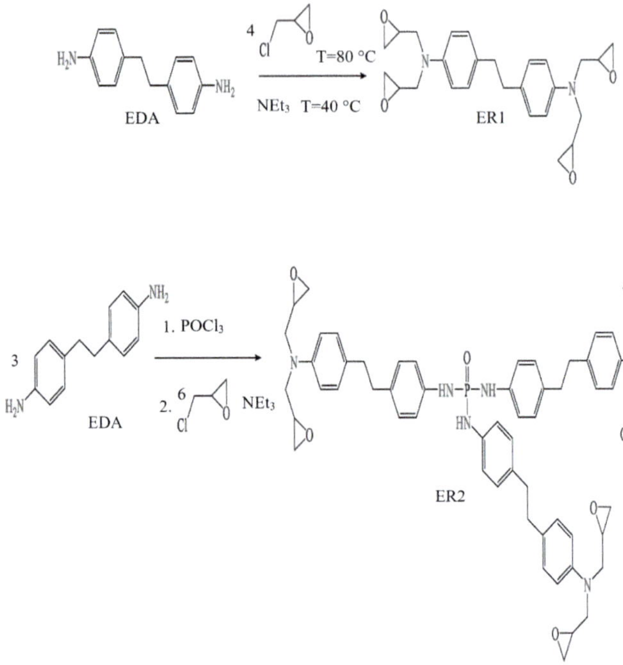

FIGURE 8.10 Epoxy resin ER1, along with ER2 synthesis methods.

Reprinted with permission from [21] © 2020 Elsevier Publications.

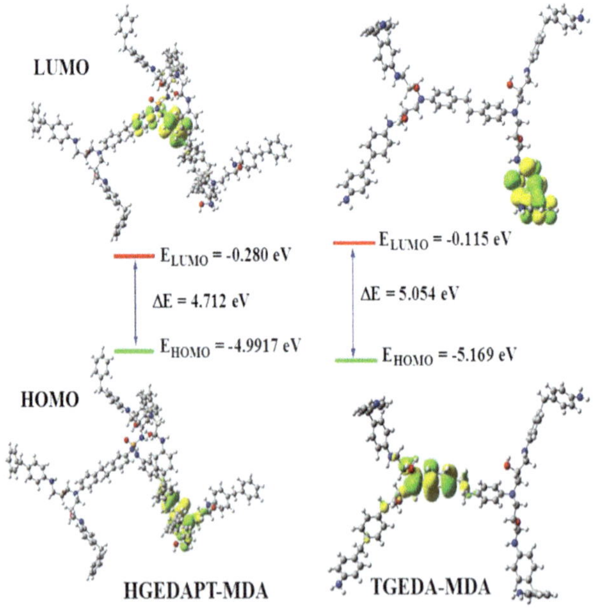

FIGURE 8.11 The ER2 (HGEDAPT-MDA) along with ER1 (TGEDA-MDA), HOMO, as well as LUMO surfaces.

Reprinted with permission from [21] © 2020 Elsevier Publications.

moieties, the HOMO is primarily distributed throughout the midsection of the ER1 molecule. A higher E_{HOMO} value and a lower E_{LUMO} value are associated with strong meta-inhibitor interaction [23]. Specifically, the E_{HOMO} values of the investigated EPs followed the sequence ER2 > ER1, whereas the E_{LUMO} values showed ER1 > ER2. This discovery showed that ER2 molecules connect with Cu (111) surfaces more forcefully than ER1 ones.

The ER1 as well as ER2 coatings on the Cu(111) surface, as well as their 5% Zn composites (ER3 along with ER4), were studied using MD modeling to illustrate how they adsorb to the surface in a NaCl medium. **Figure 8.12** shows the lowest energy equilibrium configurations of the ER1 and ER2 coatings and their 5% Zn composite (ER3, along with ER4) coatings adsorbed on the metal surface.

Figure 8.12a and **8.12b** show that ER1 (in **Figure 8.12a**) and ER2 (in **Figure 8.12b**) obtain flat or horizontal orientations on the Cu(111) surface. It is crucial

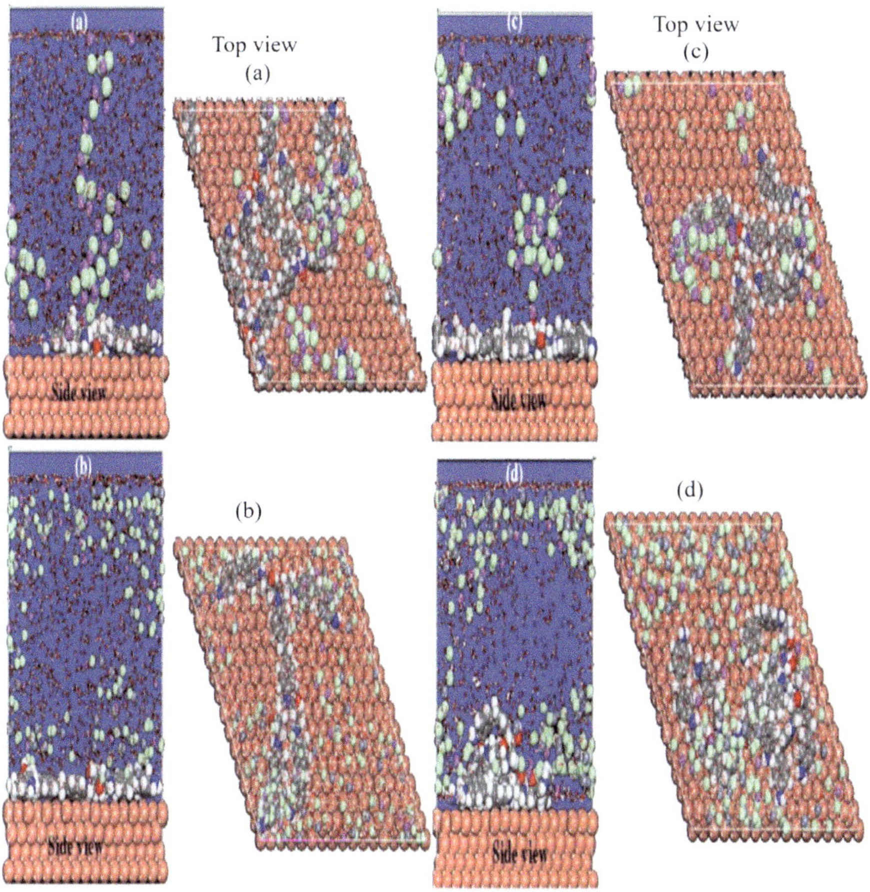

FIGURE 8.12 The most stable low-energy combinations for the adsorption of Cu (111)/ER1, along with Cu (111)/ER3 (111)/ER2, as well as Cu (111)/ER4, can be seen in side and top views.

Reprinted with permission from [21] © 2020 Elsevier Publications.

to understand that ER1 and ER2 are effectively adsorbed by aromatic rings and heteroatoms. In particular, C-O, C-N, as well as P-O atoms increased the adsorption ability of ER2 when O, N, as well as P heteroatoms were present. The ER3 (**Figure 8.12c**) and ER4 (**Figure 8.12d**) epoxy resins both showed better adsorptive properties in their composite forms, according to a careful analysis. Zn ions can also be seen near the ER1 along with ER2 molecules. The adsorption capability of ER1 and ER2, as well as their 5% Zn composites (ER3 as well as ER4) on Cu (111) surface, was further examined using the RDF. All molecules interact with metallic surfaces by chemisorption, according to an interaction peak at 2.5. This discovery complements the findings of DFT research very well.

This study evaluates the anti-corrosive effectiveness of an EP for CS corrosion in a 3% NaCl solution based on DGEDDS, MDA, and both with and without TiO_2 [24]. The composite constructed of DGEDDS-MDA as well as TiO_2 is known by the name DGEDDS-MDA-TiO_2. Computational methods were used to illustrate the anti-corrosive effectiveness of the DGEDDS-MDA along with its TiO_2 composite.

Figure 8.13 shows the creation of the DGEDDS-MDA, including its TiO_2 composite compositions.

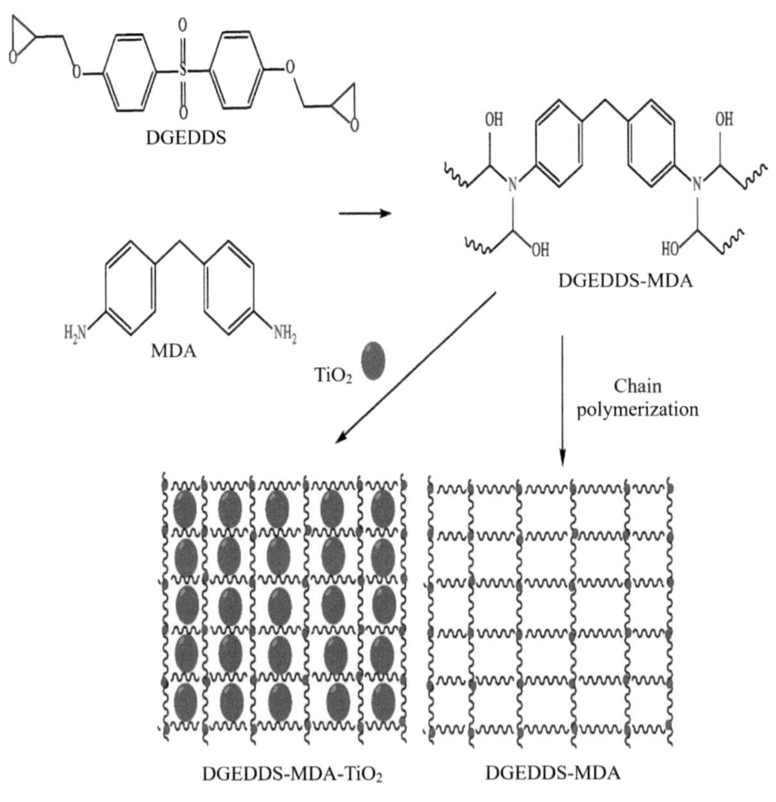

FIGURE 8.13 DGEDDS-MDA as well as its TiO_2 composite anti-corrosive compositions are synthesized.

Reprinted with permission from [24] © 2020 Elsevier Publications.

The LUMO and HOMO orbitals for the DGEDDS-MDA are shown in **Figure 8.14**. It is generally known that one of the key factors in determining an inhibitor's reactivity, along with stability, as well as predicting the course of chemical processes, is the spreading of the electron density along the HOMO and LUMO surfaces. The molecular regions responsible for charge transfer (donation) are designated as HOMO, and the molecular positions responsible for charge acceptance (retro-donation) are designated as LUMO [25]. The FMO image of the DGEDDS-MDA molecule (**Figure 8.14**) confirms that the HOMO is predominantly distributed above the middle portion of the molecule, which includes the diphenylmethane (**Figure 8.14a**), as well as the LUMO is predominantly distributed over one of the four sulfonyldiphenol portions, suggesting that this portion of the molecule is in charge of accepting electrons from the metallic surface (**Figure 8.14b**). The DGEDDS-MDA molecule's comparatively high HOMO (E_{HOMO} = 4.749 eV), along with low LUMO (E_{LUMO} = 2.276 eV) values, reveals the molecule's tendency to engage in donation and absorption interactions with the metallic surface [25]. Similar to that, ESP map shows the distribution of all charges in three dimensions [26]. On the molecule's ESP map, the positive area (electrophilic sites) can be identified by the blue regions, while the low electron density (nucleophilic sites) is illustrated by the red regions. As

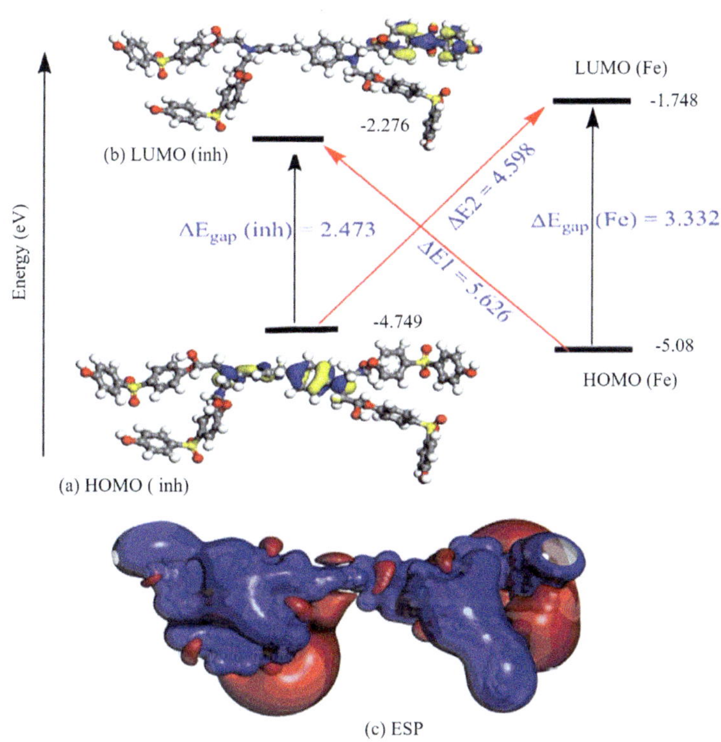

FIGURE 8.14 (a) HOMO; (b) LUMO, ΔE, ΔE1, and ΔE2; and (c) ESP of the DGEDDS-MDA molecule.

Reprinted with permission from [24] © 2020 Elsevier Publications.

shown in **Figure 8.14c**, the areas that are most positively orientated are the ones that include the Sulfonyldiphenol region, while the regions that are most negatively orientated are ones that surround the O as well as N atoms in the middle of the molecule.

The most reliable along with low-energy DGEDDS-MDA orientations either with or without TiO_2 on the metal surface are shown in **Figure 8.15a**. As can be observed, during adsorption, DGEDDS-MDA obtains flat orientations by using heteroatoms and aromatic rings. In the presence of TiO_2, TiO_2 particles may be detected close to the DGEDDS-MDA inhibitor molecule (**Figure 8.15b**), and they are essential for increasing the system's overall adsorption capability. This type of adsorption leads to the development of an effective barrier. The DGEDDS-MDA's E_{ads} values were −649.5 kJ/mol and −952.3 kJ/mol, respectively, with and without TiO_2. E_{ads} tests with negative magnitudes show that epoxy resin can spontaneously adsorb with or without TiO_2.

In a research by Rahmani et al. [27], graphene oxide (GO), a platform based on meso/microporous graphene, was suggested as a potential multifunctional filler for epoxy coating. Phytic acid molecules were used to change the GO layers (Phy-rGO), and subsequently ZIF-9 MOFs were used to completely adorn them (ZIF-9@Phy-rGO).

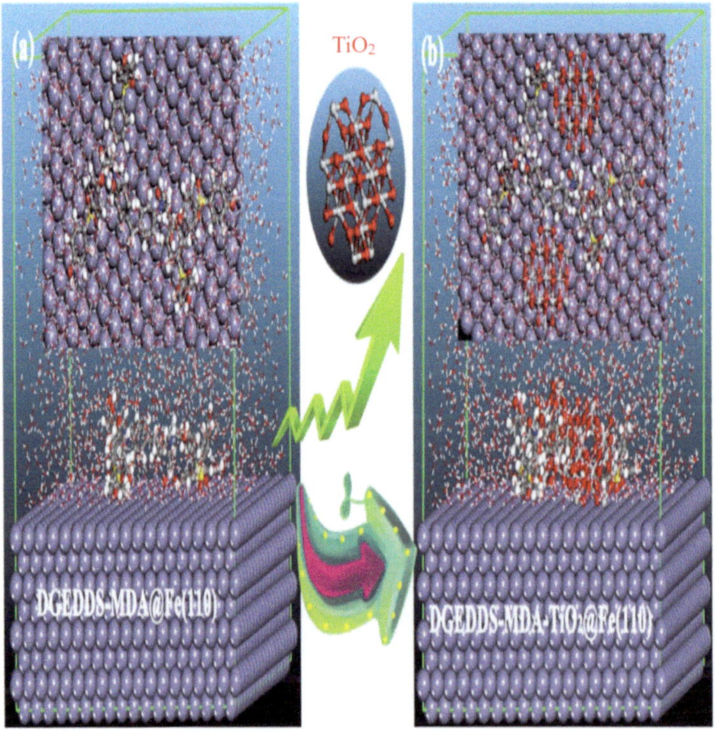

FIGURE 8.15 Side and top views of the stable adsorption configurations of (a) DGEDDS-MDA and (b) DGEDDS-MDA-TiO_2 on Fe (110) surface.

Reprinted with permission from [24] © 2020 Elsevier Publications.

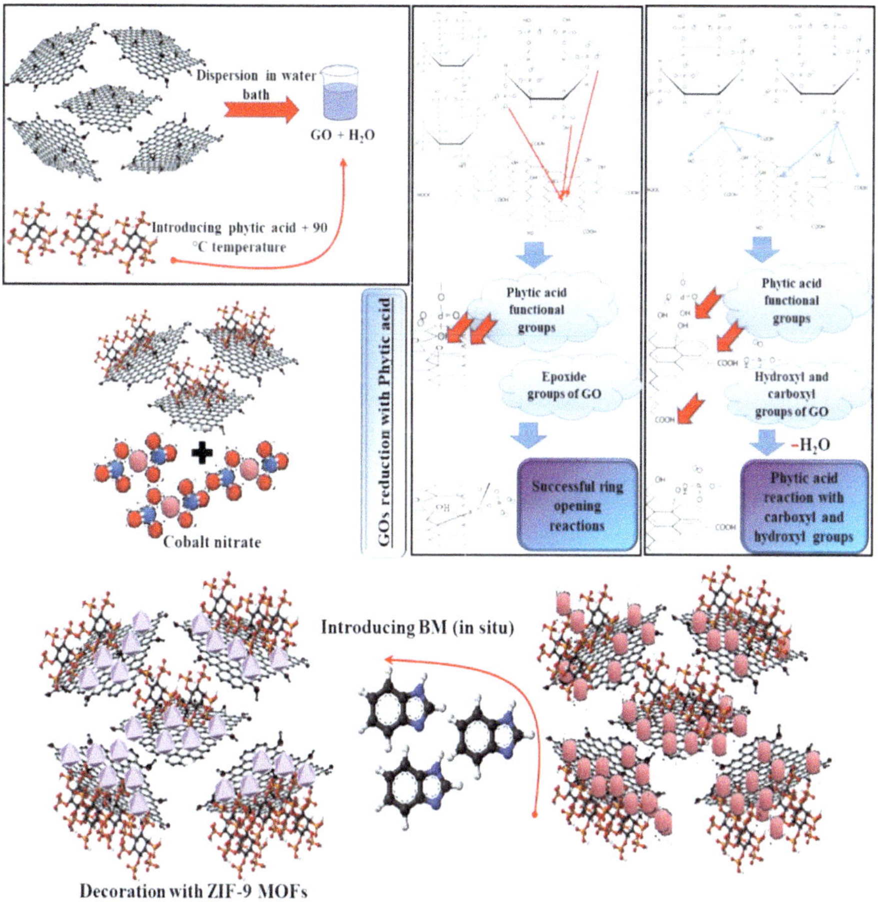

FIGURE 8.16 Diagram showing the steps in the creation of ZIF-9@Phy-rGO nanocomposites. Reprinted with permission from [27] © 2022 Elsevier Publications.

Purple nanoparticles (ZIF-9@Phy-rGOs) were first filtered and dried (**Figure 8.16**). The affinity of phytic acid for GO adsorbents was assessed using an electronic-scale calculation that utilized DFT-D.

Figure 8.17 depicts the definitive configurations of phytic acid over GO surfaces with a single active fragment and oxygenated surfaces with various groups. In order to demonstrate its preference for intermolecular interactions, the phytic acid molecule chose to stabilize near the single- or multifunctionalized GO surfaces. Further analysis of the final clusters revealed that phytic acid stabilization at oxygen-containing surfaces resulted in the formation of H-bonds that linked the OH or carbonyl moiety of phytic acid and different surface-decorated GO groups. The relationships between components of phytic acid were further investigated by calculating the binding energy [28]. Phytic acid had binding energies of −42.88, −51.15, as well as −48.08 kcal/mol over the GO(O), GO(COOH), and GO(OH) adsorbent

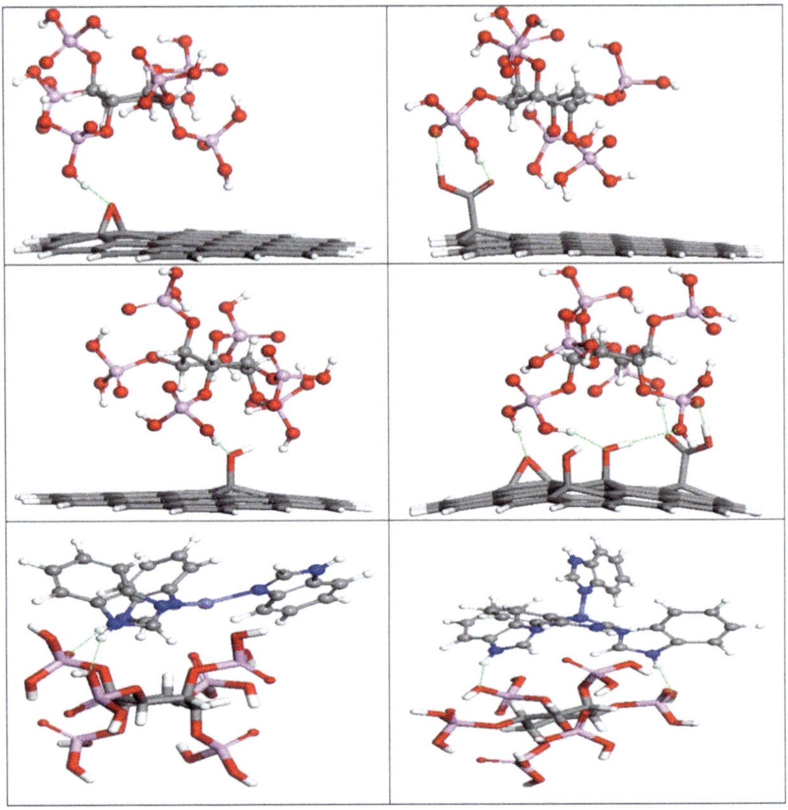

FIGURE 8.17 The phytic acid's final DFT-D refined interactions with several GO surface models, as well as with $[Co(Benzimidazole)_3]^{+2}$ along with $[Co(Benzimidazole)_4]^{+2}$ complex combinations.

Reprinted with permission from [27] © 2022 Elsevier Publications.

models, respectively. Additionally, phytic acid interacted with a multifunctionalized GO model in a 59.13 kcal/mol energy-dense manner. The negative magnitudes calculated for binding energies serve to quantitatively increase the phytic acid affinity for GO-based surfaces.

Mustard seed extract was introduced by Dehghani et al. [29] as a novel, environmentally friendly, cost-effective reductant for GO layers because it contains active molecules such as allyl isothiocyanate, arabinose, diallyltrisulfide, xylose, rhamnose, and galactose. The rGOnano-sheets were doped with Zn (II) ions (Zn@rGO) as the inorganic inhibitor to increase the active protective property (**Figure 8.18**).

The DFT-D methods were selected to assess the GO interactions between them and the aforementioned green compounds.

The divalent cations $[Zn(Allyl\ isothiocyanate)n]^{+2}$ as well as $[Zn(Arabinose)n]^{+2}$ are demonstrated in **Figure 8.19**. Complexes with the final configurations of allyl isothiocyanate, along with arabinose, were produced at the lowest energies. The

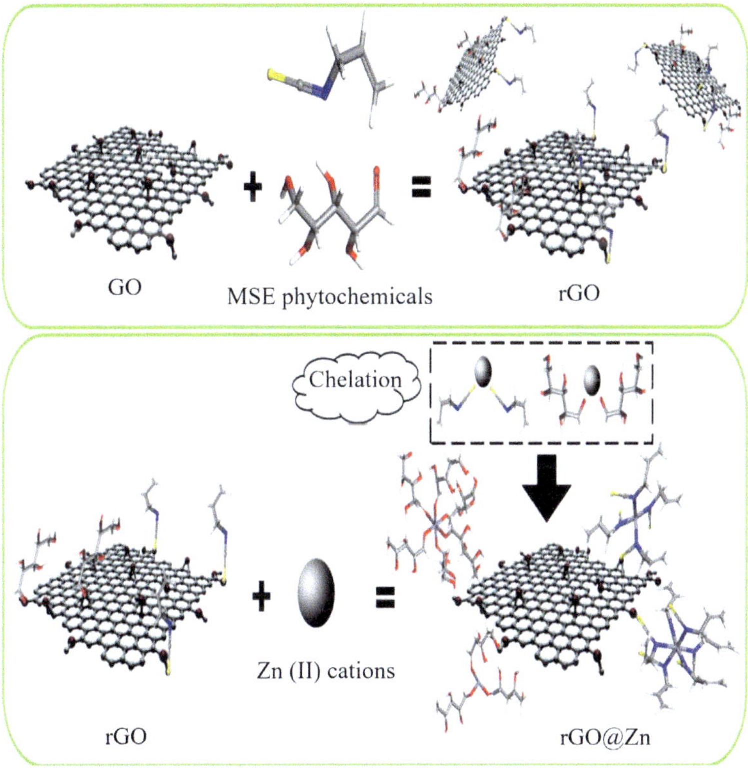

FIGURE 8.18 Diagram showing the reduction of GOs by mustard extract as well as Zn cation doping over the thinner layers of decreased GOs.

Reprinted with permission from [29] © 2022 Elsevier Publications.

shown configurations demonstrate that pure forms of chemical compounds, such as allyl isothiocyanate and arabinose, as well as complexes of green molecules and Zn (II), persisted on the oxygenated surface, demonstrating graphically the affinity of pure/complex chemical compounds against the GO adsorbent. It was discovered that the inorganic centers of the [Zn(Allyl isothiocyanate)2]$^{+2}$ and [Zn(Allyl isothiocyanate)3]$^{+2}$ compounds aligned with the GO. Zn(II) cations align in this manner because of electrostatic attraction to GO heteroatoms. In the adsorbed configurations of the other compounds, the allyl isothiocyanate ligands were situated closer to the surface. Additionally, it should be noted that during interactions of complexes of [Zn(Allyl isothiocyanate)5]$^{+2}$ and [Zn(Allyl isothiocyanate)6]$^{+2}$, some organic ligands separated from inorganic cations with a single hydroxyl moiety detached from the surface of GO. The resulting structures were further studied by studying binding energy [28]. Binding energies (kcal/mol) of −77.82, −131.54, −121.92, −97.35, −104.68, including −69.55 have been calculated for allyl isothiocyanate, [Zn(Allyl isothiocyanate)$_2$]$^{+2}$, [Zn(Allyl isothiocyanate)$_3$]$^{+2}$, [Zn(Allyl isothiocyanate)$_4$]$^{+2}$, [Zn(Allyl isothiocyanate)$_5$]$^{+2}$, as well as [Zn(Allyl isothiocyanate)$_5$]$^{+2}$, respectively.

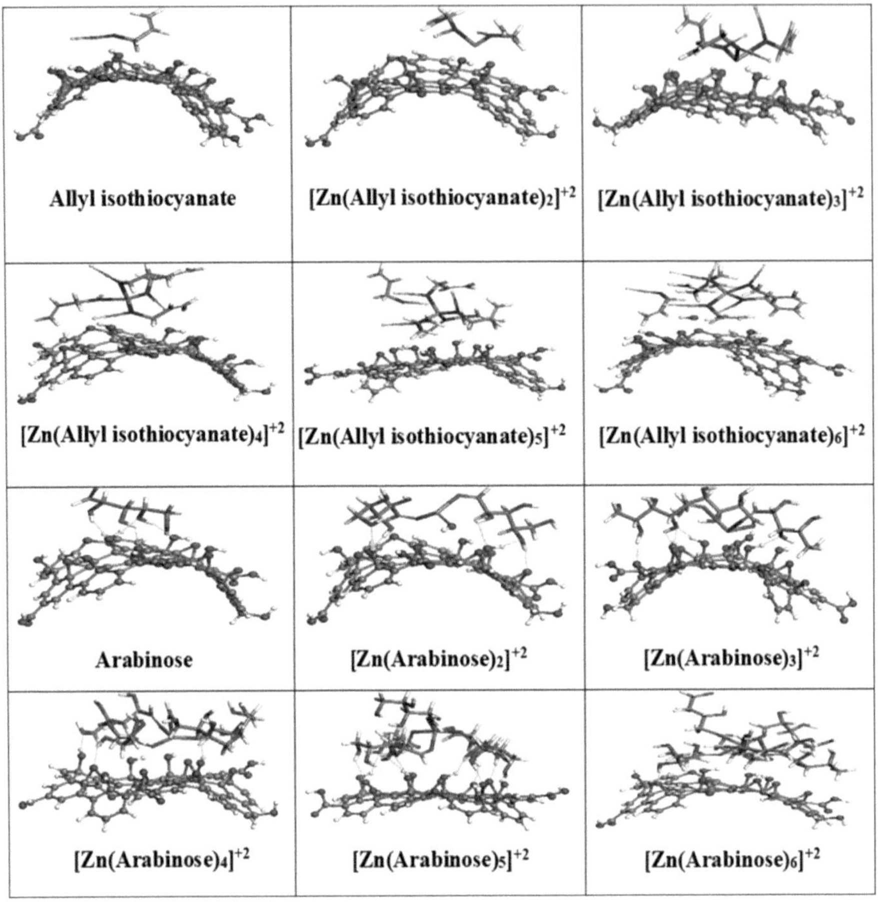

FIGURE 8.19 The allyl isothiocyanate, five $[Zn(Allyl\ isothiocyanate)_n]^{+2}$, arabinose, and five $[Zn(Arabinose)_n]^{+2}$ (n = 2–6) complexes over the GO surface were all optimized using the GGA/PBE/DNP method [29].

The negative binding energies provide additional evidence for the interactions with GO that were shown.

By integrating a graphene-based pH-sensitive nanofiller featuring self-control releasing ability, Dehghani et al. [30] focused on improving the epoxy polymeric coating's active/passive protective properties. In order to obtain long-term self-healing, the release of the encapsulated L-His active agents (L-His@ β-CD) was controlled by a β-CD nano-container. The APTES-amended graphene oxide (AGO) nanoparticles (β-CD@L-His-AGO) were given inclusion complexes for improved targeted release and considerable passive protection. The particle that was produced was called L-His@ β-CD-AGO (**Figure 8.20**).

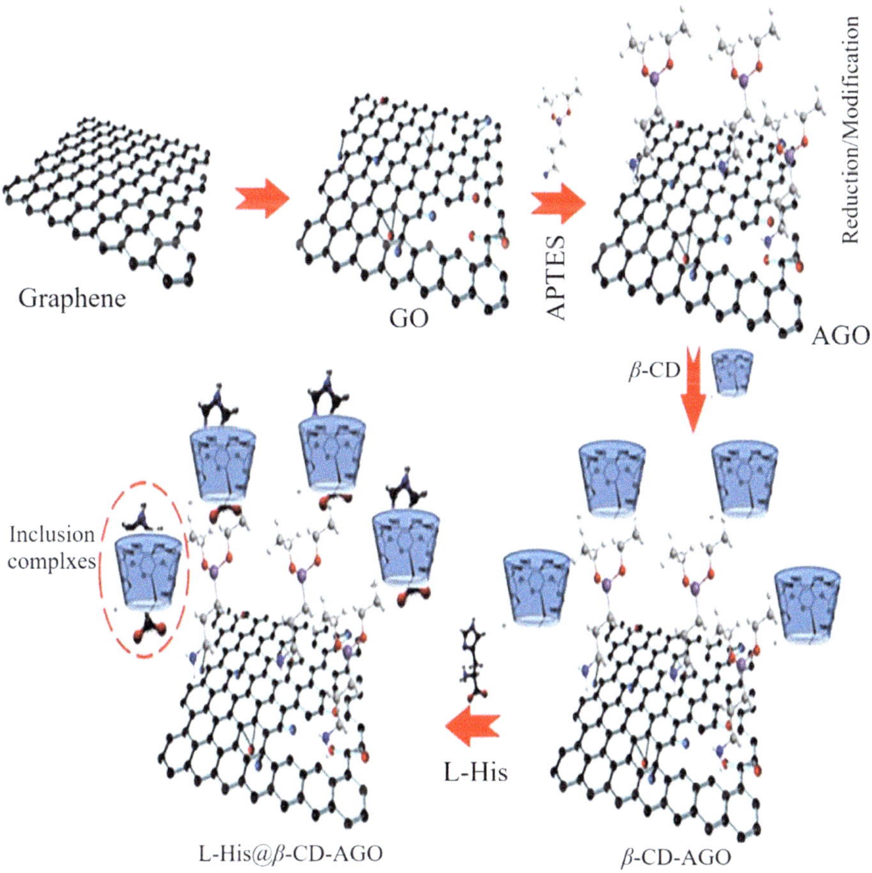

FIGURE 8.20 Construction of L-His@-CD-AGI nanocomposites schematic.

DFT-D calculations were used to analyze the theoretical effects of histidine loading in the β-CD cavity in both the gas phase and the aqueous phase. The geometrically optimized hybrid gas-phase and aqueous-phase histidine/β-CD clusters are shown in **Figure 8.21**. The side/top views of the clusters reveal that the histidine substance may be swallowed by β-CD, resulting in the development of the β-CD/histidine inclusion complex, in either the absence or presence of solvent molecules. This conclusion was confirmed by the examination of binding energy [31]. Histidine inclusion has happened at −116.77 (orientation 1), along with −120.67 (orientation 1) kcal/mol, in the gas phase. Aqueous-phase inclusion complex synthesis also took place at energies within −24.36 as well as −27.51 kcal/mol (orientation-1). The existence of negative energy and histidine complexes is another convincing argument. It is demonstrated that the binding energies in the aqueous phase are substantially lower than those in the gas phase, indicating a poorer ability for inclusion of complex

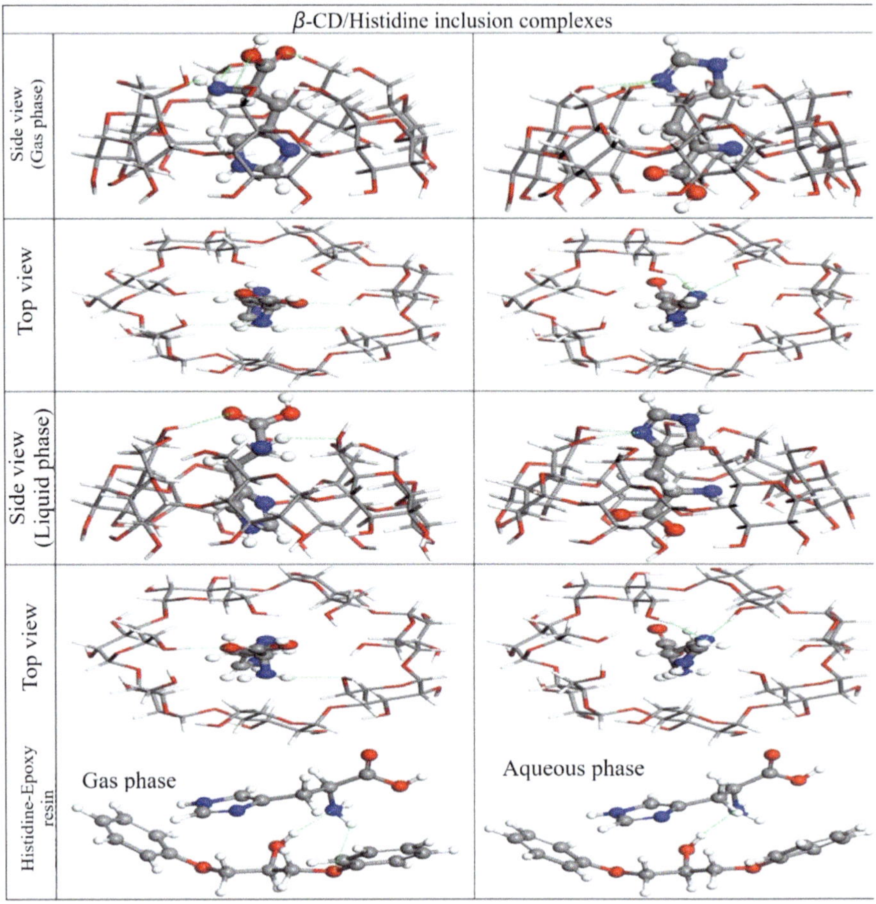

FIGURE 8.21 The optimized geometry of histidine inclusion complexes with β-CD molecules in gas and aqueous phases, as well as their interactions with the molecule of epoxy resin, were calculated using DFT-D.

Reprinted with permission from [30] © 2021 Elsevier Publications.

formation in the presence of solvent molecules. This result may be explained by the fact that histidine interacts with β-CD less favorably in the aqueous phase than in the gas phase. For instance, histidine molecules formed smaller H-bonds with the cavity in its orientation-1 model. The epoxy resin and histidine molecules interacted through intermolecular H-bonds to create a stable cluster in both the gas phase and the aqueous phase. These linkages were made between the oxygenated epoxy moieties and the histidine amine fragments. According to measurements, the strength of histidine-epoxy binding is −40.16 kcal/mol in the gas phase and −17.16 kcal/mol in the aqueous phase. The binding energies with negative indications quantitatively ensure histidine interactions and stabilization near the epoxy covering molecule.

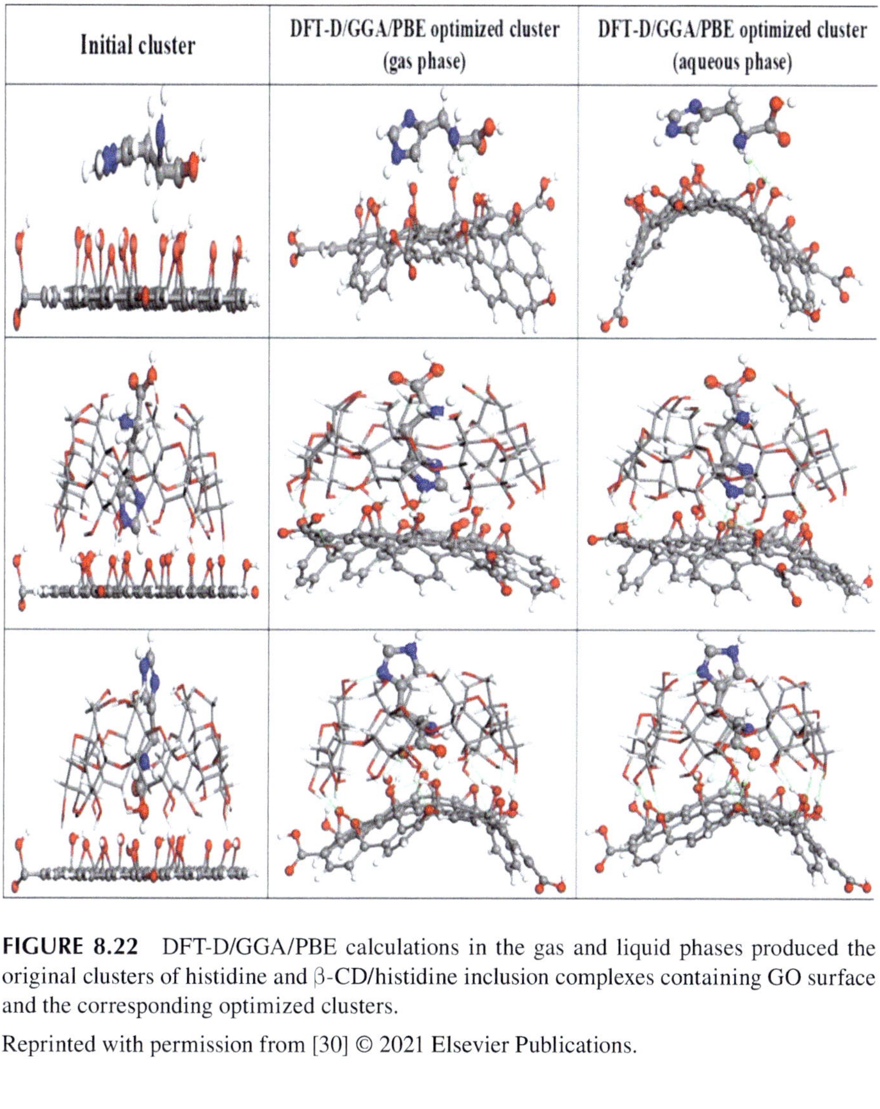

FIGURE 8.22 DFT-D/GGA/PBE calculations in the gas and liquid phases produced the original clusters of histidine and β-CD/histidine inclusion complexes containing GO surface and the corresponding optimized clusters.

Additionally, the lower (i.e., less negative) energy, which is caused by the weaker intermolecular H-bonding interactions in the presence of water molecules, suggests that the histidine-epoxy binding in the aqueous phase is worse. As shown, the histidine molecule contributed just one hydrogen link to the epoxy resin in the aqueous phase, as compared to two hydrogen links in the gas phase. These findings might be explained by the completeness of histidine molecules' interactions with epoxy resin and water solvent molecules.

The initial GO-histidine and GO-CD/histidine clusters designed for modeling in the gas alongside aqueous phases are listed in **Figure 8.22**. The best-looking related clusters from the DFT-D modeling are included in this image as well. It was found that the histidine molecule by itself, as well as histidine loaded in the β-CD cavity, were capable of geometrical stabilization in close proximity to the oxygenated surface (i.e., GO), in both the absence and presence of the solvent molecules. Histidine and β-CD/histidine were stabilized via H-bonds that connect the O/N atoms of histidine, along with the O atoms of β-CD and GO surface groups. It is clear from the thorough investigation that the histidine molecule and the neighboring β-CD molecule had hydrogen bonding interactions. In order to offer a quantitative understanding of L-His and β-CD/L-His interactions with GO, the binding energy was calculated. L-His, β-CD/L-His (first orientation), along with β-CD/L-His (second orientation), yielded energies (kcal/mol) of −0.69, −29.99, as well as −85.30, respectively, when they bound to GO in the gas phase. L-His, β-CD/L-His (first orientation), along β-CD/L-His (second orientation), all had energies of −9.83, −2.12, alongside −60.85 kcal/mol for the aqueous-phase binding. The larger negative energy levels calculated for L-His-loaded β-CD compared to histidine support the higher binding of L-His/β-CD inclusion complexes. The reduced negative binding energies in the presence of solvent molecules emphasize the weaker interactions and adsorption of the L-His/β-CD inclusion complexes on the GO surface.

8.3 CONCLUSIONS

This chapter reviews various physical properties calculated using simulation methods, particularly different-sized DFT, MC, and MD methods. It begins by introducing the fundamentals of calculating physical properties of conducting polymers used as anti-corrosive materials using simulation methods. The interactions between components in conducting polymers, such as nanofiller-nanofiller and nanofillers-polymer matrices, as well as the microstructures of nanofillers and conducting polymer matrix, determine the physical properties of conducting polymers employed as anti-corrosive materials. The acquired information can aid in understanding the connections between conducting polymers' microstructure and their key physical characteristics, enabling the application of modern MD techniques to the creation of conducting polymers with cutting-edge capabilities.

Abbreviation

CS: carbon steel
DGEDDS: diglycidyl ether 4, 4′-dihydroxydiphenylsulfone
EP: epoxy resin
ESP: electrostatic potential
HGEDPAT: hexaglycidylTris (p-ethylene dianiline) phosphoric triamide
MAC: Mulliken atomic charge
MC: Monte Carlo
MD: molecular dynamics
PAA: polyaminoamide
RDF: radial distribution function
TGEDA: tetra glycidyl of ethylene dianiline

REFERENCES

[1] H. Shokry, "Molecular dynamics simulation and quantum chemical calculations for the adsorption of some Azo-azomethine derivatives on mild steel," *Journal of Molecular Structure*, vol. 1060, pp. 80–87, 2014.
[2] A. Javidparvar, B. Ramezanzadeh, and E. Ghasemi, "The effect of surface morphology and treatment of Fe3O4 nanoparticles on the corrosion resistance of epoxy coating," *Journal of the Taiwan Institute of Chemical Engineers*, vol. 61, pp. 356–366, 2016.
[3] T. T. X. Hang, T. A. Truc, T. H. Nam, V. K. Oanh, J.-B. Jorcin, and N. Pébère, "Corrosion protection of carbon steel by an epoxy resin containing organically modified clay," *Surface and Coatings Technology*, vol. 201, pp. 7408–7415, 2007.
[4] A. Fu and Y. Cheng, "Characterization of the permeability of a high performance composite coating to cathodic protection and its implications on pipeline integrity," *Progress in Organic Coatings*, vol. 72, pp. 423–428, 2011.
[5] O. Dagdag, A. El Harfi, A. Essamri, M. El Gouri, S. Chraibi, M. Assouag, et al., "Phosphorous-based epoxy resin composition as an effective anticorrosive coating for steel," *International Journal of Industrial Chemistry*, vol. 9, pp. 231–240, 2018.
[6] O. Dagdag, A. E. Harfi, A. Essamri, A. E. Bachiri, N. Hajjaji, H. Erramli, et al., "Anticorrosive performance of new epoxy-amine coatings based on zinc phosphate tetrahydrate as a nontoxic pigment for carbon steel in NaCl medium," *Arabian Journal for Science and Engineering*, vol. 43, pp. 5977–5987, 2018.
[7] O. Dagdag, O. Hamed, H. Erramli, and A. El Harfi, "Anticorrosive performance approach combining an epoxy polyaminoamide-zinc phosphate coatings applied on sulfo-tartaric anodized aluminum alloy 5086," *Journal of Bio-and Tribo-Corrosion*, vol. 4, pp. 1–11, 2018.
[8] P. Chandrasekhar, *Conducting Polymers, Fundamentals and Applications*. Springer, 2018.
[9] J. Zhao, L. Wu, C. Zhan, Q. Shao, Z. Guo, and L. Zhang, "Overview of polymer nanocomposites: Computer simulation understanding of physical properties," *Polymer*, vol. 133, pp. 272–287, 2017.
[10] O. Dagdag, A. Berisha, V. Mehmeti, R. Haldhar, E. Berdimurodov, O. Hamed, et al., "Epoxy coating as effective anti-corrosive polymeric material for aluminum alloys: Formulation, electrochemical and computational approaches," *Journal of Molecular Liquids*, vol. 346, p. 117886, 2022.
[11] W. Daoudi, B. El Ibrahimi, O. Dagdag, E. Berdimurodov, L. Guo, E. E. Ebenso, et al., "New chlorophenyl-imidazole derivative as a novel corrosion inhibitor in the gas and oil industry," *Journal of Physics and Chemistry of Solids*, vol. 179, p. 111409, 2023.
[12] S. K. Gupta, R. K. Mehta, M. Yadav, O. Dagdag, V. Mehmeti, A. Berisha, et al., "Diazenyl derivatives as efficient corrosion inhibitors for mild steel in HCl medium: Gravimetric, electrochemical and computational approach," *Journal of Molecular Liquids*, vol. 382, p. 121976, 2023.
[13] O. Dagdag, A. Berisha, Z. Safi, S. Dagdag, M. Berrani, S. Jodeh, et al., "Highly durable macromolecular epoxy resin as anticorrosive coating material for carbon steel in 3% NaCl: Computational supported experimental studies," *Journal of Applied Polymer Science*, vol. 137, p. 49003, 2020.
[14] O. Dagdag, A. El Harfi, L. El Gana, Z. S Safi, L. Guo, A. Berisha, et al., "Designing of phosphorous based highly functional dendrimeric macromolecular resin as an effective coating material for carbon steel in NaCl: Computational and experimental studies," *Journal of Applied Polymer Science*, vol. 138, p. 49673, 2021.

[15] M. H. Abdellattif, S. H. Alrefaee, O. Dagdag, C. Verma, and M. Quraishi, "Calotropis procera extract as an environmental friendly corrosion Inhibitor: Computational demonstrations," *Journal of Molecular Liquids*, vol. 337, p. 116954, 2021.

[16] A. Berisha, F. Podvorica, V. Mehmeti, F. Syla, and D. Vataj, "Theoretical and experimental studies of the corrosion behavior of some thiazole derivatives toward mild steel in sulfuric acid media," *Macedonian Journal of Chemistry and Chemical Engineering*, vol. 34, pp. 287–294, 2015.

[17] M. Rbaa, P. Dohare, A. Berisha, O. Dagdag, L. Lakhrissi, M. Galai, et al., "New Epoxy sugar based glucose derivatives as ecofriendly corrosion inhibitors for the carbon steel in 1.0 M HCl: Experimental and theoretical investigations," *Journal of Alloys and Compounds*, vol. 833, p. 154949, 2020.

[18] O. Dagdag, A. El Harfi, Z. Safi, L. Guo, C. Verma, E. Ebenso, et al., "Fabrication on designing of a macromolecular epoxy resin as anti-corrosive coating material for electrocatalytically deposited cadmium on 15CDV6 steel in 3% NaCl solution," *Journal of Materials Research and Technology*, vol. 9, pp. 5549–5563, 2020.

[19] O. Dagdag, R. Hsissou, A. El Harfi, A. Berisha, Z. Safi, C. Verma, et al., "Fabrication of polymer based epoxy resin as effective anti-corrosive coating for steel: Computational modeling reinforced experimental studies," *Surfaces and Interfaces*, vol. 18, p. 100454, 2020.

[20] S. Shahabi and P. Norouzi, "Electrochemical and quantum chemical assessment of some thiosemicarbazide derivatives as carbon steel corrosion inhibitors in HCl; continuously monitoring the current change by FFT voltammetry," *International Journal of Electrochemical Science*, vol. 12, pp. 2628–2646, 2017.

[21] O. Dagdag, R. Hsissou, A. El Harfi, Z. Safi, A. Berisha, C. Verma, et al., "Epoxy resins and their zinc composites as novel anti-corrosive materials for copper in 3% sodium chloride solution: Experimental and computational studies," *Journal of Molecular Liquids*, vol. 315, p. 113757, 2020.

[22] I. Obot and N. Obi-Egbedi, "Anti-corrosive properties of xanthone on mild steel corrosion in sulphuric acid: Experimental and theoretical investigations," *Current Applied Physics*, vol. 11, pp. 382–392, 2011.

[23] F. Zhang, Y. Tang, Z. Cao, W. Jing, Z. Wu, and Y. Chen, "Performance and theoretical study on corrosion inhibition of 2-(4-pyridyl)-benzimidazole for mild steel in hydrochloric acid," *Corrosion Science*, vol. 61, pp. 1–9, 2012.

[24] O. Dagdag, L. Guo, Z. Safi, C. Verma, E. E. Ebenso, N. Wazzan, et al., "Epoxy resin and TiO2 composite as anticorrosive material for carbon steel in 3% NaCl medium: Experimental and computational studies," *Journal of Molecular Liquids*, vol. 317, p. 114249, 2020.

[25] O. Dagdag, A. Berisha, Z. Safi, S. Dagdag, M. Berrani, S. Jodeh, et al., "Highly durable macromolecular epoxy resin as anticorrosive coating material for carbon steel in 3% NaCl: Computational supported experimental studies," *Journal of Applied Polymer Science*, p. 49003, 2020.

[26] Y. S. Mary, C. Y. Panicker, M. Sapnakumari, B. Narayana, B. Sarojini, A. A. Al-Saadi, et al., "FT-IR, NBO, HOMO-LUMO, MEP analysis and molecular docking study of 1-[3-(4-Fluorophenyl)-5-phenyl-4, 5-dihydro-1H-pyrazol-1-yl] ethanone," *Spectrochimica Acta Part A: Molecular and Biomolecular Spectroscopy*, vol. 136, pp. 483–493, 2015.

[27] M. H. Rahmani, A. Dehghani, G. Bahlakeh, and B. Ramezanzadeh, "Introducing GO-based 2D-platform modified via phytic acid molecules decorated by zeolite imidazole ZIF-9 MOFs for designing multi-functional polymeric anticorrosive system; DFT-D computations and experimental studies," *Journal of Molecular Liquids*, vol. 364, p. 119945, 2022.

[28] B. Ramezanzadeh, P. Kardar, G. Bahlakeh, Y. Hayatgheib, and M. Mahdavian, "Fabrication of a highly tunable graphene oxide composite through layer-by-layer assembly of highly crystalline polyaniline nanofibers and green corrosion inhibitors: Complementary experimental and first-principles quantum-mechanics modeling approaches," *The Journal of Physical Chemistry C*, vol. 121, pp. 20433–20450, 2017.
[29] A. Dehghani, A. H. Mostafatabar, G. Bahlakeh, and B. Ramezanzadeh, "Metal-doped 2D rGO nano-sheets fabrication utilizing plant source bio-molecules and application in the epoxy anti-corrosive coating: Combined experimental and DFT-D modeling investigations," *Progress in Organic Coatings*, vol. 170, p. 106938, 2022.
[30] A. Dehghani, G. Bahlakeh, B. Ramezanzadeh, and A. H. J. Mofidabadi, "2D reduced-graphene oxide (rGO) nanosheets decorated with l-histidine loaded-β-cyclodextrin for efficient epoxy nano-composite anti-corrosion properties; DFT-D modeling/experimental assessments," *FlatChem*, vol. 30, p. 100309, 2021.
[31] B. Ramezanzadeh, G. Bahlakeh, M. M. Moghadam, and R. Miraftab, "Impact of size-controlled p-phenylenediamine (PPDA)-functionalized graphene oxide nanosheets on the GO-PPDA/Epoxy anti-corrosion, interfacial interactions and mechanical properties enhancement: Experimental and quantum mechanics investigations," *Chemical Engineering Journal*, vol. 335, pp. 737–755, 2018.

9 Micro/nano containers as self-healing anti-corrosion materials

Amir Hossein Mostafatabar, Amir Hossein Jafari Mofidabadi, and Ali Dehghani

9.1 INTRODUCTION

Corrosion, the gradual deterioration of materials caused by chemical reactions with their surrounding environment, poses significant challenges and disadvantages in various industries [1, 2]. The impact of corrosion extends beyond mere cosmetic or superficial concerns, as it can have far-reaching consequences that affect operational efficiency, safety, and economic viability. One of the major disadvantages of corrosion is the substantial financial burden it places on industries. The cost of corrosion-related maintenance, repairs, and replacements of equipment, structures, and infrastructure is staggering [3]. The expenses associated with combating corrosion can consume a significant portion of an industry's budget, diverting resources that could have been invested in research, development, and expansion. Moreover, corrosion compromises the structural integrity of materials, leading to reduced performance and reliability. When metals corrode, their strength and load-bearing capacity can be significantly diminished, posing risks to the safety of personnel and the public. This can be particularly critical in industries such as aerospace, automotive, and oil and gas, where failure of corroded components can have catastrophic consequences [4–7]. Corrosion also affects operational efficiency by impeding the smooth functioning of equipment and processes. The accumulation of corrosion products can obstruct flow paths, restrict the movement of mechanical components, and impair the efficiency of heat exchange systems. Consequently, industries may experience reduced productivity, increased energy consumption, and compromised product quality, ultimately impacting their competitiveness in the market [8]. Furthermore, corrosion-related failures and disruptions can result in unplanned downtime and production losses. Unscheduled maintenance and repairs to address corrosion issues can lead to production delays, decreased output, and missed delivery deadlines [9]. These disruptions can have ripple effects throughout the supply chain, affecting customer satisfaction, contractual obligations, and overall business performance. Environmental concerns also arise from corrosion. The chemical reactions involved in corrosion processes often result in the release of harmful substances into the environment. This not only poses risks to ecosystems

 DOI: 10.1201/9781003456025-9

and human health but also subjects industries to potential regulatory compliance issues and legal liabilities [10–12].

Given the multifaceted disadvantages of corrosion, it is imperative for industries to adopt effective corrosion protection strategies. As mentioned earlier, corrosion poses a significant threat to the integrity, functionality, and longevity of materials used in various industries. Recognizing the importance of corrosion protection is paramount for safeguarding assets, minimizing financial losses, ensuring operational reliability, and promoting sustainable practices [13]. One of the primary reasons why corrosion protection is crucial is the economic impact it has on industries. The direct costs associated with corrosion, including maintenance, repairs, and replacements, are staggering. By implementing effective corrosion protection measures, industries can substantially reduce these expenses and allocate resources toward other critical areas such as research, development, and innovation [14]. Furthermore, corrosion protection plays a vital role in maintaining the structural integrity of materials. By implementing robust corrosion protection strategies, industries can prevent or mitigate the weakening of materials, ensuring safe and reliable operations. Operational efficiency is another key aspect where corrosion protection proves its importance. Corrosion can impede the performance and functionality of equipment, resulting in reduced productivity and increased downtime. The accumulation of corrosion products can restrict fluid flow, hinder mechanical movement, and compromise the efficiency of heat exchange systems [15]. By implementing effective corrosion protection measures, industries can minimize disruptions, enhance operational efficiency, and optimize production processes. In addition to financial and operational considerations, corrosion protection is also vital from an environmental perspective. Corrosion-related failures and leaks can result in the release of hazardous substances into the environment, posing risks to ecosystems and human health. Implementing robust corrosion protection measures can help prevent or minimize such incidents, thereby reducing the environmental impact of industrial operations and promoting sustainable practices [16, 17]. Moreover, effective corrosion protection contributes to the longevity and sustainability of assets and infrastructure. By extending the lifespan of materials and structures, industries can reduce the need for frequent replacements and minimize waste generation. This not only optimizes resource utilization but also reduces the carbon footprint associated with the manufacturing, transportation, and disposal of materials [18, 19].

In the ongoing battle against corrosion, a wide range of methods have been employed to protect materials from degradation. These corrosion protection methods aim to impede or prevent the chemical reactions that cause corrosion, thereby preserving the integrity and functionality of industrial assets. One of the commonly employed approaches is the use of corrosion inhibitors [20, 21]. While inhibitors have shown promise in mitigating corrosion, they are not without their disadvantages. Corrosion inhibitors are chemicals that are designed to inhibit the corrosion process by forming a protective layer on the metal surface or altering the electrochemical reactions involved [22, 23]. They can be classified into two main types: organic inhibitors and inorganic inhibitors. Organic inhibitors, such as organic compounds and polymers, form a protective film on the metal surface, while inorganic inhibitors, such as chromates and phosphates, react with the metal surface to form a passivating layer [24, 25]. One of the advantages of corrosion inhibitors is their ease of application. They can be easily added to coatings, paints, or other protective systems, making them convenient to use in various industries. Additionally, corrosion inhibitors can provide temporary or long-term protection, depending on their specific formulation and application conditions. They have the potential to extend the service life of materials and reduce the frequency of maintenance and repairs [26]. However, corrosion inhibitors also come with some disadvantages. One significant drawback is their limited longevity and sustainability. Inhibitors may leach out or deplete over time, especially in harsh environments or under prolonged exposure to corrosive conditions. This necessitates regular monitoring and replenishment, which can be both time-consuming and costly. Another disadvantage of inhibitors is their dependence on continuous contact with the metal surface to provide effective protection. If the protective layer formed by the inhibitor is compromised or disrupted, corrosion can resume. This is particularly problematic in situations where the protected surface is subjected to mechanical wear, abrasion, or high-velocity fluid flow [27]. Furthermore, some inhibitors can have adverse environmental and health impacts. Certain traditional inhibitors, such as chromates, have been found to be toxic and carcinogenic. Strict regulations and environmental concerns have prompted the search for alternative, environmentally friendly corrosion inhibitors [28, 29].

The limitations and challenges associated with traditional corrosion inhibitors have prompted researchers and engineers to explore innovative approaches that can overcome these drawbacks [30, 31]. One promising strategy involves the use of micro/nano carriers to deliver corrosion inhibitors, mitigating their disadvantages and enhancing their effectiveness in corrosion protection [32]. Carriers act as protective vessels that encapsulate corrosion inhibitors, shielding them from premature depletion or degradation. By incorporating inhibitors within these carrier materials, it becomes possible to control their release and ensure a sustained supply of inhibitors to the corroding surface [33]. This controlled release mechanism offers several advantages over the direct application of inhibitors and addresses the limitations associated with their use [34]. One of the primary advantages of using carriers is the improved longevity and sustainability of corrosion protection. By encapsulating inhibitors within carriers, their leaching and depletion can be significantly reduced [35]. The carriers act as reservoirs, releasing inhibitors gradually and steadily over time, extending the protective effects and minimizing the need for frequent replenishment. This ensures long-term corrosion inhibition and reduces the maintenance burden on industries [36]. Moreover, carriers provide physical protection to the encapsulated inhibitors, shielding them from environmental factors and preventing their premature degradation. The carrier materials act as barriers, preventing direct contact between the inhibitors and the corrosive environment. This protection enhances the stability and effectiveness of inhibitors, increasing their durability and prolonging their lifespan [37]. The use of carriers also allows for targeted and controlled delivery of inhibitors to the corroding surface. The carriers can be designed with specific release mechanisms, such as diffusion or pH-dependent triggers, enabling the precise and selective release of inhibitors at the desired locations [38]. This targeted delivery minimizes the wastage of inhibitors and ensures optimal utilization, optimizing the overall efficiency of corrosion protection strategies.

Furthermore, carriers offer opportunities for customization and optimization of corrosion inhibition. Different types of carriers can be tailored to specific application requirements, such as desired release rates, compatibility with coatings or materials, and environmental conditions. This flexibility allows for the development of corrosion protection systems that are optimized for specific industries, environments, and materials [38]. In recent years, micro/nano containers have emerged as a particularly promising type of carrier for corrosion inhibitors. These containers, with their small size and high surface area-to-volume ratio, offer unique advantages in terms of encapsulation, controlled release, and targeted delivery of inhibitors. Micro/nanocontainers can be engineered from various materials, such as polymers, nanoparticles, or mesoporous structures, providing versatility in terms of properties and functionality [39, 40].

As previously stated, in the pursuit of innovative corrosion protection strategies, a variety of carrier materials have been explored to encapsulate corrosion inhibitors and enhance their performance. These carriers, including micro/nano containers, offer unique properties and functionalities that contribute to the sustained release and targeted delivery of inhibitors, revolutionizing the field of corrosion protection. Some prominent types of micro/nano containers that were applied as self-healing anti-corrosion materials are reviewed next:

1. **Polymer-based Carriers**: Polymers have gained significant attention as carriers due to their versatility, tuneability, and compatibility with various corrosion inhibitors. Polymers can be engineered to form encapsulating matrices that provide physical protection to inhibitors and control their release through diffusion or degradation mechanisms. Polymer-based carriers offer excellent flexibility in terms of design, allowing for tailoring of carrier properties, such as porosity, permeability, and degradation rate, to meet specific corrosion protection requirements [41, 42].
2. **Nanoparticles**: Nanoparticles have emerged as another class of carriers that hold great potential in corrosion protection applications. Their small size, large surface area, and unique surface properties make them ideal candidates for encapsulating and delivering corrosion inhibitors. Nanoparticles can be functionalized to provide controlled release mechanisms, such as pH-responsive or stimuli-responsive behaviors, enabling targeted delivery of inhibitors to the corroding surface. Additionally, the incorporation of nanoparticles into coatings or composites can enhance their barrier properties and provide localized corrosion protection [43, 44].
3. **Mesoporous Materials**: Mesoporous materials, such as mesoporous silica, Covalent-organic frameworks (COF), or metal-organic frameworks (MOFs), offer a highly ordered network of interconnected pores with controllable pore sizes and surface areas. These materials provide an ideal platform for encapsulating and releasing corrosion inhibitors due to their large internal surface area and tunable pore structures. The high loading capacity of mesoporous materials enables the efficient encapsulation of a significant amount of inhibitors, while their controlled release properties ensure sustained protection against corrosion [45, 46].

4. **Microcapsules**: Microcapsules, consisting of a shell surrounding a liquid or solid core, have been extensively investigated as carriers for corrosion inhibitors. The shell materials can range from polymers to inorganic compounds, providing protection to the encapsulated inhibitors and controlling their release. Microcapsules offer advantages such as easy incorporation into coatings or paints, protection against premature inhibitor depletion, and triggered release mechanisms in response to specific environmental conditions or stimuli [47, 48].
5. **Self-Healing Materials**: Self-healing materials represent a unique class of carriers that possess the ability to autonomously repair damage and restore functionality. These carriers can encapsulate corrosion inhibitors along with other healing agents, such as encapsulated monomers or catalysts. When damage occurs, such as the rupture of the carrier or exposure of the metal surface, the healing agents are released and initiate self-healing processes, including the formation of protective films or the restoration of damaged coatings. Self-healing materials offer the potential for long-lasting and self-sustaining corrosion protection [49, 50].

The choice of carrier material depends on various factors, including the nature of the corrosion inhibitor, compatibility with the substrate or coating, release requirements, and the targeted application. By exploring the unique properties and capabilities of these carrier materials, it is possible to advance the field of micro/nano-containers as self-healing anti-corrosion materials, providing tailored and effective corrosion protection solutions in diverse industries.

As referenced earlier, porous materials have garnered significant interest in the realm of corrosion protection due to their unique structural characteristics and functionalities. These materials, characterized by interconnected voids or pores, offer a high surface area-to-volume ratio, controllable pore sizes, and diverse surface chemistries [51]. In the context of micro/nano containers as self-healing anti-corrosion materials, porous materials have emerged as promising candidates for encapsulating corrosion inhibitors and facilitating their controlled release. Porous materials provide an ideal platform for encapsulating corrosion inhibitors within their pore structures. These materials can be engineered to have tailored porosity, pore size distribution, and surface chemistry, allowing for efficient loading and release of inhibitors. The high surface area of porous materials offers a large contact interface with the surrounding environment, enhancing the availability and reactivity of the encapsulated inhibitors [52, 53]. One of the key advantages of porous materials is their ability to sustainably release corrosion inhibitors over an extended period. The interconnected pore network facilitates the diffusion of inhibitors out of the material, enabling a controlled and gradual release. This controlled release mechanism ensures a continuous supply of inhibitors to the corroding surface, providing long-term protection against corrosion [54, 55]. Moreover, porous materials can act as physical barriers, protecting the encapsulated inhibitors from premature depletion or environmental degradation. The encapsulation within the porous structure shields the inhibitors from direct contact with corrosive agents, preventing their leaching or degradation. This protection preserves the integrity and effectiveness of the inhibitors, prolonging their lifespan and enhancing their corrosion inhibition performance [56]. The versatility of porous materials allows for the incorporation of different types of corrosion inhibitors, including organic compounds, inorganic compounds, and environmentally friendly alternatives [57, 58]. By choosing the appropriate type of inhibitor and tailoring the porous material's properties, it is possible to develop corrosion protection systems that are effective against specific types of corrosion and compatible with different environmental conditions. Furthermore, porous materials can offer additional functionalities beyond corrosion inhibition. For example, some porous materials possess catalytic properties that can accelerate corrosion inhibition reactions or promote self-healing processes. The combination of corrosion inhibitors with catalytic porous materials can lead to synergistic effects, enhancing the overall corrosion protection performance [58, 59]. Porous materials have found wide applications in various industries, including aerospace, automotive, oil and gas, and infrastructure [60, 61]. Their versatility and compatibility with different coating systems and substrates make them suitable for integration into existing corrosion protection technologies. Additionally, the popularity of porous materials is attributed to their ability to accommodate different loading techniques and deposition methods, ensuring ease of application and scalability [62].

As mentioned before, porous materials have gained significant popularity and found widespread application in numerous industries due to their unique properties and versatile functionalities. The structural characteristics of porous materials, including their high surface area, interconnected porosity, and tunable pore sizes, make them highly desirable for a range of industrial applications. In addition, porous materials spread their popularity and application across various industries, including:

1. **Aerospace Industry**: The aerospace industry demands lightweight materials with exceptional corrosion resistance. Porous materials, such as porous metals or metal foams, have gained traction in this sector due to their low density, high strength-to-weight ratio, and excellent corrosion resistance. These materials find applications in aircraft components, structural elements, and fuel tanks, where they provide corrosion protection and structural integrity while reducing weight [63–66].
2. **Automotive Industry**: The automotive industry faces challenges related to corrosion, especially in regions with harsh climates and exposure to road salts. Porous materials integrated into coatings, paints, and surface treatments offer improved corrosion protection for automotive components, such as body panels, chassis, and exhaust systems. The high surface area of porous materials facilitates the incorporation of corrosion inhibitors, enhancing their efficiency and longevity [67, 68].
3. **Oil and Gas Industry**: The oil and gas industry operates in aggressive environments that can accelerate corrosion processes. Porous materials, including mesoporous silica and porous ceramics, find application as corrosion inhibitor carriers and protective coatings for pipelines, storage tanks, and offshore structures. These materials provide sustained release of inhibitors, protection against harsh conditions, and enhanced durability, ensuring the integrity and safety of oil and gas infrastructure [69, 70].

4. **Infrastructure and Construction**: Infrastructure, such as bridges, buildings, and pipelines, is susceptible to corrosion due to exposure to environmental factors and harsh conditions. Porous materials incorporated into coatings, admixtures, or surface treatments act as effective barriers against corrosive agents, extending the service life of structures. These materials contribute to the durability, maintenance reduction, and sustainability of infrastructure assets [71].
5. **Marine Industry**: The marine environment poses significant challenges to corrosion protection, with structures exposed to saltwater, humidity, and constant moisture. Porous materials, such as nanoporous coatings or composites, offer enhanced resistance to corrosion and biofouling in marine applications. These materials provide efficient encapsulation and controlled release of corrosion inhibitors, preventing the deterioration of marine structures, ship hulls, and offshore equipment [72, 73].
6. **Electronics and Semiconductor Industry**: Porous materials find applications in the electronics and semiconductor industry for corrosion protection of delicate components and circuits. The high surface area and controlled release properties of porous materials enable the encapsulation and sustained delivery of corrosion inhibitors, safeguarding electronic devices against environmental factors and prolonging their lifespan [74].
7. **Renewable Energy Industry**: The renewable energy sector, including solar panels and wind turbines, requires corrosion protection solutions to ensure the longevity and efficiency of energy systems. Porous materials, with their tailored porosity and encapsulation capabilities, offer corrosion inhibitor delivery systems that can withstand harsh environmental conditions, such as humidity, UV exposure, and temperature fluctuations. These materials contribute to the reliability and performance of renewable energy infrastructure [75, 76].

The popularity of porous materials in these industries stems from their ability to address corrosion challenges while providing tailored solutions to meet specific application requirements. As micro/nano containers, porous materials offer enhanced corrosion protection through the controlled release of inhibitors, physical barrier properties, and self-healing functionalities. The widespread adoption of porous materials in various industries highlights their effectiveness and potential to revolutionize corrosion protection strategies.

Considering the previously stated cases, porous materials have emerged as promising candidates for serving as corrosion inhibitor carriers, revolutionizing the field of corrosion protection. These materials offer unique advantages in terms of encapsulation, controlled release, and enhanced performance of corrosion inhibitors. One of the primary advantages of porous materials as corrosion inhibitor carriers is their ability to encapsulate a wide range of inhibitors. Porous materials, such as mesoporous silica, metal-organic frameworks (MOFs), or porous polymers, provide a host matrix for incorporating various types of corrosion inhibitors, including organic compounds, inorganic compounds, and environmentally friendly alternatives. This versatility allows for the selection of inhibitors that are specifically tailored to combat different corrosion processes and environmental conditions [77–79]. The controlled release of corrosion inhibitors is another key advantage offered by porous materials. The interconnected pore structure of these materials enables the gradual diffusion and sustained release of encapsulated inhibitors over an extended period [80]. This controlled release mechanism ensures a continuous supply of inhibitors to the corroding surface, effectively inhibiting corrosion processes and providing long-term protection. The ability to modulate the release rate and duration makes porous materials ideal for applications where a prolonged and consistent inhibitor presence is required. Porous materials also serve as physical barriers, protecting the encapsulated inhibitors from premature depletion or environmental degradation. The encapsulation within the porous structure shields the inhibitors from direct contact with corrosive agents, preventing their leaching or degradation. This protective barrier preserves the integrity and effectiveness of the inhibitors, extending their lifespan and enhancing their corrosion inhibition performance. Furthermore, the high surface area and tailored porosity of porous materials enhance the efficiency of corrosion inhibitor loading. The large contact interface between the porous material and the inhibitor allows for high loading capacity, ensuring a significant amount of inhibitor can be encapsulated within a relatively small volume. This high loading efficiency leads to improved inhibitor utilization and reduces the frequency of inhibitor replenishment, contributing to cost-effectiveness and operational convenience [81]. Additionally, porous materials can be functionalized or modified to enhance their compatibility with specific corrosion inhibitors or target corrosive environments. Surface modifications, such as functional groups or coatings, can promote stronger interactions between the porous material and the inhibitor, enhancing loading capacity and controlled release properties. By tailoring the surface chemistry of porous materials, their affinity for certain inhibitors can be optimized, ensuring improved encapsulation and performance. The use of porous materials as corrosion inhibitor carriers has demonstrated significant potential in addressing the limitations of traditional corrosion protection methods [82]. Their unique advantages, including versatile encapsulation, controlled release, physical barrier properties, and enhanced loading efficiency, make them attractive options for developing self-healing anti-corrosion materials. By harnessing the benefits of porous materials, it is possible to enhance the effectiveness, durability, and sustainability of corrosion protection systems in various industries [83–85].

In recent years, numerous porous materials such as COFs, MOFs, SiO_2, and Metal Oxides have gained prominence and found widespread use in various industries, owing to their unique properties and functionalities. These materials, with their tailored porosity, high surface area, and controllable pore sizes, have opened up new possibilities in corrosion protection and self-healing applications. In this chapter, some of the common porous materials that are widely utilized today in the context of micro/nano-containers as self-healing anti-corrosion materials will be explored.

9.2 RECENT APPLIED MATERIALS AS ADVANCED POROUS CARRIERS

9.2.1 Metal Organic Frameworks (MOFs): The Porous Materials Family Hero

When talking about porous structures, MOFs can be placed at the top of the list. With diverse structures, they are applicable in myriad industries, that is, water treatment, energy storage, and CO_2 capturing. However, with precise exploration of MOFs, it is possible to realize that the crystal-like structures are potential carriers for both organic and inorganic compounds [86]. In regular circumstances, organic ligands are occupied with heteroatoms, including S-, O-, and N-elements, which can remarkably tune the anti-corrosion function of MOFs, while metal clusters can behave as inorganic corrosion inhibitors upon MOFs degradation and ingredients release in corrosive environments [52].

Fortunately, assessment of MOFs' anti-corrosion impact was the final goal of many scientific studies [87–89], and with precise explorations, acceptable performance of these porous nanomaterials in corrosive acidic and neutral environments has been proven. With respect to the type of MOF and the sub-family, the porosity and crystal structure changes, and nowadays ZIFs [90], MILs [91], and UIOs [92] are the most applied MOFs for anti-corrosion targets. Discoveries about the MOFs' characteristics revealed that sensitivity against pH is responsible for the structure's degradation in corrosive atmospheres. Once they come into contact with acidic media, protonation of organic compounds reduces the MOFs' stability, and eventually they release in free forms and, simultaneously, the metal clusters release in the media as a function of time. Since the harsh media is now protected by both organic and inorganic corrosion inhibitors, upon metal exposure, the chemical and physical reactions protect the substrate with the development of a thin organic-inorganic inhibitive layer. In brief, organic inhibitors tend to react with anodic sites by electron donation mechanism (chemical reactions) or metal-cation exchange phenomenon, whereas the cathodic sites would be covered by oxidation/hydroxation of the freely available MOFs-metal clusters.

Confirmation of the anti-corrosion potential of MOFs in the solution phase convinced researchers to enlist them as active fillers for polymeric coatings [52, 93]. With a brief scientific history, albeit moisture accelerates coatings degradation in the long term, fillers can substantially control the deterioration by filling the solution pathways that conclusively decline the penetration volume of water/moisture. Literature clarified that MOFs are one of the potential stars for incorporation in different polymeric coatings to promote the coatings' barrier performance [94, 95]. However, the MOFs' responsibility will not be limited to coating barrier promotion duty. Undoubtedly, the fabricated coatings' long life is threatened by mechanical damage. Subsequently, in a very short period, moisture passes through the coating, and corrosion occurs. For such cases, researchers proposed smart coatings with active-inhibition capability. With a simple theory, the inhibitive compounds are incorporated in the polymeric matrix, and in response to environmental stimuli, they release, and function recovery happens by the development of an inhibitive layer over the damaged zones. Fortunately, the MOFs' pH-sensitivity paves the way for an active inhibition mechanism and design of smart coatings. With a similar inhibition mechanism in the solution phase, the damaged areas will be covered by organic and inorganic components, and the anti-corrosion criteria will be enhanced.

But it is not the whole story of the MOF application. Till here, we explained about the pristine MOF application, while MOFs' porosity opened a new gate as potential carriers for inhibitor encapsulation. In demand of reaching a higher anti-corrosion index, researchers realized pristine MOFs cannot meet expectations, but with the encapsulated inhibitors release, the nanocomposites can donate outstanding functions for anti-corrosion aims [95, 96]. Recently, contemporary researchers have tried to encapsulate organic compounds or inorganic materials like metal ions in a variety of MOF classes. Assessments in both solution and coating phases demonstrated that the incomplete function of MOFs has been perfected with the encapsulated inhibitors, and the inhibition yield has made a significant jump.

In one step further, pioneer researchers tried to design complicated MOF-based hybrid materials [96]. Although at first glance, the construction of hybrid nanocomposites is limited only to laboratory studies, with a deep look at the coating structures, we can understand that the design of these structures is vital and practical. Frankly, the gaps and defects in polymeric matrices are far broader than they can be completely filled by MOFs, and water/moisture can easily pass through the MOFs. In usual circumstances, researchers rather grab materials with high surface to volume ratios (low-dimensional materials), including h-BN, Graphene Oxide (GO), Graphene, and Carbon Nano Tube (CNT). The conducted studies reflected that the nominated low-dimensional nanoparticles can remarkably promote the barrier function of polymeric coatings. However, due to the lack of active protection capability, these materials must be hybridized with other anti-corrosion materials. Combining the barrier improvement theory with the recently designed active MOFs leads to the design of multifunctional nanocomposites with extreme barrier function and active inhibition ability.

9.2.2 Covalent organic frameworks (COFs): potential brand-new substitutes

Nowadays, COFs have been nominated as a new class of porous materials that are successfully applied in diverse industries. To realize the differences between MOFs and COFs, we should zoom in on clusters. Actually, metals are available as clusters in MOFs, whereas in COFs, no specific cluster is available, and a porous network expands by integral connection of organic compounds [97]. As a result, the COFs' inhibition mechanism is only based on organic agents and mainly anodic protection since there is no sign of inorganic compounds exhibiting.

Unfortunately, owing to complicated and expensive synthesis, COFs were not widely applied in corrosion science, and the available literature is quite limited. Liu et al. published the only available work about designing an epoxy-based smart coating by the incorporation of a 2D-COF [98]. Interestingly, the same work tried to tune the fabricated COF anti-corrosion by encapsulation of benzotriazole.

As formerly explained, hybridizing is a desirable way to develop multifunctional active/barrier fillers for polymeric coatings. That is why researchers attempted to design COF-based hybrid materials. To the best of our knowledge, COFs have been hybridized by carbonaceous materials (CNT [99] and GO [97]) and MXenes [100] only. Outcomes from the literature clarified that the following strategy practically promoted the polymeric coatings' performance in both intact and damaged conditions. In other words, without sacrificing the barrier function of intact polymeric coatings, the decorated COFs successfully degrade, and with the development of an organic inhibitive layer, active inhibition promoted the damaged coating function.

9.2.3 Mesoporous silica: only representative of the inorganic compound family

Silicon is one of the elements with an atomic number of 14, which is a brittle crystalline solid (blue-gray metallic luster), hard, a semiconductor, and a tetravalent metalloid. In the earth's crust, more than 90% consists of silicate and its family, which, after oxygen, makes it the second-most abundant element on the earth.

By mass, it is the eighth-most common element in the world, but most of it is not pure form in the crust of the earth. In the space or planetoids, and planets, there is most silicon dioxide (silica) or silicates. Due to the high chemical affinity of silicon to oxygen, in 1823, for the first time, Berzelius was able to generate a pure form of silicon. The utilization of silicon in cosmetics is widespread due to good light diffusion and natural absorbency. Silicon dioxide, which is known as silica or SiO_2, is naturally found on our planet (the main constituent of sand) and commonly found in nature as quartz. In the world, it is one of the most abundant and the most complex synthetic product, which consists of various minerals. According to these points, the usage of SiO_2 is widely spread. The SiO_2 is utilized in various applications, like microelectronics as electrical insulators, food industries, structural materials, and pharmaceutical industries. In the structure of $SiO_{2,}$ each silicon atom displayed the tetrahedral coordination with four oxygen atoms, with the silicon atom at the center and the oxygen surrounding the Si atom. According to these, the SiO_2 had a three-dimensional structure in which the Si atoms were covalently bonded with four oxygen atoms in a tetrahedral manner.

The crystalline forms of SiO_2 are the main factor of its solubility in a water environment, which is more than four or five times greater than that of quartz. The mesoporous SiO_2 is one of the silicon dioxide family, which resulted from (synthesized) the reaction of tetraethyl orthosilicate with a template made of micellar rods. The outcomes lead to the generation of nano-sized rod or sphere collections, which are filled by orderly serialization of pores. The template can be removed by washing with a solvent (with proper pH). According to these points and due to the properties of SiO_2, among the inorganic nano-containers, the utilization of mesoporous SiO_2 is widely investigated and studied. There are some important properties that lead to the spread of a nano container, including high loading capacity, controlled pore diameter, and large surface area.

In some investigations, the pure SiO_2 nanoparticles were mixed with the coating matrix, and the functionalities of the designed coating were studied, which displayed the fact that the SiO_2 nanoparticles can provide active corrosion protection. As an example, in one research, Yin et al. investigated the effect of SiO_2 on the corrosion protection and self-healing of organic composite coating (PVB) on the Zn [101]. Their results displayed that the presence of SiO_2 nanoparticles in the PVB layer can lead to self-healing of the delaminated layer of PVB and prevention of organic coating disbandment. The protection mechanism consisted of the degradation of mesoporous SiO_2 in the high alkaline pH at the delamination area, which leads to the formation of a silicate layer. The formed silicate layer can block the under film corrosion and reestablishes of protective and delamination-resistant interface. In addition, in another research, Zhou et al. surveyed the effect of SiO_2/Polybenzoxazine on the mild steel corrosion [102]. Their results displayed that the corrosion protection of mild steel was enhanced due to the covalent bonding of SiO_2 and Polybenzoxazine, which can lead to improving interfacial interaction at the coating/nanoparticles interfaces. Zhou et al. studied the influence of interfacial interaction of Polybenzoxazine/SiO_2 (SNP) nanocomposite coatings in 3.5% harsh media [103]. The results of their study illustrated that the presence of an SNP leads to a reduction in coating porosities, and the coating, including an SNP, had excellent corrosion resistance with more than 1GΩ. cm^2 charge transfer resistance after 31 days of immersion into the 3.5% salinity solution, which was three orders of magnitude higher than pure coating. Also, Alam et al. surveyed the effect of various contents of SiO_2 on the anti-corrosion performance of epoxy coating at different immersion times in 3.5% NaCl media [104]. The outcomes showed that among various contents of SiO_2 (1%, 3% and 5%), the 5% displayed higher anti-corrosion efficiency up to 30 days of immersion.

However, the results of these studies affirm good performance and even self-healing of SiO_2, but this point should be noted that the self-healing ability of pure SiO_2 is not high enough, and in the long-term connection with corrosive agents cannot be acceptable. Thus, the researchers made decisions to utilize SiO_2 as a nano-container due to the pointed important properties of SiO_2 nanoparticles. In addition, the large specific surface area, large pore volumes, and good stability, which can lead to loading and releasing of many corrosion inhibitors via a diffusion-controlled process, the SiO_2 nanoparticles attracted more attention from the researchers. According to the past, by mixing nano-containers, which were loaded with corrosion inhibitors or other anti-corrosive agents in the coating matrix, the coating lifetime and anti-corrosion properties were enhanced. So, by mixing loaded SiO_2 with organic/inorganic coating, the coating barrier properties and lifetime are enhanced due to the filling of the porosities or cracks (which are naturally present in the coating structure) and the self-healing properties, respectively. Various studies that investigated the effect of loaded silica on the coating's anti-corrosion properties and surveyed the self-healing ability of the coating are pointed out next:

As an example, Sun et al. surveyed the functionalities of novel coating (SiO_2-loaded tannic acid, which was anchored on the MXene) on the anti-corrosion performance of carbon steel in harsh media [105]. The outcomes of their study displayed that the designed coating showed higher resistance compared with a pure one, even after 132 h of continuous immersion, and the efficiency of electrochemical self-healing was acceptable. In addition, in another study, Nguyen et al. surveyed the anti-corrosion protection of carbon steel with the help of silica nanoparticle-loaded cerium in the epoxy nanocomposite coating [106]. Their results demonstrated that both coating and polarization resistance, along with the barrier performance of the coating, were enhanced. In addition, their results displayed the important role of Ce inhibitor in improving coating performance. In another research, Skorb et al. studied the effect of silica-loaded inhibitor (named 2-(benzothiazol-2-ylsulfanyl)-succinic acid)-zirconia for the corrosion protection of aluminum alloys [107]. The technique of scanning vibrating electrodes affirmed the effective healing ability of containers to cure the corrosion defects. In another study, the effect of a hybrid organic/inorganic film on the corrosion resistance of Mg alloys was investigated by Kartsonakis et al. [108]. The coating consisted of cross-linked polymers based on Bisphenol

A diglycidyl ether, and in the organic section, the modified silicate was loaded with cerium molybdate. Their results confirm that for the samples with artificial defects after immersion in 1mM NaCl after 72 h for the coated sample, the charge transfer resistance was enhanced which displayed better self-healing ability of coated sample with the designed coating. Also, Gharagozlou et al. investigated the novel anti-corrosive nanoparticles ($NiFe_2O_4$), which were dispersed into the SiO_2 matrix to improve the anti-corrosive performance of epoxy coating [109]. The results of their study displayed that introducing the $NiFe_2O_4$-SiO_2 to the epoxy matrix helped the enhancement of coating protection properties before and after the scratch, which was applied intentionally. Their results affirm the good self-healing ability and anti-corrosion performance of the designed nanoparticles against the corrosive agent. Li et al. showed the positive effect of rGO/SiO_2-loaded benzotriazole on the corrosion protection of coating (epoxy) [110]. The results of their study displayed that for scratched and unscratched samples, the corrosion resistance reached the highest, which displayed the acceptable self-healing and barrier properties.

Nowadays, according to the results of these studies, the utilization of mesoporous silica, which is economical and one of the more effective nano-containers (leading from the noted properties) for the production of more usable coatings with acceptable lifetime, resistance, and active protection properties, is widespread. SiO_2 displayed good functionalities as a nano-container, especially alongside other two-dimensional materials such as MXene and/or rGO, which can result from the high surface area, controlling the pore diameter, and great absorbency.

9.2.4 Current application and future suggestions about metal oxides

The metal is categorized into two groups, which are chemical elements (like iron) and alloys (such as stainless steel). The alloys were produced by mixing various metallic elements with different contents. There are more metallic elements in the periodic table of the elements, which have various properties such as good electrical and heat conductive, alongside their malleability and ductility. These properties resulted from the metallic bond between the atoms or molecules of the metal. As chemical elements, they are contained in 25% of the earth's crust, which has an important role in modern life. For example, due to their strength and resiliency, some of them could be used in bridge construction. Alongside these applications, some metals are historically valuable and utilized as coinage in the past.

So, due to good properties of metal, the utilization of these materials as nano-filler in the coating to enhance coating barrier properties was paid attention to. Niroumandrad et al. surveyed the influence of three different modified Al nanoparticles (Ce, APTES, and Si) on the corrosion resistance of composite coating [111]. The results demonstrated better performance of the nanocomposite coating containing Si-Ce-Al in comparison with others, which can be attributed to the good dispersion of Ce-Al with modified APTES. In one other study, Jalili et al. assessed the modified Al nanoparticles with amino trimethylenephosphonic acid influence on the Zinc-rich epoxy coating corrosion resistance [112], illustrating that the galvanic action and corrosion resistance improved due to the enhancement in the barrier action and performance of the coating.

One of the most common forms of metal in nature is the metal oxide, which is the result of metal reactions with at least one oxygen atom. It is good to know that most of the earth's crust consists of oxides. Some pure elements have a natural thin oxide layer over themselves, which is called a passivation layer. There are various properties known for metal oxide, such as mechanical, surface, chemical, and optical, which are good reasons for metal oxide utilization in various research areas. According to these, in this part, we decided to survey the effect of metal oxide (zinc oxide (ZnO), zirconium oxide (ZrO_2), cerium oxide (CeO_2), etc.) on the barrier and self-healing properties of the coating. These inorganic nanofillers can improve the barrier properties of the coating by filling the porosities and enhancing the barrier properties of the coating against corrosive against diffusion to the metal/coating interface. Each of the metal oxide nanoparticles had various properties, such as titanium oxide, which is nontoxic, has good electrical properties, and UV light adsorption. As an example, Wang et al. surveyed the influence of three different structures of TiO_2 on the corrosion resistance of epoxy coating [113]. The electrochemical results displayed that the meso TiO_2had better function on the anti-corrosion efficiency of epoxy coating due to the large pore size of TiO_2 (the large pore size of TiO_2 helps to absorb more resin in the surface and channel, leading to better interaction of TiO_2 with the polymeric matrix). In addition, Tourknezhad et al. studied the influence of various amounts and types of metal oxides on the corrosion resistance of epoxy coating on carbon steel [114], demonstrating that both of these two points can affect the corrosion resistance and performance of the coating. Also, they found a direct relation between the nanoparticle content and the anti-corrosion efficiency. On the other hand, the results displayed better functionalities of Al_2O_3 in comparison with other nanoparticles. Ramezanzadeh and Attar investigated the influence of micro- and nano-sized ZnO particles on the anti-corrosion performance of epoxy coating [115]. Their results affirm the enhancement of barrier properties of the designed composite coating, resulting from the high specific surface area of ZnO nanoparticles. Alongside their study, Haddai et al. surveyed the influence of silane-modified and unmodified ZrO_2 on the corrosion performance of epoxy coating in corrosive media [116]. Their results showed that higher corrosion resistance was achieved after the usage of silane-modified ZrO_2/epoxy in comparison with unmodified and pure epoxy coating.

There are various studies on the enhancement of the barrier properties of coatings with the help of metal oxide nanoparticles. Nevertheless, it is good to know that according to the agglomeration of additional concentration of metal oxide into the polymeric matrix, which resulted in the creation of diffusion ways for the corrosive agents' diffusion into the metal/coating interface, the modification of these nanoparticle surfaces can be a great way to resolve this problem. For this aim, the researchers considered the utilization of some other organic/inorganic materials (such as corrosion inhibitors, silanes, surfactants, and other chemical agents) to modify the surface of nanoparticles. As an example, Wang et al. illustrated the effect of meso-TiO_2 (which was modified with polyethylenimine with various Mw (600, 70,000, and 700,000)) on the corrosion resistance of waterborne epoxy coating [117]. Their outcomes showed that the higher corrosion resistance was demonstrated after utilization of modified TiO_2 with a lower Mw of polyethylenimine as a nanofiller in the epoxy coating. On the other hand, Radoman et al. prepared the TiO_2 nanoparticles modified with the help of gallic acid ester on the epoxy coating and then investigated the influence of this nanofiller on the vapor transmission in the composite coating [118]. The outcomes displayed that by enhancing the length of the gallic acid ester chain and the corrosion resistance of the epoxy coating was improved. Also, Duraibabu et al. studied the functionalized (APES) and unfunctionalized zinc oxide on anti-microbial and anti-corrosion performance of 4-amine-phenoxy benzene (TGBAPB) epoxy coating [119]. Their results displayed that the corrosion resistance of coating with functionalized zinc oxide is more than other coatings. However, their outcomes affirm that the corrosion resistance of pure TGBAPB is more than the coating, which contained non-functionalized zinc oxide nanoparticles. This can result from the inappropriate dispersion of non-functionalized zinc oxide particles in the coating matrix, leading to the creation of cracks and porosities in the coating matrix. Mahulikar et al. studied the effect of modified TiO_2 with the help of polyaniline on the epoxy coating corrosion resistance [120]. The results of their research confirm the better corrosion resistance of the designed composite coating in comparison with pure epoxy, which can result from the enhancement of the barrier properties of the designed composite coating.

Besides these good properties of metal oxide nanoparticles, they can be utilized to help with the self-healing application of coating, which can be a good research context for the future. Nevertheless, it is good to know that alongside these metal oxide and their good properties, there is another metal oxide, which was surveyed as a nanofiller in the composite coating, and the results displayed the impressive enhancement in the barrier properties of the coating. Also, after the utilization of this nanofiller, it can play as role of a nano-container and improve the self-healing ability of the composite coating.

9.3 CONCLUSION

With all sound studies, it can be shown that porous nanocarriers containing inhibitors play a significant role in the self-healing of coatings. By following different mechanisms and responding to different environmental stimuli, the released inhibitor covers the damaged areas, preventing the coating deterioration. For example, the sensitivity of MOFs and COFs to pH changes will lead to the destruction of the coordinative network, and the result of this destruction will be the release of active inhibitors. Or, in some cases, carriers such as metal oxide and SiO_2 do not have inhibitory capabilities by themselves and only play the role of carriers for active inhibitors. However, all these nanoparticles induce self-healing ability, and the obtained results prove the continued efficiency of the damaged coating.

REFERENCES

[1] P.R. Roberge, *Handbook of Corrosion Engineering*, McGraw-Hill Education, 2019.

[2] A.H. Mostafatabar, G. Bahlakeh, M. Ramezanzadeh, B. Ramezanzadeh, Eco-friendly protocol for zinc-doped amorphous carbon-based film construction over steel surface using nature-inspired phytochemicals: Coupled experimental and classical atomic/molecular and electronic-level theoretical explorations, *Journal of Environmental Chemical Engineering*, 9 (2021) 105487.

[3] P.R. Roberge, *Corrosion Inspection and Monitoring*, John Wiley & Sons, 2007.
[4] V. Golubović-Bugarski, S. Petković, G. Globočki-Lakić, The effect of corrosion on a structural integrity and vehicle safety, *Mobility and Vehicle Mechanics*, 47 (2021) 49–62.
[5] S. Hertelé, A. Cosham, P. Roovers, Structural integrity of corroded girth welds in vintage steel pipelines, *Engineering Structures*, 124 (2016) 429–441.
[6] C.A. Apostolopoulos, D. Michalopoulos, P. Koutsoukos, The corrosion effects on the structural integrity of reinforcing steel, *Journal of Materials Engineering and Performance*, 17 (2008) 506–516.
[7] A.A. Basheer, Advances in the smart materials applications in the aerospace industries, *Aircraft Engineering and Aerospace Technology*, 92 (2020) 1027–1035.
[8] B. Cao, Z. Li, D. Yang, L. Mu, P. Liu, L. Wang, L. Yang, Y. Wu, Formation and influence analysis of electrolytic corrosion products on insulator surface, *High Voltage*, 6 (2021) 1024–1030.
[9] E. Shekari, F. Khan, S. Ahmed, Economic risk analysis of pitting corrosion in process facilities, *International Journal of Pressure Vessels and Piping*, 157 (2017) 51–62.
[10] S. Zehra, M. Mobin, J. Aslam, *An Overview of the Corrosion Chemistry*, Environmentally Sustainable Corrosion Inhibitors, 2022: 3–23.
[11] R.W. Revie, *Corrosion and Corrosion Control: An Introduction to Corrosion Science and Engineering*, John Wiley & Sons, 2008.
[12] K.R. Trethewey, J. Chamberlain, *Corrosion for science and engineering*, 1995. https://www.osti.gov/biblio/378108
[13] M. Aliofkhazraei, *Developments in Corrosion Protection*, BoD-Books on Demand, 2014.
[14] N.G. Thompson, M. Yunovich, D. Dunmire, Cost of corrosion and corrosion maintenance strategies, *Corrosion Reviews*, 25 (2007) 247–262.
[15] K. Li, Y. Zeng, Corrosion of heat exchanger materials in co-combustion thermal power plants, *Renewable and Sustainable Energy Reviews*, 161 (2022) 112328.
[16] M.R. Schock, Internal corrosion and deposition control, *Water Quality and Treatment*, 4 (1999).
[17] R. Heidersbach, *Metallurgy and Corrosion Control in Oil and Gas Production*, John Wiley & Sons, 2018.
[18] M. Iannuzzi, G. Frankel, The carbon footprint of steel corrosion, *NPJ Materials Degradation*, 6 (2022) 101.
[19] L. Huang, Q. Liao, J. Yan, Y. Liang, H. Zhang, Carbon footprint of oil products pipeline transportation, *Science of the Total Environment*, 783 (2021) 146906.
[20] A. Kadhim, A. Al-Amiery, R. Alazawi, M. Al-Ghezi, R. Abass, Corrosion inhibitors. A review, *International Journal of Corrosion and Scale Inhibition*, 10 (2021) 54–67.
[21] G. Gece, Drugs: A review of promising novel corrosion inhibitors, *Corrosion Science*, 53 (2011) 3873–3898.
[22] C.G. Dariva, A.F. Galio, Corrosion inhibitors-principles, mechanisms and applications, *Developments in Corrosion Protection*, 16 (2014) 365–378.
[23] A.H. Mostafatabar, A. Dehghani, P. Ghahremani, G. Bahlakeh, B. Ramezanzadeh, Molecular-dynamic/DFT-electronic theoretical studies coupled with electrochemical investigations of the carrot pomace extract molecules inhibiting potency toward mild steel corrosion in 1 M HCl solution, *Journal of Molecular Liquids*, 346 (2022) 118344.
[24] A.H. Mostafatabar, G. Bahlakeh, B. Ramezanzadeh, A. Dehghani, M. Ramezanzadeh, A comprehensive electronic-scale DFT modeling, atomic-level MC/MD simulation, and electrochemical/surface exploration of active nature-inspired phytochemicals based on Heracleum persicum seeds phytoextract for effective retardation of the acidic-induced corrosion of mild steel, *Journal of Molecular Liquids*, 331 (2021) 115764.

[25] G. Palanisamy, *Corrosion inhibitors*, Intech Open, 2019: 1–24.
[26] A.H. Mostafatabar, G. Bahlakeh, B. Ramezanzadeh, Novel bi-functional RGO-HPSE-Zn@ epoxy nanocomposite with superior corrosion protection potency, *Journal of Industrial and Engineering Chemistry*, 108 (2022) 28–46.
[27] M. Askari, M. Aliofkhazraei, R. Jafari, P. Hamghalam, A. Hajizadeh, Downhole corrosion inhibitors for oil and gas production—a review, *Applied Surface Science Advances*, 6 (2021) 100128.
[28] R. Zuo, Biofilms: Strategies for metal corrosion inhibition employing microorganisms, *Applied Microbiology and Biotechnology*, 76 (2007) 1245–1253.
[29] P. Ghahremani, A.H. Mostafatabr, A. Dehghani, G. Bahlakeh, B. Ramezanzadeh, Apple pomace extract: A potent renewable source of active biomolecules for suppressing mild steel aggression in aquatic solution, *Biomass Conversion and Biorefinery*, (2022) 1–18.
[30] A. Dehghani, A.H. Mostafatabar, B. Ramezanzadeh, Synergistic anticorrosion effect of Brassica Hirta phytoconstituents and cerium ions on mild steel in saline media: Surface and electrochemical evaluations, *Colloids and Surfaces A: Physicochemical and Engineering Aspects*, 656 (2023) 130503.
[31] A. Dehghani, P. Ghahremani, A.H. Mostafatabar, B. Ramezanzadeh, Plant extracts: Probable alternatives for traditional inhibitors for controlling alloys corrosion against acidic media: A review, *Biomass Conversion and Biorefinery* (2022) 1–20.
[32] A. Dehghani, A.H. Mostafatabar, G. Bahlakeh, B. Ramezanzadeh, Poppy-leaf extract-derived biomolecules adsorption on the rGO-nanoplatforms and application as smart self-healing material for epoxy coating, *Journal of Molecular Liquids*, 370 (2023) 120931.
[33] C. Ding, J. Fu, Smart anticorrosion coatings based on nanocontainers, In *Smart Nanocontainers*, Elsevier, 2020: 413–429.
[34] Y. Qian, Y. Li, S. Jungwirth, N. Seely, Y. Fang, X. Shi, The application of anti-corrosion coating for preserving the value of equipment asset in chloride-laden environments: A, *International Journal of Electrochemical Science*, 10 (2015) 10756–10780.
[35] A. Dehghani, A.H. Mostafatabar, G. Bahlakeh, B. Ramezanzadeh, Metal-doped 2D rGO nano-sheets fabrication utilizing plant source bio-molecules and application in the epoxy anti-corrosive coating: Combined experimental and DFT-D modeling investigations, *Progress in Organic Coatings*, 170 (2022) 106938.
[36] A. Dehghani, G. Bahlakeh, B. Ramezanzadeh, A.H.J. Mofidabadi, A.H. Mostafatabar, Benzimidazole loaded β-cyclodextrin as a novel anti-corrosion system; Coupled experimental/computational assessments, *Journal of Colloid and Interface Science*, 603 (2021) 716–727.
[37] Z. Chen, N. Scharnagl, M.L. Zheludkevich, H. Ying, W. Yang, Micro/nanocontainer-based intelligent coatings: Synthesis, performance and applications—A review, *Chemical Engineering Journal*, 451 (2023) 138582.
[38] P. Ghahremani, A.H. Mostafatabar, G. Bahlakeh, B. Ramezanzadeh, Rational design of a novel multi-functional carbon-based nano-carrier based on multi-walled-CNT-oxide/polydopamine/chitosan for epoxy composite with robust pH-sensitive active anti-corrosion properties, *Carbon*, 189 (2022) 113–141.
[39] X. Zhou, Q. Dong, D. Wei, J. Bai, F. Xue, B. Zhang, Z. Ba, Z. Wang, Smart corrosion inhibitors for controlled release: A review, *Corrosion Engineering, Science and Technology*, 58 (2023) 190–204.
[40] E. Alibakhshi, B. Ramezanzadeh, M. Mahdavian, Self-healing materials in corrosion protection, *Self-Healing Smart Materials and Allied Applications* (2021) 247–296.
[41] M.H. Nazari, X. Shi, Polymer-based nanocomposite coatings for anticorrosion applications, in *Industrial Applications for Intelligent Polymers and Coatings*, Springer, 2016: 373–398.

[42] M. Zhang, C. Li, X. Wang, J. Peng, S. Yuan, H. Geng, Y. Zhou, Y. Gao, H. Wang, Ultra-high anti-corrosion performance of polymer-based coating filled with a novel micro network nanofiller, *Corrosion Science*, 190 (2021) 109685.
[43] A.H. Mostafatabar, G. Bahlakeh, B. Ramezanzadeh, Designing a novel anti-corrosion metal-organic platform based on dual-action epoxy coating, *Progress in Organic Coatings*, 170 (2022) 107007.
[44] S. Habib, R. Shakoor, R. Kahraman, A focused review on smart carriers tailored for corrosion protection: Developments, applications, and challenges, *Progress in Organic Coatings*, 154 (2021) 106218.
[45] A. Dehghani, P. Ghahremani, A.H. Mostafatabar, B. Ramezanzadeh, Plant extracts: Probable alternatives for traditional inhibitors for controlling alloys corrosion against acidic media—A review, *Biomass Conversion and Biorefinery*, 14 (2024) 7467–7486.
[46] R. Majidi, M. Ramezanzadeh, B. Ramezanzadeh, Developing a dual-functional self-healing nanocomposite utilizing oxidized-multiwall carbon nanotube/highly-porous metal-organic framework (OCNT/ZIF-8) nano-hybrid, *Applied Materials Today*, 32 (2023) 101830.
[47] P. Ghahremani, A.A. Sarabi, S. Roshan, Cerium containing pH-responsive microcapsule for smart coating application: Characterization and corrosion study, *Surface and Coatings Technology*, 427 (2021) 127820.
[48] S. An, M.W. Lee, A.L. Yarin, S.S. Yoon, A review on corrosion-protective extrinsic self-healing: Comparison of microcapsule-based systems and those based on core-shell vascular networks, *Chemical Engineering Journal*, 344 (2018) 206–220.
[49] N.K. Guimard, K.K. Oehlenschlaeger, J. Zhou, S. Hilf, F.G. Schmidt, C. Barner-Kowollik, Current trends in the field of self-healing materials, *Macromolecular Chemistry and Physics*, 213 (2012) 131–143.
[50] D. Bekas, K. Tsirka, D. Baltzis, A.S. Paipetis, Self-healing materials: A review of advances in materials, evaluation, characterization and monitoring techniques, *Composites Part B: Engineering*, 87 (2016) 92–119.
[51] S. Kitagawa, Future porous materials, *Accounts of Chemical Research*, 50 (2017) 514–516.
[52] M. Ramezanzadeh, B. Ramezanzadeh, M. Mahdavian, G. Bahlakeh, Development of metal-organic framework (MOF) decorated graphene oxide nanoplatforms for anti-corrosion epoxy coatings, *Carbon*, 161 (2020) 231–251.
[53] P. Choudhary, S.K. Ola, I. Chopra, V. Dhayal, D.S. Shekhawat, Metal-organic framework (MOF)/graphene-oxide (GO) nanocomposites materials: A potential formulation for anti-corrosive coatings-a review, *Materials Today: Proceedings*, 79 (2023) 172–178.
[54] C. Yang, W. Xu, X. Meng, X. Shi, L. Shao, X. Zeng, Z. Yang, S. Li, Y. Liu, X. Xia, A pH-responsive hydrophilic controlled release system based on ZIF-8 for self-healing anticorrosion application, *Chemical Engineering Journal*, 415 (2021) 128985.
[55] C. Zhang, W. Li, C. Liu, C. Zhang, L. Cao, D. Kong, W. Wang, S. Chen, Effect of covalent organic framework modified graphene oxide on anticorrosion and self-healing properties of epoxy resin coatings, *Journal of Colloid and Interface Science*, 608 (2022) 1025–1039.
[56] S.B. Ulaeto, R. Rajan, J.K. Pancrecious, T. Rajan, B. Pai, Developments in smart anti-corrosive coatings with multifunctional characteristics, *Progress in Organic Coatings*, 111 (2017) 294–314.
[57] A.H. Lu, F. Schüth, Nanocasting: A versatile strategy for creating nanostructured porous materials, *Advanced Materials*, 18 (2006) 1793–1805.
[58] I.K. Nwokolo, H. Shi, A.I. Ikeuba, N. Gao, J. Li, S. Ahmed, F. Liu, Synthesis, Characterization and investigation of anticorrosion properties of an innovative metal-organic framework, ZnMOF-BTA, on carbon steel in HCl solution, *Coatings*, 12 (2022) 1288.

[59] S. Swetha, B. Janani, S.S. Khan, A critical review on the development of metal-organic frameworks for boosting photocatalysis in the fields of energy and environment, *Journal of Cleaner Production*, 333 (2022) 130164.

[60] Z. Haeri, B. Ramezanzadeh, M. Ramezanzadeh, Recent progress on the metal-organic frameworks decorated graphene oxide (MOFs-GO) nano-building application for epoxy coating mechanical-thermal/flame-retardant and anti-corrosion features improvement, *Progress in Organic Coatings*, 163 (2022) 106645.

[61] P. Silva, S.M. Vilela, J.P. Tome, F.A.A. Paz, Multifunctional metal-organic frameworks: From academia to industrial applications, *Chemical Society Reviews*, 44 (2015) 6774–6803.

[62] I. Ahmed, S.H. Jhung, Covalent organic framework-based materials: Synthesis, modification, and application in environmental remediation, *Coordination Chemistry Reviews*, 441 (2021) 213989.

[63] K.L. Rubow, AEROSPACE, *Advanced Materials & Processes* (2009) 27.

[64] L.F. Dumée, L. He, B. Lin, F.-M. Ailloux, J.-B. Lemoine, L. Velleman, F. She, M.C. Duke, J.D. Orbell, G. Erskine, The fabrication and surface functionalization of porous metal frameworks—a review, *Journal of Materials Chemistry A*, 1 (2013) 15185–15206.

[65] C. Ensarioglu, A. Bakirci, H. Koluk, M.C. Cakir, Metal foams and their applications in aerospace components, in *Materials, Structures and Manufacturing for Aircraft*, Springer, 2022: 27–63.

[66] C.-J. Yu, H.H. Eifert, J. Banhart, J. Baumeister, Metal foams, *Advanced Materials & Processes*, 154 (1998) 45–45.

[67] A. Bisht, V.K. Patel, B. Gangil, Future of metal foam materials in automotive industry, *Automotive Tribology*, (2019) 51–63.

[68] W. Kim, S. Ahn, H. Kim, J. Kim, I. Ozdemir, Y. Tsunekawa, Corrosion performance of plasma-sprayed cast iron coatings on aluminum alloy for automotive components, *Surface and Coatings Technology*, 200 (2005) 1162–1167.

[69] S.R.S. Rodrigues, V. Dalmoro, J.H.Z. dos Santos, *Mesoporous Silica-Based Systems for Anti-Biofouling Applications, Supramolecular Chemistry in Corrosion and Biofouling Protection*, CRC Press, 2021: 347–358.

[70] M.A. Zavareh, A.A.D.M. Sarhan, B.B. Abd Razak, W.J. Basirun, Plasma thermal spray of ceramic oxide coating on carbon steel with enhanced wear and corrosion resistance for oil and gas applications, *Ceramics International*, 40 (2014) 14267–14277.

[71] Z. Chen, D. Koleva, K. van Breugel, A review on stray current-induced steel corrosion in infrastructure, *Corrosion Reviews*, 35 (2017) 397–423.

[72] C. Zhou, M. Pan, S. Li, Y. Sun, H. Zhang, X. Luo, Y. Liu, H. Zeng, Metal organic frameworks (MOFs) as multifunctional nanoplatform for anticorrosion surfaces and coatings, *Advances in Colloid and Interface Science*, (2022) 102707.

[73] P. Sambyal, G. Ruhi, R. Dhawan, S.K. Dhawan, Designing of smart coatings of conducting polymer poly (aniline-co-phenetidine)/SiO2 composites for corrosion protection in marine environment, *Surface and Coatings Technology*, 303 (2016) 362–371.

[74] M. Usman, S. Mendiratta, K.L. Lu, Semiconductor metal-organic frameworks: Future low-bandgap materials, *Advanced Materials*, 29 (2017) 1605071.

[75] X. Li, X. Yang, H. Xue, H. Pang, Q. Xu, Metal-organic frameworks as a platform for clean energy applications, *EnergyChem*, 2 (2020) 100027.

[76] U. Farooq, T. Ahmad, F. Naaz, S.U. Islam, Review on metals and metal oxides in sustainable energy production: Progress and perspectives, *Energy & Fuels*, 37 (2023) 1577–1632.

[77] G. Cai, P. Yan, L. Zhang, H.-C. Zhou, H.-L. Jiang, Metal-organic framework-based hierarchically porous materials: Synthesis and applications, *Chemical Reviews*, 121 (2021) 12278–12326.

[78] R.V. Dennis, V. Patil, J.L. Andrews, J.P. Aldinger, G.D. Yadav, S. Banerjee, Hybrid nanostructured coatings for corrosion protection of base metals: A sustainability perspective, *Materials Research Express*, 2 (2015) 032001.

[79] M.F. Montemor, Functional and smart coatings for corrosion protection: A review of recent advances, *Surface and Coatings Technology*, 258 (2014) 17–37.

[80] J. Zeng, B. Tan, S. Zhang, W. Li, The behavior of two indazole derivatives on the copper/sulfuric acid interface in terms of adsorption and corrosion inhibition, *Journal of the Taiwan Institute of Chemical Engineers*, 140 (2022) 104567.

[81] D.G. Shchukin, H. Möhwald, Self-repairing coatings containing active nanoreservoirs, *Small*, 3 (2007) 926–943.

[82] X. Fu, L. Tian, Y. Fan, W. Ye, Z.-A. Qiao, J. Zhao, L. Ren, W. Ming, Stimuli-responsive self-healing anticorrosion coatings: From single triggering behavior to synergetic multiple protections, *Materials Today Chemistry*, 22 (2021) 100575.

[83] S. Zehra, M. Mobin, R. Aslam, Nanocontainers: A comprehensive review on their application in the stimuli-responsive smart functional coatings, *Progress in Organic Coatings*, 176 (2023) 107389.

[84] S.-Y. Ding, W. Wang, Covalent organic frameworks (COFs): From design to applications, *Chemical Society Reviews*, 42 (2013) 548–568.

[85] A.U. Czaja, N. Trukhan, U. Müller, Industrial applications of metal-organic frameworks, *Chemical Society Reviews*, 38 (2009) 1284–1293.

[86] S. Zafari, M. Niknam Shahrak, M. Ghahramaninezhad, New MOF-based corrosion inhibitor for carbon steel in acidic media, *Metals and Materials International*, 26 (2020) 25–38.

[87] K. Cao, Z. Yu, D. Yin, Preparation of Ce-MOF@ TEOS to enhance the anti-corrosion properties of epoxy coatings, *Progress in Organic Coatings*, 135 (2019) 613–621.

[88] H. Chen, F. Wang, H. Fan, R. Hong, W. Li, Construction of MOF-based superhydrophobic composite coating with excellent abrasion resistance and durability for self-cleaning, corrosion resistance, anti-icing, and loading-increasing research, *Chemical Engineering Journal*, 408 (2021) 127343.

[89] Y. Zhang, J. Wang, S. Zhao, M. Serdechnova, C. Blawert, H. Wang, M.L. Zheludkevich, F. Chen, Double-ligand strategy to construct an inhibitor-loaded Zn-MOF and its corrosion protection ability for aluminum alloy 2A12, *ACS Applied Materials & Interfaces*, 13 (2021) 51685–51694.

[90] S.M. Lashgari, H. Yari, M. Mahdavian, B. Ramezanzadeh, G. Bahlakeh, M. Ramezanzadeh, Application of nanoporous cobalt-based ZIF-67 metal-organic framework (MOF) for construction of an epoxy-composite coating with superior anti-corrosion properties, *Corrosion Science*, 178 (2021) 109099.

[91] M.Y. Zorainy, M. Sheashea, S. Kaliaguine, M. Gobara, D.C. Boffito, Facile solvothermal synthesis of a MIL-47 (V) metal-organic framework for a high-performance Epoxy/MOF coating with improved anticorrosion properties, *RSC Advances*, 12 (2022) 9008–9022.

[92] J. Abdi, M. Izadi, M. Bozorg, Improvement of anti-corrosion performance of an epoxy coating using hybrid UiO-66-NH2/carbon nanotubes nanocomposite, *Scientific Reports*, 12 (2022) 10660.

[93] F. Seidi, M. Jouyandeh, M. Taghizadeh, A. Taghizadeh, H. Vahabi, S. Habibzadeh, K. Formela, M.R. Saeb, Metal-organic framework (MOF)/epoxy coatings: A review, *Materials*, 13 (2020) 2881.

[94] L. Chen, Z. Yu, D. Yin, K. Cao, C. Xie, L. Zhu, Y. Jiang, Preparation and anticorrosion properties of GO-Ce-MOF nanocomposite coatings, *Journal of Applied Polymer Science*, 139 (2022) 51571.

[95] K. Cao, Z. Yu, D. Yin, L. Chen, Y. Jiang, L. Zhu, Fabrication of BTA-MOF-TEOS-GO nanocomposite to endow coating systems with active inhibition and durable anticorrosion performances, *Progress in Organic Coatings*, 143 (2020) 105629.

[96] A. Dehghani, Z. Sanaei, M. Fedel, M. Ramezanzadeh, M. Mahdavian, B. Ramezanzade, Fabrication of an intelligent anti-corrosion silane film using a MoO42– loaded Micro/mesoporous ZIF67-MOF/multi-walled-CNT/APTES core-shell nano-container, *Colloids and Surfaces A: Physicochemical and Engineering Aspects*, 656 (2023) 130511.

[97] M. Zhang, X. Yu, Y. Lin, J. Liu, J. Wang, Anti-corrosion coatings with active and passive protective performances based on v-COF/GO nanocontainers, *Progress in Organic Coatings*, 159 (2021) 106415.

[98] T. Liu, W. Li, C. Zhang, W. Wang, W. Dou, S. Chen, Preparation of highly efficient self-healing anticorrosion epoxy coating by integration of benzotriazole corrosion inhibitor loaded 2D-COF, *Journal of Industrial and Engineering Chemistry*, 97 (2021) 560–573.

[99] N. Keshmiri, P. Najmi, M. Ramezanzadeh, B. Ramezanzadeh, G. Bahlakeh, Ultrastable porous covalent organic framework assembled carbon nanotube as a novel nanocontainer for anti-corrosion coatings: Experimental and computational studies, *ACS Applied Materials & Interfaces*, 14 (2022) 19958–19974.

[100] P. Najmi, N. Keshmiri, M. Ramezanzadeh, B. Ramezanzadeh, M. Arjmand, Porous 2D Ti3C2 MXene nanosheets sandwiched between imine-based covalent organic frameworks (COFs) for excellent corrosion protective coatings, *Chemical Engineering Journal*, 456 (2023) 141001.

[101] Y. Yin, H. Zhao, M. Prabhakar, M. Rohwerder, Organic composite coatings containing mesoporous silica particles: Degradation of the SiO2 leading to self-healing of the delaminated interface, *Corrosion Science*, 200 (2022) 110252.

[102] C. Zhou, X. Lu, Z. Xin, J. Liu, Y. Zhang, Polybenzoxazine/SiO2 nanocomposite coatings for corrosion protection of mild steel, *Corrosion Science*, 80 (2014) 269–275.

[103] C. Zhou, M. Tao, J. Liu, T. Liu, X. Lu, Z. Xin, Effects of interfacial interaction on corrosion resistance of polybenzoxazine/SiO2 nanocomposite coatings, *ACS Applied Polymer Materials*, 1 (2019) 381–391.

[104] M.A. Alam, U.A. Samad, E.-S.M. Sherif, A.M. Poulose, J.A. Mohammed, N. Alharthi, S.M. Al-Zahrani, Influence of SiO2 content and exposure periods on the anticorrosion behavior of epoxy nanocomposite coatings, *Coatings*, 10 (2020) 118.

[105] X. Sun, C. Ma, F. Ma, T. Wang, C. Feng, W. Wang, A novel coating with SiO2 anchored on MXene loading tannic acid for self-healing anticorrosive performance, *Journal of Alloys and Compounds*, 928 (2022) 167202.

[106] T.V. Nguyen, D.L. Tran, T.A. Nguyen, T.T.H. Nguyen, P.H. Dao, V.P. Mac, M.T. Do, T.M. Nguyen, T.M.L. Dang, Ce-loaded silica nanoparticles in the epoxy nanocomposite coating for anticorrosion protection of carbon steel, *Anti-Corrosion Methods and Materials*, 69 (2022) 514–523.

[107] E.V. Skorb, D. Fix, D.V. Andreeva, H. Möhwald, D.G. Shchukin, Surface-modified mesoporous SiO2 containers for corrosion protection, *Advanced Functional Materials*, 19 (2009) 2373–2379.

[108] I. Kartsonakis, A. Balaskas, E. Koumoulos, C. Charitidis, G. Kordas, Evaluation of corrosion resistance of magnesium alloy ZK10 coated with hybrid organic-inorganic film including containers, *Corrosion Science*, 65 (2012) 481–493.

[109] M. Gharagozlou, B. Ramezanzadeh, Z. Baradaran, Synthesize and characterization of a novel anticorrosive cobalt ferrite nanoparticles dispersed in silica matrix (CoFe2O4-SiO2) to improve the corrosion protection performance of epoxy coating, *Applied Surface Science*, 377 (2016) 86–98.

[110] D. Li, H. Peng, Z. Lin, J. Zhu, J. Yu, J. Liu, R. Chen, Q. Liu, J. Wang, Corrosion protection coatings embedded with silane-functionalized rGO/SiO2 nanocontainers: Enhancing dispersive and corrosion-inhibitor loading capabilities, *Surface and Coatings Technology*, 427 (2021) 127850.

[111] S. Niroumandrad, M. Rostami, B. Ramezanzadeh, Effects of combined surface treatments of aluminium nanoparticle on its corrosion resistance before and after inclusion into an epoxy coating, *Progress in Organic Coatings*, 101 (2016) 486–501.

[112] M. Jalili, M. Rostami, B. Ramezanzadeh, An investigation of the electrochemical action of the epoxy zinc-rich coatings containing surface modified aluminum nanoparticle, *Applied Surface Science*, 328 (2015) 95–108.

[113] N. Wang, W. Fu, M. Sun, J. Zhang, Q. Fang, Effect of different structured TiO2 particle on anticorrosion properties of waterborne epoxy coatings, *Corrosion Engineering, Science and Technology*, 51 (2016) 365–372.

[114] Y. Torknezhad, M. Khosravi, M. Assefi, Corrosion protection performance of nanoparticle incorporated epoxy paint assessed by linear polarization and electrochemical impedance spectroscopy, *Materials and Corrosion*, 69 (2018) 472–480.

[115] B. Ramezanzadeh, M. Attar, Studying the effects of micro and nano sized ZnO particles on the corrosion resistance and deterioration behavior of an epoxy-polyamide coating on hot-dip galvanized steel, *Progress in Organic Coatings*, 71 (2011) 314–328.

[116] S. Haddadi, M. Mahdavian, E. Karimi, Evaluation of the corrosion protection properties of an epoxy coating containing sol-gel surface modified nano-zirconia on mild steel, *RSC Advances*, 5 (2015) 28769–28777.

[117] N. Wang, W. Fu, J. Zhang, X. Li, Q. Fang, Corrosion performance of waterborne epoxy coatings containing polyethylenimine treated mesoporous-TiO2 nanoparticles on mild steel, *Progress in Organic Coatings*, 89 (2015) 114–122.

[118] T.S. Radoman, J.V. Džunuzović, K.B. Jeremić, B.N. Grgur, D.S. Miličević, I.G. Popović, E.S. Džunuzović, Improvement of epoxy resin properties by incorporation of TiO2 nanoparticles surface modified with gallic acid esters, *Materials & Design (1980–2015)*, 62 (2014) 158–167.

[119] D. Duraibabu, T. Ganeshbabu, R. Manjumeena, P. Dasan, Unique coating formulation for corrosion and microbial prevention of mild steel, *Progress in Organic Coatings*, 77 (2014) 657–664.

[120] P.P. Mahulikar, R.S. Jadhav, D.G. Hundiwale, Performance of polyaniline/TiO2 nanocomposites in epoxy for corrosion resistant coatings, (2011) 367–376.

10 Nanostructured carbon allotropes as self-healing anti-corrosive coating

Amir Hossein Jafari Mofidabadi and Ali Dehghani

10.1 INTRODUCTION

Regarding the extreme demand for metal utilization in various industries, the improvement of metal durability in harsh environments is one of the important subjects from a scientific perspective. Without a doubt, corrosion is one of the responsible issues leading to the metal's lifetime decrement and over the years carries significant damages and costs. As an example, we can point to the corrosion damage in oil pipelines and storage tanks, which brings noticeable financial and time losses. In such regard, corrosion prevention became an interesting title that targeted substantial improvement of the applied metallic equipment function in upstream and downstream industries.

In the first step, to control metal corrosion, it is better to understand how corrosion occurs. There are various definitions of corrosion, and after the conclusion of these definitions, we can briefly state that corrosion is the chemical and/or electrochemical phenomena between metal and the environment that converts a refined metal into chemically stable oxides.

For controlling such conversion, significant attempts were made with the orientation of corrosion retardation, and up to now, significant methods have been developed. In regular circumstances, the main way of corrosion control is cutting off the connection of metal and environment, leading to the chemical and/or electrochemical reactions control between metal substrates and surrounding environments with known techniques, including corrosion inhibitors application, anodic and/or cathodic protection, and coatings fabrication, which improve the performance of the metallic substrate through this mechanism.

Recent advances in a series of corrosion-oriented studies clarified that the utilization of coatings has been at the center of attention, thanks to superior benefits, including ease of fabrication, durability, and high protection efficiency. Upon surface coverage, the free contact area between metal and the surrounding environment prosperously declines, which concludes corrosion retardation and metal life-long enhancement. However, alongside all these advantages, there are some basic issues for specific aims.

According to the literature and research, there are different kinds of coatings, but the most applied ones are organic coatings like epoxy, which are so valid due

 DOI: 10.1201/9781003456025-10

to their exceptional properties. However, against our desire after fabrication, some unwanted, identifiable micro-cracks or porosities appear unintentionally in the matrices. The porosities regularly rise in coating structures during the curing process owing to internal stresses or polymerization failure, which are known as inherited natural defects in coating structures. In such conditions, these micro porosities and/or cracks are potential zones for the corrosive agent diffusion into the coating structures. By the diffusion of these corrosive agents, the coating adhesion bond deteriorates, and the corrosion initiates at the metal/coating interfaces. Eventually, the corrosion product accumulation at the substrate-coating interface, alongside the appearance of superior osmotic pressure, diminishes the applied coating's proper performance.

To control the hazards caused by water-ion penetration, researchers offer the utilization of fillers. By passing conventional fillers, which require a large volume to control water penetration, thanks to the emergence of the nano era, a low percentage of nanofillers can significantly control water penetration. Basically, the filler is responsible for the coating defects occupation, and with substantial diffusion control, the coating barrier properties are enhanced subsequently.

Although the initial performance of the coating has been significantly improved, the effect of stimuli on the applied coatings has been neglected. Undoubtedly, in the long term, the coating will not remain intact and will suffer various environmental damages. One of the most important of them is mechanical damage, which is extremely common. The damaged coating is highly susceptible to water and ion penetration from the damaged areas. In this case, the destructive agent will easily reach the substrate without encountering the polymer entanglements that control the penetration and passing through the damaged area, resulting in corrosion initiation. This is where the need for the emergence of self-healing coatings is felt. In this case, the scratched area will be protected with the help of the induced repair agents, and as a result, performance will increase significantly in the long term.

10.1.1 Coatings' self-healing mechanisms and reasons for carbonaceous materials application

In general, the self-healing mechanism is based on two modes: autonomous and non-autonomous. In the state of autonomous self-healable coatings, the broken bonds (damage area) are connected through the induction of dynamic bonds, or thermal, ion stimuli, etc., causing the damage area coverage. In addition to the great advantages of instinctive self-healing, the need for catalysts or inducing complex polymers increases the cost of developing this class of coatings.

For this reason, in many studies, researchers have preferred to use non-autonomous coatings in order to achieve self-healing capability. Among all the methods used to develop non-autonomous self-healing anti-corrosion coatings, incorporating an inhibitor into the coating body is a facile way. In this way, in the scratch area and upon contact with the corrosive solution, the inhibitor release and the development of the inhibitive layer prevent the rapid destruction of the damaged coating. Although the results obtained for the introduction of the first generation of self-healing anti-corrosion coatings seemed acceptable, with the passage of time, significant problems have appeared in this generation of coatings. Unfortunately, the uncontrolled release of the inhibitor has only been able to delay the corrosion for a short period of time, and then protection decreases rapidly.

A reliable solution to this problem is the controlled release of the inhibitor from the coating in response to environmental stimuli. To date, various types of carriers have been revealed, including MOFs, LDHs, SiO_2, metal oxides, MXenes, and the carbon-based nanoparticles family.

In response to environmental stimuli, these carriers release the incorporated inhibitor from their structure over time, and as a result, the performance remains acceptable for a remarkable time period. But among all the applied carriers, the carbon nano-family has been the focus of a wide range of research. In fact, carbonaceous nanomaterials, including graphene and its derivatives, CNT, etc., have been considered as potential carriers with barrier capabilities thanks to their high surface-to-volume ratio. As a result, this class of carriers can be utilized as multi-functional fillers with the purpose of improving the barrier function of polymer coatings in addition to inducing self-healing ability.

10.2 GRAPHENE AND ITS COMPOSITES

Graphene is the pure form of carbon, which is thin and nearly transparent, has about an atom thickness, and is interestingly strong (about 100 times harder than steel). In this allotrope, the atoms of carbon are compactly stacked on an atomic scale of regular sp^2 with a hexagonal pattern. The unique property of the regularly arranged bonded carbon atoms is the superior conduction of both heat and electricity. Nowadays, graphene is a helpful material to promote many physical properties of composites and has affected the future of nanotechnology, electronics, and material sciences. Graphene is a thick layer of graphite with one atom thickness, which can be brought up as a large and flat polycyclic aromatic molecule, and has been at the center of attention since the first study in 2004. Each single carbon atom in the graphene molecular structure is exposed to chemical reactions from two sites (the graphene has a 2D structure). Thus, the carbon atoms that are at the edge of graphene sheets have particular chemical reactivity. Regarding the highest ratio of edgy carbon, graphene has special chemical reactivity compared with other similar materials (such as carbon nanotubes). In addition, the presence of different types of defects in the sheet structure commonly resulted in chemical reactivity enhancement. This material's surface volume ratio is high, and its mass density is low.

However, alongside the coating fabrication, graphene can be added as a filler into an organic coating structure to improve protection efficiency. The high surface-to-volume ratio of graphene can fill up the micro porosities, which were introduced into the polymeric matrix of the coating. As was mentioned in the past, by filling up the porosities, the way corrosive agents diffuse into the coating matrix was closed, resulting in the coating lifetime, adhesion bond strength, and protection efficiency increment. But the researchers have realized that the inter-plane forces of graphene prevent the proper dispersion of the plates within the polymer matrix. Alongside, due to a lack of active functional groups on the pure graphene, it is necessary to modify the graphene surface before utilization as a filler. Therefore, most researchers turned to using graphene oxide (GO) as a potential substitute for pristine graphene in most applications.

Surface oxidation causes the appearance of oxygen groups on the surface and can substantially control the interfacial forces to some extent. Thus, the graphene oxide (GO) in water and other solvents is dispersible as there is presented oxygen functionalities. There are various methods in manufacturing for GO generation, which consist of Staudenmaier, Hofmann, Brodie, and Hummers' methods. Due to the good chemical and physical properties of GO, some researchers utilized graphene and GO as corrosion repressors or as agents that could help the anti-corrosion properties of coatings. The uniform and stable dispersion of GO into the coating matrix is an important factor that can lead to an increase in coating protection efficiency. The large surface area of the GO leads to an enhancement of attraction toward a great tendency to agglomerate in the coating matrix. For resolving this problem, the researchers tendered the utilization of doping or the functionalized with functional groups methods.

Nevertheless, when we look at the lack of graphene oxide, we can see that graphene oxide can be used in water-processable environments only, and its inhomogeneity with the non-polar polymer phase is undeniable. As a result, again, the problem of unbalanced surface tension between polymer and graphene oxide causes GO plate agglomeration in the polymer matrix. Therefore, the return to the graphene state is sensitive to dispersion control. This state, known as reduction, removes oxygen groups from the surface and achieves a pure carbon sheet. As a result, the phase difference between polymer and filler disappears, and the performance improves. The graphene was modified with the help of silane coupling agents, in which the silane molecules were hydrolyzed, leading to the production of active Si-OH. The active Si-OH can react chemically with the active, edgy groups of graphene and graphene oxide, such as carboxyl and hydroxyl. The presented functional group in the silane molecules can create active sites for future coupling modification. It is good to know that if there was presented epoxy group in the silane coupling agent, the composites of graphene would have better compatibility with the epoxy resin and dispersibility. There are various studies that displayed the effect of reduced graphene oxide on the anti-corrosion, resistance of different coatings. In an interesting research, Lijuan Zhu et al. studied the corrosion behavior of epoxy coating, which was reinforced by reduced graphene oxide (RGO) in saline media [1]. The results showed that the adhesion, toughness, and corrosion resistance of the epoxy coating after reinforcement with reduced graphene oxide were enhanced. In addition, in another research, M. B. Hegde et al. investigated the effect of reduced graphene oxide and epoxidized linseed oil combination on the mild steel corrosion in neutral 3.5% saline solution [2]. The outcomes of morphological and electrochemical studies displayed the excellent protection efficiency of the coated specimen in salinity media.

Besides the good effect of rGO and/or GO on the anti-corrosion performance of the coating after the combination, there was one problem. Alongside their acceptable barrier action (high barrier properties), corrosion deteriorates performance after the creation of defects or scratches (if they have no active inhibition). However, since there are various oxygen-containing functional groups, such as hydroxyl on the GO, leading to an enhancement of the tendency to interact with organic compounds by covalent or non-covalent bonding. Therefore, they can play the role of an impressive platform for loading a high amount of organic corrosion inhibitors. There was a lot of researches that confirm the higher adsorption capacity of GO in comparison with common adsorbents. It can be attributed to the presence of a high amount of oxidized zone over the GO sheets, which can be a result of the electrostatic attraction and/or hydrogen bonding of GO with organic/inorganic corrosion inhibitors.

Nevertheless, nowadays, the incorporation of GO-modified corrosion inhibitors (such as green, metal cations, and hybrid inhibitors) and coating is widely used due to the enhancement of anti-corrosion efficiency, adhesion bond strength, and the coating lifetime improvement targets. The presence of corrosion inhibitors/GO plays the role of a barrier against the diffusion of corrosive agents to the coating defects or mechanical damage. Also, the rGO-modified corrosion inhibitor, due to its high adsorption properties, can play the role of a nano container (corrosion inhibitor), leading to the enhancement of coating lifetime and anti-corrosion efficiency (the presence of corrosion inhibitor leads to the active inhibition properties). Besides these points, the modification of rGO with corrosion inhibitors can lead to the self-healing ability of the protection coating, due to the presence of corrosion inhibitors in the defects or scratch area. Various researches demonstrate the influence of rGO/corrosion inhibitor/coating. Yangmin Wu et al. surveyed the effect of rGO-CeO_2 nanofiller on the corrosion protection of waterborne epoxy on mild steel [3]. The outcomes of EIS and salt spray confirm that after the usage of the synthesized nanofiller, the corrosion protection performance was enhanced, especially in long term. In another study, Ziwei Zhu et al. studied the influence ofther GO/titanium dioxide/polyvinyl butyral composite coating on the stainless steel corrosion in various corrosive media [4]. The results of their study displayed that the composite coating could protect steel effectively for over 430 days in 3.5% salinity (NaCl) solution. In one other research, Fei Zhong et al. investigated the effect of rGO/aspartic acid intercalated layer double hydroxide on the self-healing and barrier properties of epoxy coating [5]. The electrochemical outcomes of their study displayed that the impedance of epoxy coating was changed impressively after adding nanofiller into the epoxy coating matrix. In addition, the epoxy life (coating failure time) was enhanced to 40 days. In addition, A. Madhan Kumar et al. studied the corrosion protection efficiency of epoxy coating on mild steel after adding RGO/polyaniline and chitosan nanocomposites [6]. Their research outcome illustrated that even after 45 days of exposure to the chloride (NaCl) solution, the impedance value at the low frequency remained constant, demonstrating the higher corrosion protective performance of epoxy coating/nanocomposites.

In a step forward, researchers have turned to the hybridization of modified/reduced GO with metal ions. In fact, despite the active protection ability of GO-modified inhibitors, their negligible performance has caused metal ions to be doped on them. Dehghani and his colleagues revealed that the simultaneous release of metal ions and inhibitors remarkably improves the defective coating performance, and the coating efficiency is impressively acceptable in the long term [7–9]. The main focus of this research team is devoted to investigating the effect of zinc metal ion on graphene modified with plant inhibitors, and the results of their studies have shown that the plant inhibitor was often responsible for anodic protection, while the release of zinc ions caused corrosion control in the cathodic areas.

But with the passage of time, the criteria of self-healing have changed a lot. In fact, today, researchers are looking for the design of intelligent systems with controlled release capability to achieve long-term self-healing. For this reason, graphene and its derivatives are only used as coating fillers, and inhibitive carriers are introduced for achieving the controlled release function. Fortunately, until today, different carriers have been introduced to achieve this feature. MOFs can be mentioned among the most common ones. These particles with their organic-inorganic coordination have been used as an active inhibitor in corrosion studies. By looking closely at their structure, it can be seen that these structures are made of organic ligands and metallic clusters. Due to the sensitivity to pH changes, they are degraded over time, and by releasing organic and inorganic inhibitors from their structure, they cause active protection of the coating. Although in many cases, MOF has been used only to improve the self-healing of the coating [10–12], in many studies, researchers have decorated MOF on GO for designing advanced systems, and by including it in polymer systems, it has led to the emergence of a multi-function passive-active composite. As an example, Ramezanzadeh et al. surveyed the influence of MOF (ZIF-8) decorated GO on the epoxy coating anti-corrosion properties [13]. The outcomes displayed that the wet adhesion strength, cathodic delamination, corrosion resistance, and the inhibition efficiency were enhanced by about 63%, 70%, 70%, and 79%, respectively. In another study, Keshmiri et al. investigated the effect of lanthanide-based MOF (Ce-MOF) assembled GO in epoxy composite on the mild steel corrosion in 3.5% saline media [14]. The results demonstrated the impressive improvement in the corrosion resistance of the scratched sample in comparison with the epoxy-coated sample. Pan et al. investigated the effect of the growth of MOF (derived from layered double hydroxide) on the graphene to fabricate multifunctional epoxy composites [15]. The outcomes demonstrated that the electrical resistivity of epoxy-treated nanocomposite was acceptable, and the thermal conductivity of the treated sample was higher than pure epoxy coating.

Recently, researchers have switched to COFs by passing MOF systems. Unlike MOFs, which are the result of the interaction of organic particles with metal clusters, COFs are the result of a collision between organic particles. Like MOFs, they are sensitive to pH, and after the decomposition of the organic particles released by adsorption on the naked areas, they induce self-healing capability. Chunfeng Zhang et al. constructed a mesoporous nanocomposite (mesoporous silica/2D-COF) and combined it with epoxy coatings to improve the self-healing ability and anti-corrosion performance of protective coating [16]. The results displayed that even after 90 days of immersion in corrosive media, the anti-corrosion efficiency of the designed coating was still high. In addition, Tengfei Liu et al. investigated the effect of a constructed nanocomposite (benzotriazole corrosion inhibitor-loaded 2D-COF) on the epoxy coating protection efficiency [17]. The results showed that the impedance value of the defective sample (pure coating and coating designed with 2D COF) decreased over time, but for the sample that was protected by benzotriazole-loaded 2D-COF/epoxy, the impedance value increased over time, demonstrating self-healing performance. But in a series of more advanced studies, the hybridization of graphene with COF has been considered. Zhang et al. studied the protective performance of coating based on v-COF/GO nanocontainers [18]. They showed that the impedance value at low frequency, even after 28 days of exposure for the designed protective coating, was 1.3 times higher than the pure epoxy coating.

But the use of these complex structures has always been far from basic and so-called straightforward studies, and many researchers have been looking for easier methods to achieve self-healing capability. One of the methods is to use prefabricated carriers. One of the most notable carriers used is Beta-Cyclodextrin. Due to its significant sensitivity to pH changes and its different behavior at different pHs, this carrier can control the release of the inhibitor well and provide long-term inhibition. In fact, at low pH, this carrier dissolves in acid, and the encapsulated inhibitor enters the environment directly, while in the basic environment, repulsive forces are responsible for releasing the inhibitor and injecting it into the system. Dehghani et al. surveyed the influence of rGO, which was decorated by L-Histidine-loaded β-Cyclodextrin, on the anti-corrosion properties of epoxy on mild steel in saline environment [7]. The results of their investigation displayed that the corrosion attacks of epoxy coating were reduced by more than 50 percent, and the coating resistance was raised to higher than $10^4 M\Omega$. In addition, the outcomes affirmed that by the utilization of β-Cyclodextrin, the L-Histidine was released to the environment slowly, demonstrating the enhancement of active inhibition time.

10.2.1 Carbon nanotubes (CNT)

The elemental carbon in the SP^2 hybridization can produce various interesting forms, one of which is carbon nanotubes [19]. The carbon nanotube is another allotrope of carbon, which can be used a lot due to its structure: single, double, triple, and multi-walled. In 1991, carbon nanotubes were discovered and came to the forefront of industrial research. The tubular carbon molecules with significant thermal, electrical, and chemical properties lead to the high popularity of these materials. There are three different methods for CNT synthesis called fixed or fluidized bed catalytic, laser ablation, and arc-discharge [20]. The use of CNT alongside graphene is so common due to their super mechanical strength and high chemical and thermal stability. Therefore, CNTs drew significant attention in the field of corrosion. Due to these good properties of CNT, it can be utilized as a filler in coating matrix to improve the anti-corrosion efficiency, the adhesion bond strength, conductivity, thermal, and mechanical properties of organic coating. Nowadays, the utilization of these materials as a nano-carrier has expanded dramatically. Therefore, there are more researches that surveyed the effect of these nanocomposites as a filler or the filler-loaded corrosion inhibitor. The outcomes of the researches displayed the impact of electrical current repatriation on the protection of metal surface against the positional attacks. They can also fill the porosities, microcracks, and flaws to compact the coating layer, leading to the reduction of cracks in the coating structure. Therefore, they can induce the suitable barrier properties of coating. Nevertheless, it is good to know that the conventional CNTs had a strong orientation to aggregation due to their weak interfacial interaction with the polymeric matrix and high aspect ratio. The CNTs aggregation resulted in weak corrosion protection and low mechanical properties of coatings. The right solution to resolve this problem was the chemical or physical functionalization of CNTs, which leads to improving the interfacial interaction (in the form of chemical bonding or physical adsorption) of nanoparticles and polymeric matrix. The physical interaction and chemical bonding consisted of the polymer adsorption to nanoparticles and the chemical bonding of the CNTs surface with polymers, respectively. In addition, it was found that there are some mechanical methods that dominate the aggregation tendency of CNTs, such as high shear mixing and ultrasonic dispersion. These mechanical methods can help to disperse CNTs uniformly in a polymeric matrix. Various researches have displayed the utilization of CNT on the corrosion resistance and barrier properties of coating.

The utilization of organic modifiers is another way to solve this challenge. With the help of the covalent bonding of the functional groups (which were presented at the ends or midsection), the polymer chain can react with the nanoparticles. As an example for introducing the groups, which consisted of oxygen (like hydroxyl and carbonyl) on the surface of CNTs, we can use the strong acids. The results demonstrated the fact that the oxidized CNTs improved the interfacial interaction between particles and different polymeric matrices [21]. Therefore, the nanoparticle dispersion state and the crosslink density of the polymer structure were enhanced. The selection of appropriate organic molecules is the main factor that can lead to favored covalent functionalization to cause well reactivity of the matrix and suitable bonding to nanotubes [22]. Alongside these methods, the designing or cladding of CNTs with the help of inorganic particles is widely attention due to the significant effects on the CNTs dispersion and the combination of CNTS with high-performance clads.

With all the efforts made, the main problem of self-healing remains. In fact, like the challenge faced by researchers with graphene/GO, all the modifications made have been to improve the dispersibility of CNT, and the studies have demonstrated that none of them induce self-healing capability to the system. As a result, the path of studies conducted on CNT has been exactly the same as graphene/graphene oxide, and researchers have tried to induce active protection by spreading inhibitory substances on the surface of CNT.

In an easy idea, the modification of CNT with the help of hybrid organic polymers has been given much attention. Since most of these polymers have the ability to passivate surfaces, they can be used as a potential inhibitor. Ghahremani et al. surveyed the influence of novel nano-carrier based on carbon (oxidized MWCNTs/polydopamine/chitosan) for the epoxy coating on the mild steel corrosion in 3.5% chloride solution [23]. Their research results displayed that by introducing the novel carbon-based nano-carrier, the corrosion was retarded for almost nine weeks, and the total resistance of scratched epoxy after adding this nanocomposite was enhanced significantly (up to 3.6 times). Moreover, Cai et al. surveyed the polydopamine-wrapped CNTs for the enhancement of polyurethane coating anti-corrosion properties against Al alloys corrosion [24]. The experiments indicated that by introducing polydopamine@CNTs, the adhesion strength of polyurethane was enhanced, and corrosion resistance of polyurethane was improved even after seven cycles (504 h) of aging tests. In addition, the morphological outcomes demonstrated that the defects of PU coating were decreased, and in the long term, the dissemination of defects on the PU coating was prevented.

Undoubtedly, MOFs have found their place in these studies as well. The simultaneous carrying of organic and inorganic inhibitors is one of the capabilities of

MOFs, which, by decorating CNTs, can lead to a hybrid nanoparticle with both barrier-active protection capabilities. Also, Hu et al. investigated the synergistic influence of the functionalized CNT and GO on the anti-corrosion performance of epoxy coating in corrosive media (3.5% NaCl solution) [25]. The outcomes of their study displayed that the best anti-corrosion performance was achieved (the impedance value was largest) after the usage of the designed composite coating. In another research, Guangyi Cai et al. surveyed the effect of cerium oxide grafted CNT via polydopamine wrapping to enhance the anti-corrosion performance of polyurethane coating [26]. The results showed better function of the designed composite coating in corrosive media, and the coating resistance was acceptable even after 7 cycles.

But with the passage of time, researchers have gone beyond the use of MOF alone. In fact, MOFs themselves can act as inhibitor carriers due to their porous structure. In this way, the inhibitory ability of the MOF structure is doubled by encapsulating the inhibitor, and the self-healing performance is greatly increased. In a study, after the decoration of CNTs by ZF67MOF, molybdate ions were incorporated into the designed CNT-ZIF67 structure. Based on the presented results, despite the improvement of the coating performance after the release of MOF components, the release of molybdate ions has doubled the coating's self-healing [27].

10.2.2 Carbon dots

The carbon dots, including carbon nanodots, graphene quantum dots (GQDs), and polymer dots, are commonly called carbon dots and are abbreviated by C-dots or CDs. They are the carbon nano particles with less than 100 nm size, and they have some surface passivation. For the first time, they were discovered accidentally in 2004 by Xu et al. [28] in the CNT purification process. This detection was the main factor in the next extensive research and studies on the usage of the CDs' fluorescence properties. The CDs are known as a new class of graphene-based zero-dimensional materials, and they have recently been developed with a common size of 3 to 20 nm. These particles were synthesized by the breaking of large carbon structures, such as graphite or CNTs, with three various methods, including electrochemical, laser ablation, and arc discharge techniques, which are occupied by numerous functional groups at the edge of these particles, such as carbonyl, hydroxyl, and carboxyl. The presence of these functional groups leads to easier solubility in water and rather high light stability and low bio-toxicity. Alongside CDs chemical groups, the particle size is another important factor, which can affect the CDs' physical and chemical proficiency and functionalities, particularly when the size of particles is less than 10 nm. The size of the nanoparticles is so important, especially after the utilization of these nanoparticles as a corrosion protective coating. Alongside these points, it is good to know that various researches affirm the good influence of CDs as corrosion inhibitors for metal in different corrosive media [29–31].

Some researches confirm that the CDs had good potential for self-healing and initiating healing of polymers, and they can help to improve the coating's anti-corrosion performance. Various researches and investigations illustrated the influence of CDs on the anti-corrosion efficiency and properties of different coatings in corrosive media. Zhu et al. introduced CDs, which were synthesized initially from electrochemical etching of graphite as filler to the polymeric matrix of some coating, such as polymethacrylate and polyurethane [32]. The results displayed that the corrosion protection performance was enhanced for all nanocomposites in comparison with the pure coating.

But in addition to the use of carbon dots alone, in some studies, this particle has been used as a dispersion improver. With the hybridization of the particles, the dispersion improves, which increases the performance of the coating. Jiheng Ding et al. investigated the structure and performance of CDs functionalized Mxene/epoxy ultrathin anti-corrosion coating [33]. The impedance modules demonstrated more than four times improvement, even in the coating with about 25-μm thickness. Alongside this research, in another study, Ye et al. investigated the anti-corrosion resistance and self-healing behavior of functionalized CDs, which were intercalated with graphene-based waterborne epoxy coating [34]. They were providing that by introducing 0.5 wt.% of CDs-G to the epoxy coating, the premier protection efficiency was reached for the steel in corrosive media.

Like other carbon fillers, the tendency to carbon dot agglomeration in non-polar polymer matrices is very high, and for this reason, a high concentration must be incorporated in order to achieve high efficiency. In order to achieve proper dispersion, surface functionalization is the most straightforward but practical way. The researches displayed that the modified CDs had good effect on the metal protection mechanism due to the three points: (i) hydrophilic functional group over the CDs surface, leading to processability in aquatic environments (it can help to good filling of waterborne coating porosities motivated from solvent evaporation); (ii) amino CDs can play as corrosion inhibitor adsorbent layer, absorption over the metal substrate can generate monomolecular layer on the anodic and cathodic area of metal. By the adsorption over the anodic and/or cathodic area, the electrochemical reactions are prevented and the corrosion is suppressed; (iii) alongside, the CO–NH covalent bond between polymer chain and CDs, the hydrogen bond, and the CDs van der Waals forces lead to improvement of resin structure flexibility and self-healing.

However, the hydrophilic functional groups limit the CDs processability in the water-based media but are insoluble in the non-polar coating matrices. In order to skip this challenge and achieve the desired properties, element doping was proposed by researchers. By such a technique, the compatibility between the incorporated filler and host matrices is promoted, providing a more uniform composite with superior dispersion. For instance, Siming Ran et al. surveyed the influence of nitrogen-doped CDs on the epoxy coating anti-corrosion properties [35]. The results of the electrochemical analysis displayed that after utilization of 2.0 wt. % of N-CDs/EP is 364 times higher than pure EP, even after 800 h of immersion. The presence of lots of functional groups over the N-CDs surface can lead to the reduction of oxygen diffusion in the coating, resulting in the oxygen and metal surface connection being cut off.

But in addition to proper distribution, it should be noted that the functional group can tune the anti-corrosion performance of the carbon dot. In fact, like pristine CNT or graphene, carbon dot has no anti-corrosion ability, and it is activated by introducing active functional groups. Wang et al. studied the enhancement of anti-corrosion performance of waterborne epoxy with the help of nitrogen-doped CDs [36]. The outcomes affirm the preferable anti-corrosion performance and defect repairing after utilization of 0.5 wt% N-CDs in the waterborne epoxy.

In addition, in the smart coating structure, the CDs can play as a dopant, leading to coating some self-healing properties but not for the long term [32]. In a research, Zhu et al. surveyed the self-healing and anti-corrosion performance of carbon dots in various coatings, such as polymethacrylate (PMMA) and polyurethane (PU) [32]. The self-healing ability was evaluated by comparing of pristine and scratched coatings' tensile strength. The results displayed better function of both pristine and scratched samples of the modified coating by CDs. Alongside this critical result (self-healing ability), the modified coating displayed superior anti-corrosion performance compared to the pure coating.

Nevertheless, functionalized carbon dots by themselves have a very limited inhibition ability, and for long-term control, researchers have recommended inhibitors grafting. By releasing the inhibitor from the functionalized carbon dot surface, the inhibitor covers the surface completely, and an acceptable efficiency is obtained. Zhang et al. surveyed the resistivity of Mg alloys with the help of CDs and polydopamine composite coating [37]. The results showed that the coating anti-corrosion efficiency was enhanced after the utilization of N-CDs/PDA even by increasing the N-CDs particle size can lead to the enhancement of coating performance. Ramezanzadeh et al. surveyed the influence of polyaniline-designed graphene oxide quantum dots on the anti-corrosion performance of epoxy coating, which was applied over the steel surface [38]. The outcomes of salt spray and electrochemical impedance spectroscopy affirm the enhancement in the anti-corrosion performance of the designed novel composite coating in 3.5% salinity media (NaCl) [39]. Also, Saikia and Karak investigated the influence of epoxy nanocomposite with designed nanofiber (CNT, polyaniline)-CDs nano hybrid as anti-corrosive material on mild steel in 3.5% NaCl media [40]. The outcomes affirm the highest anti-corrosion performance of the designed coating. Alongside these researches, Ji et al. studied the experimental effect of epoxy/graphene quantum dots@TiO_2nano tubes coating [41]. The results showed that the electrochemical corrosion reaction was retarded, and the long-term corrosion resistance was enhanced.

10.2.3 Suggestions and future perspectives

According to the good chemical and physical properties, various applications can endure remarkable changes with carbonaceous materials. Although these particles do not have a high self-healing ability by themselves, this ability can be induced by modification. After passing through the first generations of self-healing coatings, the hybridization of carbonaceous materials with the help of advanced nanoparticles has received much attention.

From a simple perspective, the utilization of metal organic frameworks (MOFs) is a good idea to enhance anti-corrosion efficiency and self-healing ability of the coating. Or COFs can be widely applied as alternatives to MOFs. Although these porous particles have been considered in many studies, they are far from benchmark standards. Looking at their structure, it can be seen that in addition to their anti-corrosion ability, these particles can be used as carriers of inhibitors. By inserting the inhibitor in the structure gallery of these particles, the self-healing capability of the designed coating could be doubled. In the future, this series of studies can be viewed as a very interesting idea.

10.3 CONCLUSION

Nowadays, metal corrosion is one of the most important global problems, especially due to the expansion of metals in most industries. The ways to deal with this factor are so important. One way to prevent metal corrosion is the utilization of a coating that can play the role of a barrier against the establishment of direct connections of corrosive agents with the metal substrate to reduce metal corrosion. However, the presence of defects or porosities in the structure of the coating (lead by solvent evaporation or the application process) can reduce the protection efficiency and coating lifetime. Due to the good mechanical and chemical properties of carbon and carbon-based materials such as graphene, carbon nanotubes, and carbon dots, the utilization of these materials in various industries and research areas has expanded. So, the researchers were tendered the usage of carbon-based material as filler in the polymeric matrix to fill these porosities, but the problem was not solved completely. After the usage of these materials as filler, the porosities were filled, but the mechanical damage that was applied to the coating structure from the outcomes was another problem. The solution to solve this problem pushed the researchers to utilize another material alongside carbon-based material to generate the smart coating, which can fill the damage area with the help of an external agent (such as heat or the light shining) or the chemical changes in the damage area (e.g., the changes in pH value). In addition, there was another problem after the direct utilization of carbon-based material in polymeric matrix, which was their tendency to agglomerate, and hard dispersibility in the coating matrix. To eliminate these problems and due to the super adsorption properties of these materials, the researches offered the usage of a corrosion inhibitor to resolve both problems. On the other hand, some researches have suggested the usage of some nano materials, which can control the corrosion inhibitor release to the damaged area and play as role of a container, such as cyclodextrine. All of the research outcomes conclusively affirmed that the utilization of carbon-based materials alongside corrosion inhibitors or smart materials with sensitive responsivity against changes and/or control release capability with the coating can enhance the coating protection efficiency and lifetime.

REFERENCES

1. Zhu, L., C. Feng, and Y. Cao, Corrosion behavior of epoxy composite coatings reinforced with reduced graphene oxide nanosheets in the high salinity environments. *Applied Surface Science*, 2019. **493**: p. 889–896.
2. Hegde, M.B., et al., Reduced graphene oxide-epoxidized linseed oil nanocomposite: A highly efficient bio-based anti-corrosion coating material for mild steel. *Progress in Organic Coatings*, 2021. **159**: p. 106399.
3. Wu, Y., et al., Synthesizing a novel fluorinated reduced graphene oxide-CeO2 hybrid nanofiller to achieve highly corrosion protection for waterborne epoxy coatings. *Carbon*, 2021. **176**: p. 39–51.

4. Zhu, Z., et al., Corrosion resistance of polyvinyl butyral/reduced graphene oxide/titanium dioxide composite coatings for stainless steel in different environments. *Progress in Organic Coatings*, 2022. **173**: p. 107226.
5. Zhong, F., et al., One-step hydrothermal synthesis of reduced graphene oxide/aspartic acid intercalated layered double hydroxide for enhancing barrier and self-healing properties of epoxy coating. *Reactive and Functional Polymers*, 2019. **145**: p. 104380.
6. Kumar, A.M., J. Jose, and M.A. Hussein, Novel polyaniline/chitosan/reduced graphene oxide ternary nanocomposites: Feasible reinforcement in epoxy coatings on mild steel for corrosion protection. *Progress in Organic Coatings*, 2022. **163**: p. 106678.
7. Dehghani, A., et al., 2D reduced-graphene oxide (rGO) nanosheets decorated with l-histidine loaded-β-cyclodextrin for efficient epoxy nano-composite anti-corrosion properties; DFT-D modeling/experimental assessments. *FlatChem*, 2021. **30**: p. 100309.
8. Dehghani, A., et al., Electronic DFT-D modeling of L-citrulline molecules interactions with Beta-CD aligned rGO-APTES multi-functional nano-capsule for anti-corrosion application. *Journal of Molecular Liquids*, 2022. **354**: p. 118814.
9. Dehghani, A., G. Bahlakeh, and B. Ramezanzadeh, Designing a novel targeted-release nano-container based on the silanized graphene oxide decorated with cerium acetylacetonate loaded beta-cyclodextrin (β-CD-CeA-MGO) for epoxy anti-corrosion coating. *Chemical Engineering Journal*, 2020. **400**: p. 125860.
10. He, Z., et al., Self-healing epoxy composite coating based on polypyrrole@ MOF nanoparticles for the long-efficiency corrosion protection on steels. *Colloids and Surfaces A: Physicochemical and Engineering Aspects*, 2023. **657**: p. 130601.
11. Liu, T., et al., Self-healing and corrosion-sensing coatings based on pH-sensitive MOF-capped microcontainers for intelligent corrosion control. *Chemical Engineering Journal*, 2023. **454**: p. 140335.
12. Cao, K., Z. Yu, and D. Yin, Preparation of Ce-MOF@ TEOS to enhance the anti-corrosion properties of epoxy coatings. *Progress in Organic Coatings*, 2019. **135**: p. 613–621.
13. Ramezanzadeh, M., et al., Development of metal-organic framework (MOF) decorated graphene oxide nanoplatforms for anti-corrosion epoxy coatings. *Carbon*, 2020. **161**: p. 231–251.
14. Keshmiri, N., et al., Designing an eco-friendly lanthanide-based metal organic framework (MOF) assembled graphene-oxide with superior active anti-corrosion performance in epoxy composite. *Journal of Cleaner Production*, 2021. **319**: p. 128732.
15. Pan, Y.-T., et al., Interfacial growth of MOF-derived layered double hydroxide nanosheets on graphene slab towards fabrication of multifunctional epoxy nanocomposites. *Chemical Engineering Journal*, 2017. **330**: p. 1222–1231.
16. Zhang, C., et al., Controllable construction of mesoporous silica/2D-COF nanocomposites reinforced epoxy coatings with excellent self-repairing and long-lasting anticorrosion performances. *Progress in Organic Coatings*, 2023. **177**: p. 107441.
17. Liu, T., et al., Preparation of highly efficient self-healing anticorrosion epoxy coating by integration of benzotriazole corrosion inhibitor loaded 2D-COF. *Journal of Industrial and Engineering Chemistry*, 2021. **97**: p. 560–573.
18. Zhang, M., et al., Anti-corrosion coatings with active and passive protective performances based on v-COF/GO nanocontainers. *Progress in Organic Coatings*, 2021. **159**: p. 106415.
19. Khare, R., Carbon nanotube based composites-a review. *Journal of Minerals and Materials Characterization and Engineering*, 2005. **4**(01): p. 31.
20. Baddour, C.E. and C. Briens, Carbon nanotube synthesis: A review. *International Journal of Chemical Reactor Engineering*, 2005. **3**(1).

21. Ding, K., et al., Polyaniline and polyaniline-thiokol rubber composite coatings for the corrosion protection of mild steel. *Materials Chemistry and Physics*, 2002. **76**(2): p. 137–142.
22. Ezzeddine, A., et al., Surface modification of multiwalled carbon nanotubes with cationic conjugated polyelectrolytes: Fundamental interactions and intercalation into conductive poly (methyl methacrylate) composites. *ACS Applied Materials & Interfaces*, 2015. **7**(23): p. 12903–12913.
23. Ghahremani, P., et al., Rational design of a novel multi-functional carbon-based nano-carrier based on multi-walled-CNT-oxide/polydopamine/chitosan for epoxy composite with robust pH-sensitive active anti-corrosion properties. *Carbon*, 2022. **189**: p. 113–141.
24. Cai, G., et al., Polydopamine-wrapped carbon nanotubes to improve the corrosion barrier of polyurethane coating. *RSC Advances*, 2018. **8**(42): p. 23727–23741.
25. Hu, H., et al., Synergistic effect of functional carbon nanotubes and graphene oxide on the anti-corrosion performance of epoxy coating. *Polymers for Advanced Technologies*, 2017. **28**(6): p. 754–762.
26. Cai, G., et al., CeO2 grafted carbon nanotube via polydopamine wrapping to enhance corrosion barrier of polyurethane coating. *Corrosion Science*, 2021. **178**: p. 1090.
27. Dehghani, A., et al., Fabrication of an intelligent anti-corrosion silane film using a MoO42– loaded Micro/mesoporous ZIF67-MOF/multi-walled-CNT/APTES core-shell nano-container. *Colloids and Surfaces A: Physicochemical and Engineering Aspects*, 2023. **656**: p. 130511.
28. Xu, X., et al., Electrophoretic analysis and purification of fluorescent single-walled carbon nanotube fragments. *Journal of the American Chemical Society*, 2004. **126**(40): p. 12736–12737.
29. Cui, M., et al., Carbon dots as new eco-friendly and effective corrosion inhibitor. *Journal of Alloys and Compounds*, 2017. **726**: p. 680–692.
30. Ye, Y., et al., A feasible method to improve the protection ability of metal by functionalized carbon dots as environment-friendly corrosion inhibitor. *Journal of Cleaner Production*, 2020. **264**: p. 121682.
31. Ye, Y., et al., An effective corrosion inhibitor of N doped carbon dots for Q235 steel in 1 M HCl solution. *Journal of Alloys and Compounds*, 2020. **815**: p. 152338.
32. Zhu, C., et al., Carbon dots as fillers inducing healing/self-healing and anticorrosion properties in polymers. *Advanced Materials*, 2017. **29**(32): p. 1701399.
33. Ding, J., H. Zhao, and H. Yu, Structure and performance insights in carbon dots-functionalized MXene-epoxy ultrathin anticorrosion coatings. *Chemical Engineering Journal*, 2022. **430**: p. 132838.
34. Ye, Y., et al., Study on self-healing and corrosion resistance behaviors of functionalized carbon dot-intercalated graphene-based waterborne epoxy coating. *Journal of Materials Science & Technology*, 2021. **67**: p. 226–236.
35. Ren, S., et al., Effect of nitrogen-doped carbon dots on the anticorrosion properties of waterborne epoxy coatings. *Surface Topography: Metrology and Properties*, 2018. **6**(2): p. 024003.
36. Wang, J., et al., Novel nitrogen doped carbon dots enhancing the anticorrosive performance of waterborne epoxy coatings. *Nanoscale Advances*, 2019. **1**(9): p. 3443–3451.
37. Zhang, H., et al., Corrosion protection investigations of carbon dots and polydopamine composite coating on magnesium alloy. *Journal of Magnesium and Alloys*, 2022. **10**(5): p. 1358–1367.
38. Ramezanzadeh, B., et al., Synthesis and characterization of polyaniline tailored graphene oxide quantum dot as an advance and highly crystalline carbon-based luminescent nanomaterial for fabrication of an effective anti-corrosion epoxy system on mild steel. *Journal of the Taiwan Institute of Chemical Engineers*, 2019. **95**: p. 369–382.

39. Pourhashem, S., et al., Corrosion protection properties of novel epoxy nanocomposite coatings containing silane functionalized graphene quantum dots. *Journal of Alloys and Compounds*, 2018. **731**: p. 1112–1118.
40. Saikia, A. and N. Karak, Fabrication of renewable resource based hyperbranched epoxy nanocomposites with MWCNT-polyaniline nanofiber-carbon dot nanohybrid as tough anticorrosive materials. *eXPRESS Polymer Letters*, 2019. **13**(11).
41. Ji, X., et al., Experimental and DFT studies on corrosion protection performance of epoxy/graphene quantum dots@ TiO2 nanotubes coatings. *Construction and Building Materials*, 2022. **322**: p. 126501.

11 Functionalized carbon dots as self-healing anti-corrosive materials

Swati Sahu and Sanjay Tiwari

11.1 INTRODUCTION

Numerous sectors are in danger from metal corrosion, which is silent but just as catastrophic compared to earthquakes, tsunamis, and other natural calamities and results in substantial financial damages every year [1]. Statistics show that the financial losses brought on by metal corrosion are around six times greater than those brought on by natural calamities [2]. The protection of metals against corrosion has consequently been a critical issue that has to be solved in the research and engineering sectors. In addition, the safety issues for facilities and equipment produced by metal corrosion may involve a threat to people's lives. The best method for preventing corrosion on metal surfaces is to apply organic coatings [3–5].

The engineering materials (structural and functional) from which a wide range of modern products are made degrade with time, reducing their lifespan and eventually causing the product to fail completely. Due to efficiency or financial constraints, many future materials need an extended operational period and cannot be changed regularly. Material researchers and engineers were inspired by nature's capacity to self-heal localized damage, often to the point of total restoration, and created "self-healing" materials, in which the damage (such as bond breakage or fracture development) triggers a healing reaction [6–8]. Self-healing ought to occur naturally and without assistance from humans.

Self-healing man-made materials that can repair defects on their own or with the aid of external stimuli have garnered a lot of attention since the late 1970s when Wool first reported the temperature-dependent recovery processes of the deformed viscoelastic polymer [9–13]. The capacity for self-healing opens up a wide range of opportunities for the development of dependable and long-lasting materials, significantly extending their useful lives in real-world applications [14–16]. In recent years, a number of methods and tactics have been published for creating self-healing materials [17].

Several different (frequently expensive and complicated) concepts have been used to create self-healing materials. The repair could be permanent or reversible. The concept of microencapsulating a monomer and a catalyst to start restoration in the injured area was invented by White [18]. This can also be accomplished with nanoreservoirs, hollow fibers carrying healing fluid, and microvascular polymeric

DOI: 10.1201/9781003456025-11

composites [19–21]. Healing can be accomplished by the use of specifically constructed supermolecular systems that use either chemical or physical bonding, such as hydrogen or Diels-Alder bonds [22–28].

11.2 FUNDAMENTAL CONCEPTS OF CORROSION

In an electrochemical process known as corrosion, metallic surfaces of metals and alloys deteriorate by interacting with elements in the environment. The worst consequences of corrosion behavior include wet corrosion, which is electrochemical corrosion that occurs when moisture is present. Organic acids and hydrogen sulfide (H_2S) are the primary causes of corrosion in aqueous electrolytes, which are employed in a variety of industries, including the petroleum, gas, and oil sectors. Sodium chloride (NaCl), or seawater, is the primary electrolyte causing corrosion in the marine industries. Iron, copper, aluminum, and bronze are typically employed as the primary metal materials in various industries. In the steps of descaling, acid washing, oleation, and pickling metal, hydrogen fluoride, acids such as sulfuric acid, hydrochloric acid, and various other highly acidic chemicals are typically used as electrolytic media [29, 30]. The metal surface often degrades during the acidification process, changing into its sulfide, chloride, and oxide phases and leading to corrosion of the metal [31]. Personal safety and financial loss are both severely harmed by corrosion. To prevent corrosion, it is therefore vital to adopt and create cutting-edge technology. The technology that is most successful in preventing corrosion in metal materials is thought to be corrosion inhibitors. To prevent corrosion on metallic surfaces, several inorganic salts, including zinc salts, borates, chromates, phosphates, silicates, and nitrates, were first widely employed [32]. All of these inorganic salts interacted with the metal surface's cations and formed them there as a crystalline coating that shielded the metal from further corrosion. But important aspects, such as financial expenses, economic repercussions, and environmental consciousness, are taken into account when selecting a good anti-corrosion material (corrosion inhibitor) to preserve metallic materials from corrosion.

11.3 SELF-HEALING ANTI-CORROSIVE MATERIALS AND CORROSION INHIBITION

On metallic surfaces, self-healing anti-corrosion materials are often coated as protective sheets after interacting with the metal. The primary determinants of the adsorption mechanism of molecules that inhibit above the metal surface are the structural configuration and chemical characteristics of the molecule that inhibits [33]. Due to the toxicity of inorganic salts, research has concentrated on the discovery and synthesis of anti-corrosion compounds that are affordable, environmentally benign, and have minimal toxicity. As a result, research on extracts from plants [34], molecules of organic material, heterocyclic materials, polymers, composites, and nano materials is highly helpful. However, some of these substances are poisonous and environmentally hazardous [35–37]. Thus, researchers, academics, engineers, and chemists have concentrated their research efforts on carbon-bonding materials throughout the past few years. Due to their rigidity, versatility, affordability, lack of toxicity, and environmental friendliness, carbon dots (CDs) have been used as anti-corrosion compounds. Different CD-based materials, including nitrogen-doped CDs (N-CDs), ionic liquid CDs (IL-CDs), polymer-based CDs (poly-CDs), and SN-co-doped CDs (CDs), have mostly been employed as anti-corrosion material in past studies [38–40].

In order to shield metallic surfaces from corrosive assault, anti-corrosive materials, also known as corrosion inhibitors, are modest doses of external materials that are incorporated into aggressive media throughout descaling, and acid for the purpose of pickling as well as oil-well refining processes [41]. Over the past few decades, different kinds of corrosion inhibitors have been utilized extensively. The feature adsorption and inhibitor processes of some of the most important corrosion inhibitors are discussed here. Acyclic, cyclic, aliphatic, and aromatic carbons are the major components of organic molecules. Heterocyclic compounds are organic molecules that contain heteroatoms like—P,—S,—N, and—O. But among the most well known and potent corrosion inhibitors, organic chemicals are thought to be among the best. Molecules are suitable as corrosion inhibitors for a variety of reasons, including their chemical reactivity and structural orientation as organic substances. As self-healing anti-corrosion materials, organic compounds with functional groups like—OH,—OR,—NH_2,—NR_2,—NHR,—SR,—SH,—CN,—N=N, and—COOH,—CO,—CHO,—$CONH_2$, have been widely utilized [42–45]. Several heteroatoms, including N, O, P, and S, as well as several electron-rich bonds, such as alkenes, alkyls, and aromatic rings, are present in these functional groups. When interacting with metal cations to create stable bonds, organic molecules, particularly those with electron-rich centers, exhibit nucleophilic behavior. According to earlier research, either physisorption or chemisorption mechanism is responsible for the adsorption of inhibitor molecules on the metal surface. Through an electrostatic bond, protonated, charged organic molecules adhere to the metallic surface in the physisorption mechanism. Anions like SO_4^{2-}, Cl^-, PO_4^{3-}, and NO_3^-then interact with the metal surface. The inhibitor molecules on the metallic surface do not entirely adsorb during the physisorption process; instead, they interact with the anions already present there in the protonated form [46]. In contrast, the metallic surface and inhibitor molecules directly interact during chemisorption. The organic inhibitor molecules behave as electron-rich nucleophiles during this interaction of metal ions with functional groups and pi electrons, whereas the metal surface functions as an electron acceptor due to its unoccupied 3d orbitals [47, 48]. Due to strong connections, the inhibitor molecules build up on top of the metal surface, blocking all of the metal's active sites and defending it against further attack by an aggressive electrolyte.

11.4 FUNCTIONALIZED CARBON DOTS ARE EFFECTIVE GREEN SELF-HEALING ANTI-CORROSIVE MATERIALS

Recently, there has been an increase in the usage of carbon based materials as corrosion inhibitors, including carbon quantum dots (CDs), carbon nanotubes (CNTs), carbon nanorods (CNRs), and graphene. Due to the significant variations among carbon dots obtained using various techniques, there hasn't been a consistent regulation on the name and classification of carbon dots. Graphene quantum dots (GQDs), carbon quantum dots (CQDs), carbon nanodots (CNDs), and carbonized polymer

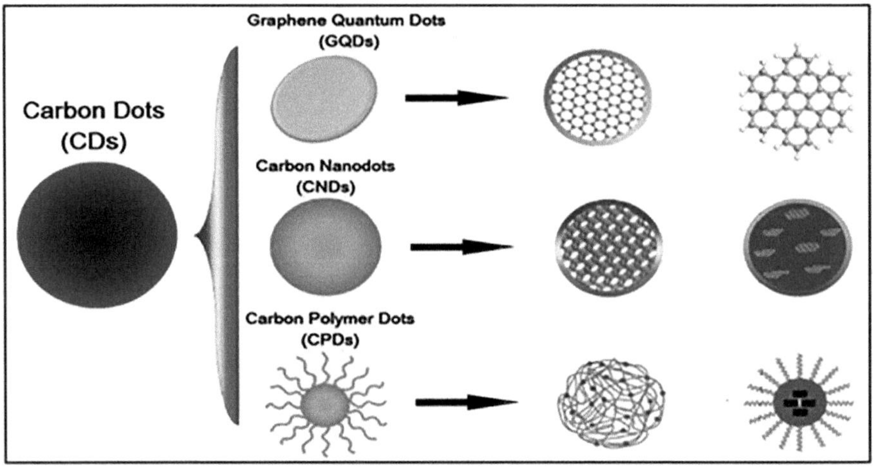

FIGURE 11.1 The three classes of carbon dots.

dots (CPDs) are the four categories of carbon dots based on the differences in their carbon cores (Figure 11.1). Additionally, CDs are nanoparticles made of carbon that have particular and distinctive properties that are employed for anti-corrosion purposes, as well as many other applications.

These techniques are used to create CDs, including polymerization, oxidation, chemical reaction, pyrolysis, and carbonization. CDs that include functional groups that are oxygen- or nitrogen-doped act very well as semiconductors. CD nanoparticles are mostly used in water contamination treatment, ion detection, capacitors, photocatalysis membrane manufacturing, heavy metal identification, and imaging of biology because of their semiconducting characteristics. Additionally, CDs are luminous materials because of the extreme quantum confinement inherent in their optoelectric and photoluminescent properties. They have chemical inertness, great thermal stability, and strong luminescent and photoluminescent capabilities as a result of their optical stability and quantum confinement characteristics [49–51]. The emissive characteristics of the raw materials' carbon-based precursors have an impact on CDs' luminous characteristics as well. Numerous carbon precursors that are widely used for the production of CDs can be found in raw materials such as carbohydrates, citric acids, and sugar. Furthermore, CDs are carbon-based nanomaterials, and as a result of their characteristics, they may be used in photocatalysis, nanosciences, photoelectrochromic technologies, detector manufacturing, and the use of nanotechnology [52–54].

Monodispersed particles of graphite with CDs that have a diameter of fewer than 10 nm. The easy and uniform dispersion of CDs in polymers is made possible by their small sizes and a wide range of alternative functional groups that may be added to their ends. Particularly, functional moieties comprising oxygen and amine that are connected to CDs promote a powerful contact (heterobonding and hydrogen bonding) with polymers. The healing/self-healing properties obtained by the incorporation of CDs into polymers are attributed to (I) the bonding of hydrogen between the surface groups that are functional of the CDs (such as amino (—NH_2), carboxyl (—COOH), and hydroxyl

(—OH) groups); (II) the covalent bonds of (C═O—NH) among CDs as well as the polymer chains; and (II) among adjacent CDs, Van der Waals bonding occurs. This innovative idea of adding terminated CDs to several popular polymers is straightforward, adaptable, and economical. The simultaneous existence of self-healing and anti-corrosive properties makes a variety of bulk polymers (such as those used as structural materials) and also numerous polymer protective coatings an attractive option [55].

Two novel classes of "green corrosion inhibitors," carbon dots (CDs) and imidazole, have drawn a lot of attention. Carbon dots have been applied to investigate the field of anti-corrosion due to their benefits of rich raw material availability, excellent water solubility, non-pollutive, cost-effectiveness, and easy synthesis [56, 57]. As an illustration, Cui et al. [58] synthesized a new CD from aminosalicylic acid and explored its inhibitory impact on steel in a hydrochloric acid condition. According to the outcomes, the strongest anti-corrosion efficacy was demonstrated by the fact that the steel corrosion rate in the 5 mg/LCDs solution was the lowest of all concentra tions. They then carried on with the synthesis of two different types of carbon dots using p- and o-phenylenediamine. According to the investigation, at a concentration of 200 mg/L, the corrosion-inhibition efficiency of p-CDs and o-CDs was 88% and 94%, respectively [59].Due to its extensive chemical characteristics, outstanding water solubility, and superior inhibitory action, imidazole has gained attention as a green, high-efficiency corrosion inhibitor [60–62]. For instance, Likhanova et al. [63] investigated how imidazole-containing ionic liquid affected mild steel's ability to be inhibited in an H_2SO_4 environment. When they tested steel in H_2SO_4 solution, they found that the imidazole compound, at 100 ppm, had a high corrosion effectiveness. In their investigation of the inhibitory effects of imidazole, benzimidazole, and its derivatives for copper in a solution containing 3 wt% sodium chloride, Milosev et al. [64] demonstrated that the copper current density for corrosionin the imidazole solution was significantly lower than that in the control solution.

Folic acid and o-phenylenediamine were used as precursors in the simple hydrothermal technique developed by Zhu et al. to synthesize N-doped CDs [65]. The produced CDs demonstrated good anodic iron dissolution inhibition performance on the Q235 steel surface in a solution of HCl and followed the Langmuir adsorption rule to produce a compact protective film. This protective film is effective in preventing both the reduction of cathodic hydrogen ions and anodic iron dissolution. In order to prevent the corrosion of carbon steel in a NaCl solution, Zeng improved CDs with melamine (Me-CDs), which were successful [66]. The physisorption with a barrier effect, the hydrophobic layer generation on the surface of Q235 steel, and the chemisorption at O and N adsorption centers were all cited as contributing factors to Me-CDs' ability to limit corrosion. Additionally, when CDs are exposed to a solution containing Fe^{3+} ions, Fe^{3+} will quench their fluorescence, enabling them to quickly and precisely probe Fe^{3+} ions [67]. CDs provide a fantastic possibility for the real-time visualization of Fe^{3+} formation during the electrochemical corrosion treatment of iron/steel materials based on the apparent variation in fluorescence intensity.

In order to address the issue of environmental contamination during the cleaning of metal, Yang et al. created green corrosion inhibitors with a size range of 3–5.5 nm using imidazole ionic liquids and carbon dots made of citric acid. By using electrochemical measurement, surface analysis, and weight loss, the inhibitory behaviors of this inhibitor for Q235 steel in 1 M HCl and 3.5 wt% NaCl solutions were investigated [35]. Liu et al. developed a variety of N, S-doped carbon dots (N, S-CDs) as an environmentally friendly corrosion inhibitor with various S/N ratios. These N, S-CDs were synthesized by precisely adjusting the ratio of L-cysteine and ammonium citrate using a hydrothermal process. The N, S-CDs inhibitor is a highly attractive replacement for conventional inhibitors because of its advantages, including excellent corrosion inhibition efficacy and low cost, which might greatly increase the lifespan of copper in corrosive conditions [68].

In order to develop coatings made of functionalized graphene/epoxy (CDs-G/EP), Yuwei et al. demonstrated that functionalized carbon dots (CDs) made from derivatives of citric acid were chosen as intercalators for transforming graphene. They also underwent in-depth analysis of their microstructure, self-healing, and corrosion resistance behaviors. The results from the investigation of the change rules of electrochemistry indicated that the addition of 0.5 wt% of CDs-G in EP coating (CDs-G0.5%/EP) displayed a higher protective behavior on steel. This is due to the chemical barrier created by highly dispersed graphene as well as the capacity of CDs to self-heal [69]. Ye et al. illustrated how to conjugate imidazole and citric acid carbon dots (CA-CDs) to yield a green and efficient inhibitor of the corrosion of functionalized carbon dots (FCDs). By using electrochemical analysis, corrosion morphological characteristics, and adsorption isotherms, a comprehensive investigation of the corrosion inhibition performance of FCDs for Q235 steel in a solution of 1 M hydrogen chloride was performed (Figure 11.2). The electrochemical results suggested that the FCDs inhibitor,

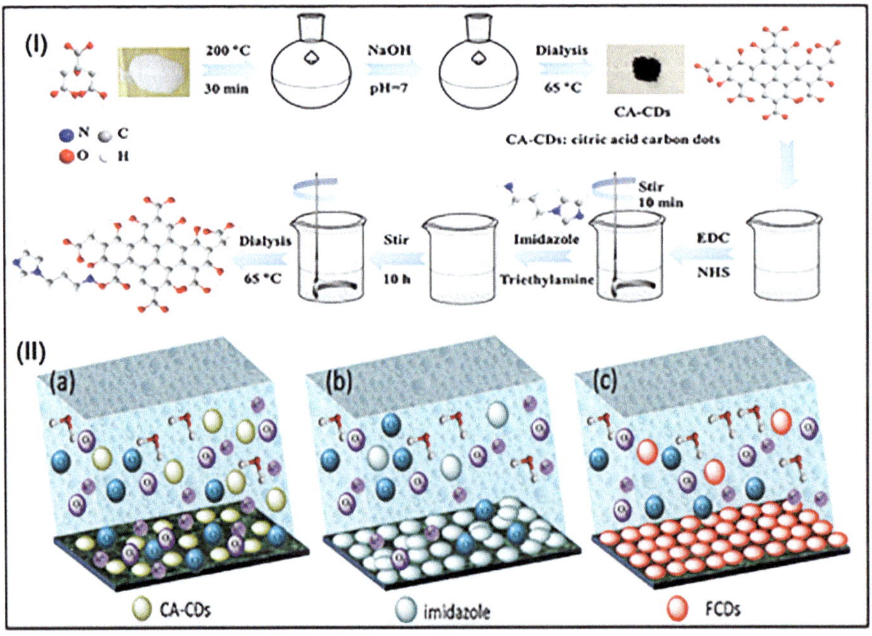

FIGURE 11.2 (I) schematic for FCD preparation, (II) various test solutions' protective mechanisms for Q235 steel: (a) CA-CDs, (b) imidazole, and (c) FCDs [70].

in its as-prepared state, could successfully prevent Q235 steel from corroding in a solution containing 1 M HCl. When the inhibitor concentration hit 100 mg/L, the inhibition efficiency of steel in 1 M HCl solution increased to more than 90% [70].

11.5 SELF-HEALING TECHNIQUES FOR FUNCTIONALIZED CARBON DOTS

Intrinsic and extrinsic self-healing are the two categories of self-healing design techniques that are currently available. The pre-embedded therapeutic compounds aid in the extrinsic self-healing's potential for recuperation. The main forces behind intrinsic self-healing strategies are polymer physical and chemical interactions. The inherent self-healing processes basically fall into two subcategories: reversible covalent bonds versus reversible noncovalent bonds. Extrinsic self-healing approaches (microcapsule or microvascular) serve as microreservoirs of healing chemicals to replenish surface defects at material interfaces. Additionally, intrinsic healing surface approaches are utilized to fix surface faults or create new interfaces by utilizing the fundamental processes of molecular mobility, redirection, and rearrangement. Recent years have seen a lot of study on intrinsic and extrinsic self-healing materials (Table 11.1).

11.5.1 Extrinsic healing

11.5.1.1 Self-healing using micro/nanocapsules

The most popular extrinsic technique, which was first put forth by White et al., uses the therapeutic chemicals contained in microcapsules [79]. When a self-healing surface material sustains damage, the microcapsules act as a micro-reservoir of healing

TABLE 11.1
Major Self-Healing Techniques

Technique	Self-Healing Anti-corrosive Material	Efficiency (%)	References
Microcapsule technique	Sodium alginate core with polyaniline shell	90	71
Microvascular technique	Alpha-aliphatic amidoamine with epoxy resin	100	72
Hybrid network technique	Dissolved epoxy	100	73
Hydrogen bonds technique	Polymethacrylate and Polyacrylate	92	74
pi-pi interactions technique	Polyimide and polyamide	95	75
Ion interactions technique	Polyelectrolytes hydrogel	95	76
Coordination bonds technique	Poly(dimethylsiloxane)	90	13
Diels-Alder (DA) reaction technique	Segmented polyurethane	94	77
Disulfide exchange reaction technique	Polyurethane	87	78

agents and feed the surface flaws. The following are the primary mechanisms by which the microencapsulated healing agent functions. Due to the spread of the crack, the microcapsules first burst, and then the healing agents are released and flow into the surface flaws of the matrix via capillary action to cause polyreaction. These two factors typically work together to repair surface flaws in the matrix, and the repaired matrix keeps its structural and functional integrity [80, 81]. In a nutshell, the healing chemistries, the suitable design of the microcapsule structure, and chemical compositions are the major drivers of the self-healing system based on microcapsules.

In situ polymerization, interfacial polymerization process, pickering emulsion templating, mini emulsion polymerization, as well as sol-gel processes are some of the more established ways for creating microcapsules in the last 10 years. Additionally, the development and production of self-healing microcapsule materials have been aided by an in-depth comprehension of chemical healing mechanisms. The core/shell substance, capsule dimensions, capsule preparation variables, the physical-chemical reaction among the matrix as well as the microcapsules, and the optimization to achieve the ideal healing function for the microcapsules are generally what determine whether healing properties based on microcapsules are appropriate [79]. Ma et al. suggested using magnetically targeted delivery microcapsules (MDTMs) based on tung oil to increase the effectiveness of self-healing coatings (SHC). In the development of microcapsules, magnetic particles served as emulsifiers and functional particles. The findings show that when the coating is mechanically damaged, magnetically driven SHC (MD-SHC) boosts the release rate of tung oil and reduces the migration path of the liberated tung oil by 50%. The healing component has excellent anti-corrosion performance; scratches were quickly coated with a significant volume of tung oil, and at a period of 24 hours, the spilled tung oil had completely established an oil layer with better polarization resistance on the scratches (Figure 11.3) [82].

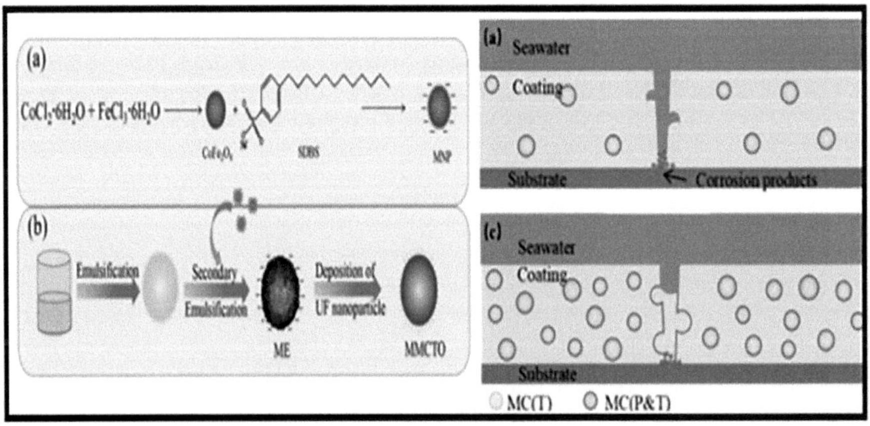

FIGURE 11.3 Schematic representation of the manufacturing of nano CoFe2O4 particles and magnetic microcapsules based on tung oil (MMCTO) [82].

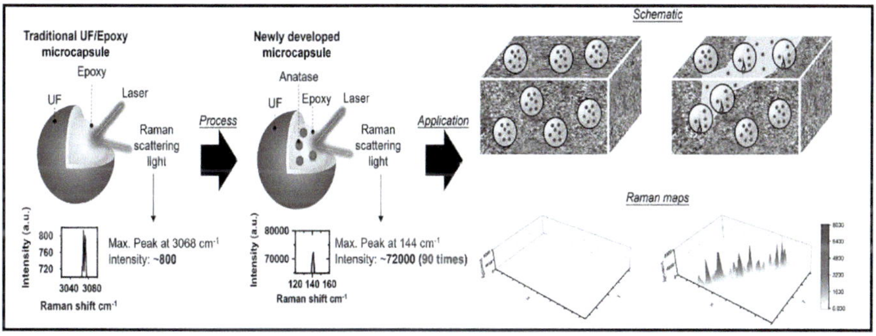

FIGURE 11.4 Schematic representation of Raman spectroscopy for monitoring microcapsule recovery in a self-healing cementitious system [83].

Raman spectroscopy was utilized by Mi et al. to monitor the development of cementitious paste repair. Additionally, a novel microcapsule with anatase incorporated in the microcapsule core that amplifies the electrical signal by around a hundred times has been developed in order to increase detection precision and effectiveness. The findings unmistakably demonstrated that Raman spectroscopy could detect the fracture of microcapsules as a result of cracking, offering a viable method for observing the functioning of self-healing microcapsules in cement-like materials (Figure 11.4) [83].

11.5.1.2 Self-healing surface using microvascular network

A microvascular structure's higher storage volume allows it to accommodate more healing agents than a self-healing technique based on microcapsules. Currently, the artificial vascular-based self-healing system has been successfully used to repair material surface flaws. In particular, the large-scale flaws on material surfaces have been successfully repaired utilizing the vascular system [19]. In order to create a network of nanofibers, Vintila et al. used coaxial electrospinning to embed the anti-inflammatory dicyclopentadiene (DCPD) in fibers made of polyacrylonitrile (PAN). The application of nano-fibers packed with a healing agent has no impact on the mechanical qualities of the matrix, unlike other encapsulation techniques, and allows for a greater healing region. Flexural testing showed that the epoxy resin had recovered 84% of its flexural strength after 48 hours, compared to 93% for the composite material (Figure 11.5) [84].

11.5.2 Intrinsic Healing

In contrast to extrinsic healing, intrinsic healing is a separate process. Embedded healing agents based on microcapsule and/or microvascular systems are frequently used in extrinsic healing. However, in intrinsic healing, damages are repaired by utilizing reversible chemical reactions powered by molecular reactions. The ability to repair injury repeatedly without the use of outside healing

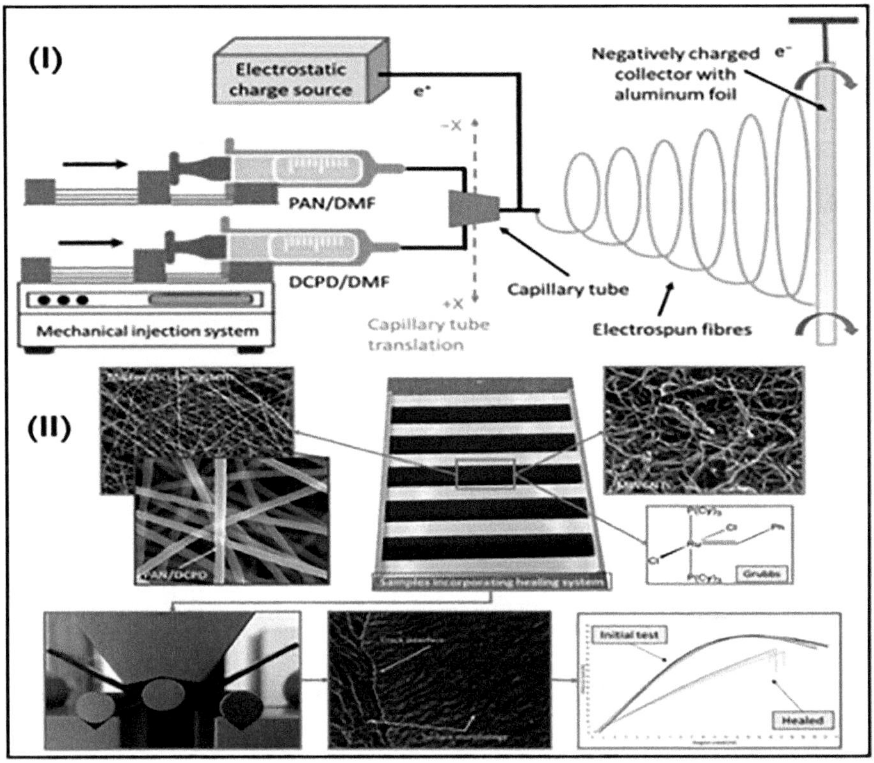

FIGURE 11.5 (I) Diagram showing the electrospinning method used to create the self-healing microvascular network and (II) using polymeric composite materials to promote microvascular system self-healing [84].

agents is a key benefit of intrinsic healing [85–87]. Reversible covalent bonds and reversible noncovalent bonds self-healing are two categories under which intrinsic healing can be classified in terms of the type of chemical reaction. Briefly stated, molecular self-migration and self-reorientation at a material's surface encourage the production of new chemical bonds, which lead to the development of new self-healing interfaces.

Carbon dots made from various aldehyde precursors reacted with branched polyethylenimine to develop multicolor, fluorescent self-healing gels, according to Bhattacharya et al. The primary amine residues in the polyethylenimine network and the aldehyde units visible on the surface of the carbon dots underwent a Schiff base reaction to produce imine linkages, which then resulted in the development of the self-healing gels. The gels have exceptional self-healing capabilities as well as high mechanical strength, thanks to the dynamic covalent imine connections between the carbon spots and polymeric matrix. Additionally, by carefully adjusting the carbon dot to polymer ratio, the viscoelastic characteristics of the gels may be precisely altered (Figure 11.6) [88].

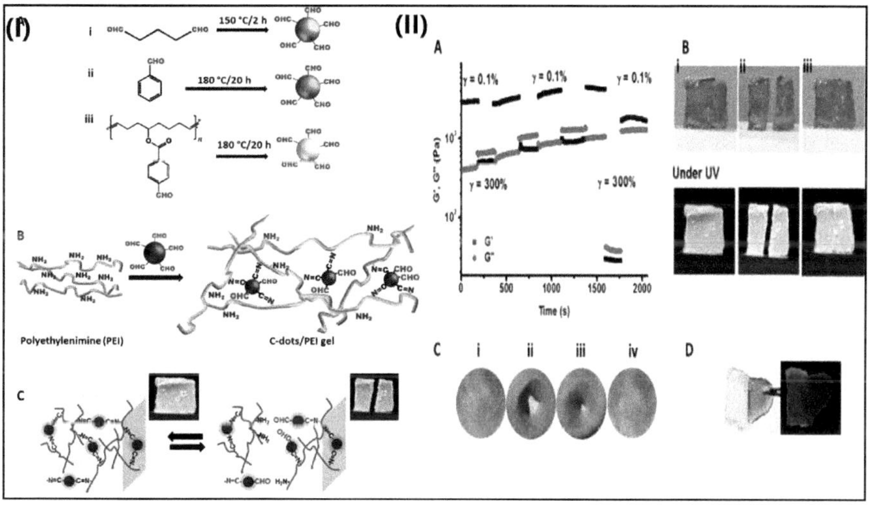

FIGURE 11.6 (I) Aldehyde-C-dot and C-dot/PEI gel synthesis, (II) C-dot/PEI gels' ability to self-heal [88].

11.5.2.1 Reversible noncovalent bond-based self-healing

Self-healing can happen because polymer chains can move and are flexible. Damage healing is enhanced by the formation of reversible noncovalent interactions among functional groups at surface defects [80]. Hydrogen bonds, π-π stacking interactions, ionic interactions, and ligand coordination are the four types of reversible noncovalent bonding. Due to their affinity and directionality, hydrogen bonds can gain significant mechanical characteristics in hydrogen bonds. Recently, significant advancements in the creation of self-healing materials based on hydrogen bonds have been made. The "π-π stacking interactions" between molecules with and without extra electrons are referred to as weak interactions. Utilizing supramolecular self-assembly as well as the reversibility of stacking interactions, a self-healing function can be created. In ion Interactions, the self-healing function induced by ion interactions has been proven to be efficient and reasonable for materials containing diametrically opposed ions or polymer chains [89]. The strength of self-healing is greatly influenced by critical variables such as electrostatic charge density, dielectric constant value, and temperature. In general, the key components of ion interaction-based self-healing materials include ionomers, polyelectrolytes, and zwitterionic polymers. The interaction of molecular ligands with metals or metal ions to generate coordination bonds, a sort of unique noncovalent link, has been defined in coordination bonds. Coordination bonds at the interface of diverse materials have received a lot of attention in order to produce self-healing surfaces [90]. The primary areas of interest in studies of self-healing materials based on coordination bonds are the molecular structural design and its interaction with other covalent bonds at surfaces.

According to Tang et al., exchangeable bonds can be engineered into the rubber-carbon nanodot (CD) interphase, utilizing CD as a high-functionality cross-linker to

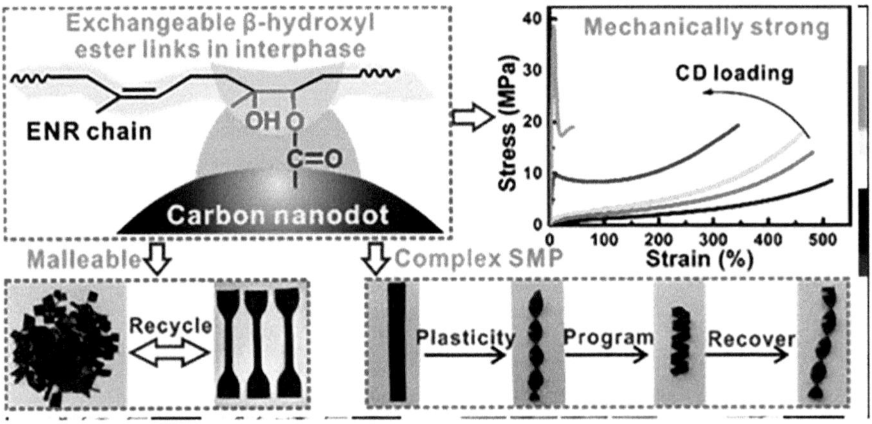

FIGURE 11.7 Illustration of adaptive elastomers with interfacial exchangeable bonds [91].

create covalently cross-linked yet recyclable, strong, and macroscopically sensitive elastomer vitrimers. The ability to recycle, reshape, and weld materials is provided by the cross-linked rubbers' ability to change the network topology through trans-esterification processes in the interphase. The mechanical characteristics are significantly improved as a result of the extensively stretched and preferentially ruptured relatively small chains connecting adjacent CDs. Additionally, the samples are able to utilize reconfigurable/multiple-shape memory effects owing to their malleable and welding capabilities (Figure 11.7) [91]. Tannic acid (TA) and graphene (Gr) were shown to self-assemble when there were π-π stacking interactions, according to Li et al. In order to enable the release of Ce^{3+} in an acidic environment, green corrosion inhibitor cerium (III) cation (Ce^{3+}) was subsequently deposited onto graphene sheets using a reversible combination with tannic acid that is pH-dependent. In comparison to bare steel, Gr-TA-Ce^{3+}nanofillers raised the impedance modulus at 0.01 Hz by an order of magnitude. This was demonstrated by an analysis of bare steel in the Gr-TA-Ce^{3+} extract. Due to the shielding ability of graphene nanosheets toward corrosive fluids and the effective self-healing capacity provided by the protective film that cerium cations form, waterborne coatings made of epoxy doped with Gr-TA-Ce^{3+} material demonstrated significant barrier effectiveness [92].

11.5.2.2 Reversible covalent bonds-based self-healing

Dynamic covalent bonds have a higher bond energy than noncovalent bonds, allowing self-healing polymers to acquire superior mechanical properties. The Diels-Alder (DA) and disulfide exchange reaction at the interface of materials are the key reversible reactions used in self-healing surfaces. DA process is a type of dynamically reversible chemical process that depends on changes in temperature and is advantageous for its moderate reaction conditions, lack of side effects, and lack of catalyst stimuli. Currently, due to their low reaction energy, the furan and maleimide groups have been frequently chosen to be used as candidates to gain self-healing ability. In general, the DA response at the interface can be controlled by temperature changes, which further activate the self-healing surface. In the disulfide exchange reaction, disulfide bonds, which can be added to crosslinked elastomers to enable the surface of the material to self-heal, are weakly covalent bonds with dynamic, reversible properties. The approach has the benefits of being inexpensive and straightforward to use. More opportunities exist to create self-healing surfaces based on disulfide bond exchange reactions, thanks to disulfide bonds, which are abundant in polymers.

Future smart materials have been rationally designed through the formation of a framework for dynamic chemistry, which gives materials dynamic function. There are numerous methods for creating dynamic polymers and materials that can adapt, react, repair, and recycle thanks to the advancement of supramolecular and dynamic covalent chemistry. With their reversible polymer formation mediated by dynamic covalent bonds composed of disulfide, the building blocks based on 1,2-dithiolanes have emerged as an essential scaffold within this toolbox. This allows for a special class of dynamic materials at the nexus of supramolecular polymerization, and able to adapt covalent networks [93]. Self-healing hydrogels have become increasingly popular in recent years due to their advantageous properties, which can increase lifespans and lower replacement costs. There have been several self-healing gels investigated up to this point using dissociation along with the recombination of either chemical or physical interactions [94]. In contrast to the chemical self-healing gels, which typically restore damaged networks through dynamic reversible covalent bonds such as imine bonds, disulfide bonds, acylhydrazone bonds, and the Diels-Alder reaction, physical self-healing gels can restore networks via such as hydrogen bonds, hydrophobic, and host-guest interactions. Additionally, these gels' morphological and mechanical characteristics are primarily used to evaluate their healing capacities. Scratched gels can self-heal using digital monitoring, optical microscopy, and scanning electron microscopy (SEM). This research provides a fresh viewpoint on creating different fluorescent assemblies through self-healing assembly, and it may encourage the use of self-healing gels in self-healing assemblies in the future.

11.6 IMPORTANCE OF A CORROSION INHIBITOR BASED ON CARBON DOTS

Through several studies, corrosion protection for metal is still being developed. Many researchers can use corrosion inhibitors because of how simple, inexpensive, and effective the synthesis process is. In various metals and corrosive conditions, a variety of compounds have been evaluated as corrosion inhibitors. The majority of them demonstrated extremely high inhibitory efficiency, making their use in the oil and gas industries quite promising. Natural materials are required, whereas chemical usage is restricted due to concerns for environmental safety. Now, carbon dots are being researched as corrosion inhibitors. Utilizing its qualities, it has already been applied to numerous applications; nevertheless, the industry is not aware of any direct applications of carbon dot-based corrosion inhibitors.

In order to safeguard the pipeline, corrosion inhibitors, especially those based on carbon dots, were primarily applied to the petroleum industry. As a basic material for pipelines, carbon steel is typically employed. Carbon steel is thus employed as a corrosion-resistant material. Aluminum and copper are cleaned with acid solutions or pickled to prevent corrosion, while the opposite is true for other metals. In order to stop further corrosion reactions, an acid solution is combined with corrosion inhibitors in this method. Along with molecular structure, CD-based corrosion inhibitors also offer data on the impact of material size on inhibitory activity. While this is happening, it is necessary to address a number of shortcomings in carbon dot-based corrosion inhibitors in order to produce, for instance, better corrosion inhibitors.

- The aqueous solubility of several corrosion inhibitors based on carbon dots is low. They have to be dissolved in an organic solvent as a result.
- Typical inhibitors are generally made up of small compounds. These inhibitors are capable of rapidly diffusing and interacting with the metal matrix on the metal surface to form a shielding coating. In contrast, nanomaterial adsorption in aqueous solution was comparatively slower. Thus, it is possible to think about researching the impact of material size on the adsorption rate.

11.7 CONCLUSION AND POTENTIAL OPPORTUNITIES

CDs were recommended as the newest generation of carbon materials because of their extraordinary qualities, which make them perfect for numerous technological applications. However, there are some limitations to concentrate on for our bright future. A thorough understanding of Photoluminescence (PL) CD's characteristics is needed for the current research. Numerous hypothesized mechanisms have been considered, including the impact of size, surface modification, and doping with other substances. Understanding the fundamental PL operations in CDs is crucial since the methodologies generated from the manufactured CDs' optical characteristics can differ. The possible advantages of functionalized CDs include strong, efficient, smart, and robust sensors that work in any environmental situation. To enable greater applications in a variety of sensors, supercapacitors, solar cells, medicinal devices, and corrosion prevention, the focus should be on manufacturing methods currently unavailable for CDs in the step-by-step CDs synthesis. Additionally, studies on CDs must be relevant to applications in the fields of energy, bioanalysis, and the environment. Considering that CDs are utilized as the electrode material, lithium-ion batteries are still in their infancy.

The self-healing surface materials have advanced greatly in production and use, as we have noted in this analysis, but there are still certain difficulties to be resolved. Extrinsic and intrinsic recovery are two categories into which the mechanisms of self-healing can currently be divided. The consistent and highly successful restoration of embedded healing agents remains difficult for the extrinsic technique. Under normal operating conditions, intrinsic self-healing surface materials can achieve rapid healable performance. However, once external stimuli, such as heat, light, humidity, or other environmental factors, which induce molecular movement at the interface, are removed, the self-healing function will deteriorate, which is not advantageous for self-healing surface materials under harsh environments, thereby limiting their practical applications.

The development of shape-memory self-healing materials has advanced significantly. The choice of proper reversible interactions is essential in the development of shape-memory self-healing polymers. Molecular entanglement, the Diels-Alder process, hydrazide bonds, disulfide bonds, hydrogen bonds, ionic interactions, metal-ligand interactions, and hydrophobic interactions are examples of common reversible interactions. Additionally, the material may acquire new transition switches or permanent dots as a result of reversible interactions. Thus, the material can be given new shape-memory properties.

CONFLICT OF INTEREST

No conflicts of interest are disclosed by the authors.

REFERENCES

[1] Li, Guidong, Heng Fang, Yang Hu, Xudong Chen, Zhuangzhuang Chu, and Zhuohong Yang. "Construction of vinyl ester resins composite coatings via introducing silane-functionalized graphene oxide for enhancing comprehensive performance." *Composites Science and Technology* 228 (2022): 109670.

[2] Nagai, Tomoyasu, Masanori Hiratsuka, Ali Alanazi, Hideki Nakamori, and Kenji Hirakuri. "Anticorrosion of DLC coating in acid solutions." *Applied Surface Science* 552 (2021): 149373.

[3] Yan, Hong, Jinxin Li, Meng Zhang, Yan Zhao, Yu Feng, and Yu Zhang. "Enhanced corrosion resistance and adhesion of epoxy coating by two-dimensional graphite-like g-C3N4 nanosheets." *Journal of Colloid and Interface Science* 579 (2020): 152–161.

[4] Luo, Xiaomin, Rong Qin, Jianyan Feng, Peng Zhang, Wenqi Wang, Huijun Ren, Hui Kong, and Wenjie Hu. "Robust, superhydrophobic, and self-healing films with hierarchical nacre mimetics: A high-efficiency and "green" route based on solvent-free polyurethane/modified graphene oxide composites." *Macromolecular Materials and Engineering* 307, no. 8 (2022): 2200127.

[5] Dehghani, Amirsalar, Mohammad Zabihi-Gargari, and Mahdi Tabatabaei Majd. "Development of a nanocomposite coating with anti-corrosion ability using graphene oxide nanoparticles modified by Echium ammonium extract." *Progress in Organic Coatings* 166 (2022): 106778.

[6] Wool, Richard P. "Self-healing materials: A review." *Soft Matter* 4, no. 3 (2008): 400–418.

[7] Syrett, Jay A., C. Remzi Becer, and David M. Haddleton. "Self-healing and self-mendable polymers." *Polymer Chemistry* 1, no. 7 (2010): 978–987.

[8] Kessler, M. R., Nancy R. Sottos, and S. R. White. "Self-healing structural composite materials." *Composites Part A: Applied Science and Manufacturing* 34, no. 8 (2003): 743–753.

[9] Wool, Richard P. "Material response and reversible cracks in viscoelastic polymers." *Polymer Engineering & Science* 18, no. 14 (1978): 1057–1061.

[10] Oh, Jin Young, Simon Rondeau-Gagné, Yu-Cheng Chiu, Alex Chortos, Franziska Lissel, Ging-Ji Nathan Wang, Bob C. Schroeder, et al. "Intrinsically stretchable and healable semiconducting polymer for organic transistors." *Nature* 539, no. 7629 (2016): 411–415.

[11] Huynh, T. P., P. Sonar, H. Haick. "Advanced materials for electronic skins: A review." *Advanced Materials* 29 (2017): 1604973.

[12] Cho, Soo Hyoun, Scott R. White, and Paul V. Braun. "Self-healing polymer coatings." *Advanced Materials* 21, no. 6 (2009): 645–649.
[13] Li, Cheng-Hui, Chao Wang, Christoph Keplinger, Jing-Lin Zuo, Lihua Jin, Yang Sun, Peng Zheng, et al. "A highly stretchable autonomous self-healing elastomer." *Nature Chemistry* 8, no. 6 (2016): 618–624.
[14] Zhang, Shiming, and Fabio Cicoira. "Water-enabled healing of conducting polymer films." *Advanced Materials* 29, no. 40 (2017): 1703098.
[15] Cao, Yue, Timothy G. Morrissey, Eric Acome, Sarah I. Allec, Bryan M. Wong, Christoph Keplinger, and Chao Wang. "A transparent, self-healing, highly stretchable ionic conductor." *Advanced Materials* 29, no. 10 (2017): 1605099.
[16] Tee, Benjamin C. K., Chao Wang, Ranulfo Allen, and Zhenan Bao. "An electrically and mechanically self-healing composite with pressure-and flexion-sensitive properties for electronic skin applications." *Nature Nanotechnology* 7, no. 12 (2012): 825–832.
[17] Cremaldi, Joseph C., and Bharat Bhushan. "Bioinspired self-healing materials: Lessons from nature." *Beilstein Journal of Nanotechnology* 9, no. 1 (2018): 907–935.
[18] White, Scott R., Nancy R. Sottos, Philippe H. Geubelle, Jeffrey S. Moore, Michael R. Kessler, S. R. Sriram, Eric N. Brown, and S. Viswanathan. "Autonomic healing of polymer composites." *Nature* 409, no. 6822 (2001): 794–797.
[19] Toohey, Kathleen S., Nancy R. Sottos, Jennifer A. Lewis, Jeffrey S. Moore, and Scott R. White. "Self-healing materials with microvascular networks." *Nature Materials* 6, no. 8 (2007): 581–585.
[20] Lanzara, G., Y. Yoon, H. Liu, S. Peng, and W. I. Lee. "Carbon nanotube reservoirs for self-healing materials." *Nanotechnology* 20, no. 33 (2009): 335704.
[21] Mauldin, Timothy C., and M. R. Kessler. "Self-healing polymers and composites." *International Materials Reviews* 55, no. 6 (2010): 317–346.
[22] Chen, Xiangxu, Matheus A. Dam, Kanji Ono, Ajit Mal, Hongbin Shen, Steven R. Nutt, Kevin Sheran, and Fred Wudl. "A thermally re-mendable cross-linked polymeric material." *Science* 295, no. 5560 (2002): 1698–1702.
[23] Cordier, Philippe, François Tournilhac, Corinne Soulié-Ziakovic, and Ludwik Leibler. "Self-healing and thermos reversible rubber from supramolecular assembly." *Nature* 451, no. 7181 (2008): 977–980.
[24] Wu, D. Y., S. Meure, and D. Solomon. "Microencapsulation of isocyanates for self-healing polymers." *Progress in Polymer Science* 33 (2008): 479–522.
[25] Banerjee, Sanjib, Ranjan Tripathy, David Cozzens, Tibor Nagy, Sandor Keki, Miklos Zsuga, and Rudolf Faust. "Photoinduced smart, self-healing polymer sealant for photovoltaics." *ACS Applied Materials & Interfaces* 7, no. 3 (2015): 2064–2072.
[26] Wilker, Jonathan J. "Sticky when wet." *Nature Materials* 13, no. 9 (2014): 849–850.
[27] Holten-Andersen, Niels, Matthew J. Harrington, Henrik Birkedal, Bruce P. Lee, Phillip B. Messersmith, Ka Yee C. Lee, and J. Herbert Waite. "pH-induced metal-ligand cross-links inspired by mussel yield self-healing polymer networks with near-covalent elastic moduli." *Proceedings of the National Academy of Sciences* 108, no. 7 (2011): 2651–2655.
[28] Krogsgaard, Marie, Manja A. Behrens, Jan Skov Pedersen, and Henrik Birkedal. "Self-healing mussel-inspired multi-pH-responsive hydrogels." *Biomacromolecules* 14, no. 2 (2013): 297–301.
[29] Verma, Chandrabhan, Eno E. Ebenso, and M. A. Quraishi. "Corrosion inhibitors for ferrous and non-ferrous metals and alloys in ionic sodium chloride solutions: A review." *Journal of Molecular Liquids* 248 (2017): 927–942.
[30] Daoud, Djamel, Tahar Douadi, Hanane Hamani, Salah Chafaa, and Mousa Al-Noaimi. "Corrosion inhibition of mild steel by two new S-heterocyclic compounds in 1 M HCl: Experimental and computational study." *Corrosion Science* 94 (2015): 21–37.

[31] De Damborenea, J., A. Conde, and M. A. Arenas. "Corrosion inhibition with rare earth metal compounds in aqueous solutions." In *Rare Earth-Based Corrosion Inhibitors*, pp. 84–116. Woodhead Publishing, 2014.
[32] Verma, Dakeshwar Kumar, Eno E. Ebenso, M. A. Quraishi, and Chandrabhan Verma. "Gravimetric, electrochemical surface and density functional theory study of acetohydroxamic and benzohydroxamic acids as corrosion inhibitors for copper in 1 M HCl." *Results in Physics* 13 (2019): 102194.
[33] Umoren, Saviour A., and Ubong M. Eduok. "Application of carbohydrate polymers as corrosion inhibitors for metal substrates in different media: A review." *Carbohydrate Polymers* 140 (2016): 314–341.
[34] Cao, Shuyun, Dan Liu, Tianxiang Wang, Aijing Ma, Chengwei Liu, Xupin Zhuang, Hui Ding, Bhekie B. Mamba, and Jianzhou Gui. "Nitrogen-doped carbon dots as high-effective inhibitors for carbon steel in acidic medium." *Colloids and Surfaces A: Physicochemical and Engineering Aspects* 616 (2021): 126280.
[35] Yang, Dongping, Yuwei Ye, Yue Su, Shuang Liu, Dirong Gong, and Haichao Zhao. "Functionalization of citric acid-based carbon dots by imidazole toward novel green corrosion inhibitor for carbon steel." *Journal of Cleaner Production* 229 (2019): 180–192.
[36] Ramesh, S., and S. Rajeswari. "Corrosion inhibition of mild steel in neutral aqueous solution by new triazole derivatives." *Electrochimica Acta* 49, no. 5 (2004): 811–820.
[37] Issa, Raafat M., Mohamed K. Awad, and Faten M. Atlam. "Quantum chemical studies on the inhibition of corrosion of copper surface by substituted uracils." *Applied Surface Science* 255, no. 5 (2008): 2433–2441.
[38] Berdimurodov, Elyor, Abduvali Kholikov, Khamdam Akbarov, Guobao Xu, Aboubakr M. Abdullah, and Morteza Hosseini. "New anti-corrosion inhibitor (3ar, 6ar)-3a, 6a-di-p-tolyltetrahydroimidazo [4, 5-d] imidazole-2, 5 (1 h, 3h)-dithione for carbon steel in 1 M HCl medium: Gravimetric, electrochemical, surface and quantum chemical analyses." *Arabian Journal of Chemistry* 13, no. 10 (2020): 7504–7523.
[39] Berdimurodov, Elyor, Abduvali Kholikov, Khamdam Akbarov, and Lei Guo. "Inhibition properties of 4, 5-dihydroxy-4, 5-di-p-tolylimidazolidine-2-thione for use on carbon steel in an aggressive alkaline medium with chloride ions: Thermodynamic, electrochemical, surface and theoretical analyses." *Journal of Molecular Liquids* 327 (2021): 114813.
[40] Berdimurodov, Elyor, Abduvali Kholikov, Khamdam Akbarov, I. B. Obot, and Lei Guo. "Thioglycoluril derivative as a new and effective corrosion inhibitor for low carbon steel in a 1 M HCl medium: Experimental and theoretical investigation." *Journal of Molecular Structure* 1234 (2021): 130165.
[41] Gece, Gökhan. "Drugs: A review of promising novel corrosion inhibitors." *Corrosion Science* 53, no. 12 (2011): 3873–3898.
[42] Berdimurodov, Elyor, Abduvali Kholikov, Khamdam Akbarov, and Lei Guo. "Experimental and theoretical assessment of new and eco-friendly thioglycoluril derivative as an effective corrosion inhibitor of St2 steel in the aggressive hydrochloric acid with sulfate ions." *Journal of Molecular Liquids* 335 (2021): 116168.
[43] Verma, Dakeshwar Kumar, Mohsin Kazi, Mohammed S. Alqahtani, Rabbani Syed, Elyor Berdimurodov, Savaş Kaya, Rajae Salim, Ashish Asatkar, and Rajesh Haldhar. "N-hydroxybenzothioamide derivatives as green and efficient corrosion inhibitors for mild steel: Experimental, DFT and MC simulation approach." *Journal of Molecular Structure* 1241 (2021): 130648.
[44] Verma, Chandrabhan, Eno E. Ebenso, Indra Bahadur, and M. A. Quraishi. "An overview on plant extracts as environmental sustainable and green corrosion inhibitors for metals and alloys in aggressive corrosive media." *Journal of Molecular Liquids* 266 (2018): 577–590.

[45] Verma, Dakeshwar Kumar, Savaş Kaya, Elhachmia Ech-Chihbi, Fadoua El-Hajjaji, Mayur Mausoom Phukan, and Hassien M. Alnashiri. "Investigations on some coumarin based corrosion inhibitors for mild steel in aqueous acidic medium: Electrochemical, surface morphological, density functional theory and Monte Carlo simulation approach." *Journal of Molecular Liquids* 329 (2021): 115531.

[46] Verma, Chandrabhan, Jiyaul Haque, M. A. Quraishi, and Eno E. Ebenso. "Aqueous phase environmental friendly organic corrosion inhibitors derived from one step multicomponent reactions: A review." *Journal of Molecular Liquids* 275 (2019): 18–40.

[47] El-Hajjaji, F., M. Messali, A. Aljuhani, M. R. Aouad, B. Hammouti, M. E. Belghiti, D. S. Chauhan, and M. A. Quraishi. "Pyridazinium-based ionic liquids as novel and green corrosion inhibitors of carbon steel in acid medium: Electrochemical and molecular dynamics simulation studies." *Journal of Molecular Liquids* 249 (2018): 997–1008.

[48] Guo, Yangyang, Zhihao Chen, Yuwei Zuo, Yun Chen, Wenzhong Yang, and Bin Xu. "Ionic liquids with two typical hydrophobic anions as acidic corrosion inhibitors." *Journal of Molecular Liquids* 269 (2018): 886–895.

[49] Xue, Bailiang, Yang, Yongchang Sun, Jinshuan Fan, Xinping Li, and Zhao Zhang. "Photoluminescent lignin hybridized carbon quantum dots composites for bioimaging applications." *International Journal of Biological Macromolecules* 122 (2019): 954–961.

[50] Liang, Zhiyu, Jun Yang, Chuxin Zhou, Qiaofei Mo, and Yuanming Zhang. "Carbon quantum dots modified BiOBr microspheres with enhanced visible light photocatalytic performance." *Inorganic Chemistry Communications* 90 (2018): 97–100.

[51] Devi, Pooja, Prachi Rajput, Anupma Thakur, Ki-Hyun Kim, and Praveen Kumar. "Recent advances in carbon quantum dot-based sensing of heavy metals in water." *TrAC Trends in Analytical Chemistry* 114 (2019): 171–195.

[52] Shen, Kui, Gui Luo, Jian Liu, Jianming Zheng, and Chunye Xu. "Highly transparent photoelectrochromic device based on carbon quantum dots sensitized photoanode." *Solar Energy Materials and Solar Cells* 193 (2019): 372–378.

[53] Ye, Yuwei, Dongping Yang, Hao Chen, Shengda Guo, Qiumin Yang, Liyong Chen, Haichao Zhao, and Liping Wang. "A high-efficiency corrosion inhibitor of N-doped citric acid-based carbon dots for mild steel in a hydrochloric acid environment." *Journal of Hazardous Materials* 381 (2020): 121019.

[54] Chatzimitakos, Theodoros, Athanasia Kasouni, Lamprini Sygellou, Apostolos Avgeropoulos, Anastasios Troganis, and Constantine Stalikas. "Two of a kind but different: Luminescent carbon quantum dots from Citrus peels for iron and tartrazine sensing and cell imaging." *Talanta* 175 (2017): 305–312.

[55] Zhu, Cheng, Yijun Fu, Changan Liu, Yang Liu, Lulu Hu, Juan Liu, Igor Bello, et al. "Carbon dots as fillers inducing healing/self-healing and anticorrosion properties in polymers." *Advanced Materials* 29, no. 32 (2017): 1701399.

[56] Jiang, Kai, Shan Sun, Ling Zhang, Yue Lu, Aiguo Wu, Congzhong Cai, and Hengwei Lin. "Red, green, and blue luminescence by carbon dots: Full-color emission tuning and multicolor cellular imaging." *Angew and Techemie* 127, no. 18 (2015): 5450–5453.

[57] Song, Yang, Chengzhou Zhu, Junhua Song, He Li, Dan Du, and Yuehe Lin. "Drug-derived bright and color-tunable N-doped carbon dots for cell imaging and sensitive detection of Fe3+ in living cells." *ACS Applied Materials & Interfaces* 9, no. 8 (2017): 7399–7405.

[58] Cui, Mingjun, Siming Ren, Qunji Xue, Haichao Zhao, and Liping Wang. "Carbon dots as new eco-friendly and effective corrosion inhibitor." *Journal of Alloys and Compounds* 726 (2017): 680–692.

[59] Cui, Mingjun, Siming Ren, Haichao Zhao, Liping Wang, and Qunji Xue. "Novel nitrogen doped carbon dots for corrosion inhibition of carbon steel in 1 M HCl solution." *Applied Surface Science* 443 (2018): 145–156.

[60] Singh, Ambrish, K. R. Ansari, Ashok Kumar, Wanying Liu, Chen Songsong, and Yuanhua Lin. "Electrochemical, surface and quantum chemical studies of novel imidazole derivatives as corrosion inhibitors for J55 steel in sweet corrosive environment." *Journal of Alloys and Compounds* 712 (2017): 121–133.

[61] El-Haddad, Mahmoud N., and A. S. Fouda. "Electroanalytical, quantum and surface characterization studies on imidazole derivatives as corrosion inhibitors for aluminum in acidic media." *Journal of Molecular Liquids* 209 (2015): 480–486.

[62] Jordan, Torsten, Susanne Schmidt, Tim Liebert, and Thomas Heinze. "Molten imidazole—a starch solvent." *Green Chemistry* 16, no. 4 (2014): 1967–1973.

[63] Likhanova, N. V. "Dom nguez-Aguilar MA, Olivares-Xometl O., Nava-Entzana N., Arce E., Dorantes H. The effect of ionic liquids with imidazolium and pyridinium cations on the corrosion inhibition of mild steel in acidic environment." *Corrosion Science* 52 (2010): 2088–2097.

[64] Milošev, Ingrid, Nataša Kovačević, Janez Kovač, and Anton Kokalj. "The roles of mercapto, benzene and methyl groups in the corrosion inhibition of imidazoles on copper: I. Experimental characterization." *Corrosion Science* 98 (2015): 107–118.

[65] Luo, Jiaxin, Xiong Cheng, Chaofa Zhong, Xinhua Chen, Y. W. Ye, H. Zhao, and H. Chen. "Effect of reaction parameters on the corrosion inhibition behavior of N-doped carbon dots for metal in 1 M HCl solution." *Journal of Molecular Liquids* 338 (2021): 116783.

[66] Zeng, Yuanxian, Lei Kang, Ying Wu, Shan Wan, Bokai Liao, Nan Li, and Xingpeng Guo. "Melamine modified carbon dots as high effective corrosion inhibitor for Q235 carbon steel in neutral 3.5 wt% NaCl solution." *Journal of Molecular Liquids* 349 (2022): 118108.

[67] Liu, Zheng, Ruonan Jia, Feng Chen, Guilong Yan, Weiguo Tian, Jinming Zhang, and Jun Zhang. "Electrochemical process of early-stage corrosion detection based on N-doped carbon dots with superior Fe3+ responsiveness." *Journal of Colloid and Interface Science* 606 (2022): 567–576.

[68] Liu, Zhixiong, Qingkai Chu, Hao Chen, Y. J. Qiang, Xianhui Zhang, and Y. W. Ye. "Experimental and molecular simulation studies of N, S-doped Carbon dots as an eco-friendly corrosion inhibitor for protecting Cu in HCl environment." *Colloids and Surfaces A: Physicochemical and Engineering Aspects* 669 (2023): 131504.

[69] Ye, Yuwei, Hao Chen, Yangjun Zou, and Haichao Zhao. "Study on self-healing and corrosion resistance behaviors of functionalized carbon dot-intercalated graphene-based waterborne epoxy coating." *Journal of Materials Science & Technology* 67 (2021): 226–236.

[70] Ye, Yuwei, Dongping Yang, and Hao Chen. "A green and effective corrosion inhibitor of functionalized carbon dots." *Journal of Materials Science & Technology* 35, no. 10 (2019): 2243–2253.

[71] Cui, Jun, Xiuqing Li, Zhiqiang Pei, and Yuansheng Pei. "A long-term stable and environmental friendly self-healing coating with polyaniline/sodium alginate microcapsule structure for corrosion protection of water-delivery pipelines." *Chemical Engineering Journal* 358 (2019): 379–388.

[72] Patrick, Jason F., Kevin R. Hart, Brett P. Krull, Charles E. Diesendruck, Jeffrey S. Moore, Scott R. White, and Nancy R. Sottos. "Continuous self-healing life cycle in vascularized structural composites." *Advanced Materials* 26, no. 25 (2014): 4302–4308.

[73] Williams, Hugo R., Richard S. Trask, and I. P. Bond. "Self-healing sandwich panels: Restoration of compressive strength after impact." *Composites Science and Technology* 68, no. 15–16 (2008): 3171–3177.

[74] Chen, Yulin, Aaron M. Kushner, Gregory A. Williams, and Zhibin Guan. "Multiphase design of autonomic self-healing thermoplastic elastomers." *Nature Chemistry* 4, no. 6 (2012): 467–472.

[75] Burattini, Stefano, Barnaby W. Greenland, Daniel Hermida Merino, Wengui Weng, Jonathan Seppala, Howard M. Colquhoun, Wayne Hayes, Michael E. Mackay, Ian W. Hamley, and Stuart J. Rowan. "A healable supramolecular polymer blend based on aromatic π-π stacking and hydrogen-bonding interactions." *Journal of the American Chemical Society* 132, no. 34 (2010): 12051–12058.

[76] Luo, Feng, Tao Lin Sun, Tasuku Nakajima, Takayuki Kurokawa, Yu Zhao, Koshiro Sato, Abu Bin Ihsan, Xufeng Li, Honglei Guo, and Jian Ping Gong. "Oppositely charged polyelectrolytes form tough, self-healing, and rebuildable hydrogels." *Advanced Materials* 27, no. 17 (2015): 2722–2727.

[77] Heo, Yunseon, Mohammad H. Malakooti, and Henry A. Sodano. "Self-healing polymers and composites for extreme environments." *Journal of Materials Chemistry A* 4, no. 44 (2016): 17403–17411.

[78] Amamoto, Yoshifumi, Hideyuki Otsuka, Atsushi Takahara, and Krzysztof Matyjaszewski. "Self-healing of covalently cross-linked polymers by reshuffling thiuram disulfide moieties in air under visible light." *Advanced Materials* 24, no. 29 (2012): 3975–3980.

[79] Suleiman, A. R., and M. L. Nehdi. "Effect of environmental exposure on autogenous self-healing of cracked cement-based materials." *Cement and Concrete Research* 111 (2018): 197–208.

[80] Zhu, Dong Yu, Min Zhi Rong, and Ming Qiu Zhang. "Self-healing polymeric materials based on microencapsulated healing agents: From design to preparation." *Progress in Polymer Science* 49 (2015): 175–220.

[81] Hillewaere, Xander K. D., and Filip E. Du Prez. "Fifteen chemistries for autonomous external self-healing polymers and composites." *Progress in Polymer Science* 49 (2015): 121–153.

[82] Ma, Yingxiang, Dan Jiang, Yuping Yang, Li Ma, Jian Zhou, Guosheng Huang, and Cunguo Lin. "Synthesis of magnetic targeted delivery microcapsules and its anti-corrosion self-healing behavior in epoxy resin coatings." *Corrosion Communications* 10 (2023): 27–37.

[83] Mi, Tangwei, Jianghong Mao, Yunhong Cai, Shuqiong Luo, Xianfeng Wang, Meilin Xiao, Haiyan Zhu, Kai Yang, and Jun Ren. "Application of Raman spectroscopy for detecting the repairing behavior of microcapsules in the self-healing cementitious system." *Construction and Building Materials* 387 (2023): 131637.

[84] Vintila, Ionut Sebastian, Jana Ghitman, HoriaIovu, Alexandru Paraschiv, Andreia Cucuruz, Dragos Mihai, and Ionut Florian Popa. "A microvascular system self-healing approach on polymeric composite materials." *Polymers* 14, no. 14 (2022): 2798.

[85] Yuan, Chan E., Min Zhi Rong, Ming Qiu Zhang, Ze Ping Zhang, and Yan Chao Yuan. "Self-healing of polymers via synchronous covalent bond fission/radical recombination." *Chemistry of Materials* 23, no. 22 (2011): 5076–5081.

[86] Zhang, Ming Qiu, and Min Zhi Rong. "Intrinsic self-healing of covalent polymers through bond reconnection towards strength restoration." *Polymer Chemistry* 4, no. 18 (2013): 4878–4884.

[87] Xiang, T. F., Min Zhang, Hisham Rabia Sadig, Z. C. Li, M. X. Zhang, C. D. Dong, Ling Yang, W. M. Chan, and Cheng Li. "Slippery liquid-infused porous surface for corrosion protection with self-healing property (vol 345, pg 147, 2018)." *Chemical Engineering Journal* 348 (2018): 1064–1065.

[88] Bhattacharya, Sagarika, Ravindra Suresh Phatake, Shiran Nabha Barnea, Nicholas Zerby, Jun-Jie Zhu, Rafi Shikler, Norberto Gabriel Lemcoff, and Raz Jelinek. "Fluorescent self-healing carbon dot/polymer gels." *ACS Nano* 13, no. 2 (2019): 1433–1442.

[89] Wu, Xiang-Fa, and Alexander L. Yarin. "Recent progress in interfacial toughening and damage self-healing of polymer composites based on electrospun and solution-blown nanofibers: An overview." *Journal of Applied Polymer Science* 130, no. 4 (2013): 2225–2237.
[90] Zhang, Luzhi, Zenghe Liu, Xueli Wu, Qingbao Guan, Shuo Chen, Lijie Sun, Yifan Guo, et al. "A highly efficient self-healing elastomer with unprecedented mechanical properties." *Advanced Materials* 31, no. 23 (2019): 1901402.
[91] Wang, Zhanhua, Xili Lu, Shaojie Sun, Changjiang Yu, and Hesheng Xia. "Preparation, characterization and properties of intrinsic self-healing elastomers." *Journal of Materials Chemistry B* 7, no. 32 (2019): 4876–4926.
[92] Li, Changhua, Yi He, Zhijun Li, Hongjie Li, and Yang Zhao. "Graphene loaded with corrosion inhibitor cerium (III) cation for enhancing corrosion resistance of waterborne epoxy coating: Physical barrier and self-healing." *Colloids and Surfaces A: Physicochemical and Engineering Aspects* 635 (2022): 128048.
[93] Zhang, Qi, Da-HuiQu, Ben L. Feringa, and He Tian. "Disulfide-mediated reversible polymerization toward intrinsically dynamic smart materials." *Journal of the American Chemical Society* 144, no. 5 (2022): 2022–2033.
[94] Liu, Chang, Qing Li, Haopeng Wang, Gefei Wang, and Haixia Shen. "Quantum dots-loaded self-healing gels for versatile fluorescent assembly." *Nanomaterials* 12, no. 3 (2022): 452.

12 Industrial anti-corrosive materials

Current challenges, future prospects, and ideal substitutes

Hüseyin Fatih Çetinkaya, Serap Çetinkaya, and Burak Tüzün

12.1 INTRODUCTION

Corrosion is a varying structural failure phenomenon observed on metal surfaces in unfavorable working environments because metals or alloys differ in their inherent corrosion sensitivity.

Industrial materials should theoretically meet certain resistance qualities against corrosive reactions in operation space and time by possessing suitable inner and/or acquired physical, mechanical, and chemical capabilities [1]. Here, the term "reaction" implies that corrosion can also be chemical or electrochemical [2, 3]. The latter involves electron transfer [3, 4]. Some other factors are oxygen and oxidizers, temperature, operation velocity, corrosive concentration, galvanic coupling, and metallurgical factors [5, 6]. A thorough insight into the corrosion phenomenon thus entails the involvement of different scientific disciplines [2]. Selection of the right material for the process could be the first essential step [2]. This should be followed by the choice of anti-corrosive coating material, for which multiple applications have been made available [7]. With coatings, many ends can be achieved, enabling efficient and cost-effective operations as well as saving in both material and pollution [8].

12.2 ANTI-CORROSION ECONOMICS AND CHALLENGES

Corrosion is the second-most metal consumer after man. Its global cost has been very high, amounting to between 1% and 3.5% of GNP [9–11]. Indirect expenses have been imagined to be much more. It thus becomes obvious that metal corrosion brings about gigantic expenditures. The worst side of corrosion is that it in fact eats up the material diversity of nature [9, 12].

Anti-corrosion measures involve numerous branches of industry. One success story has been written in the automobile industry [13], in which, by 1975, effective corrosion management programs had been implemented, as in Figure 12.1 [14].

 DOI: 10.1201/9781003456025-12

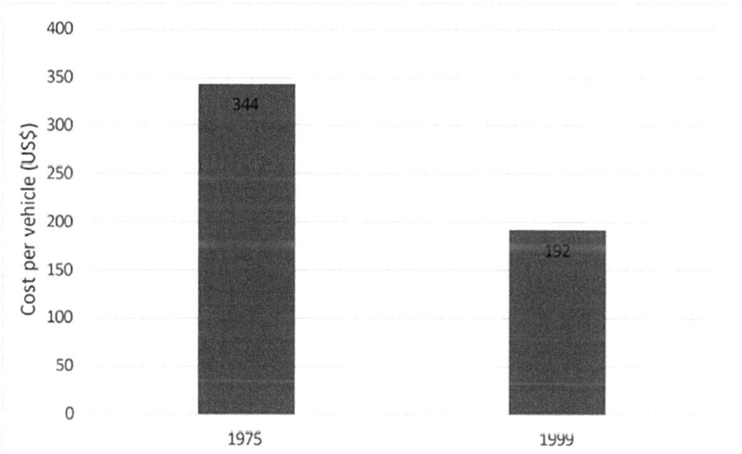

FIGURE 12.1 The change in the magnitude of corrosion cost between1975 and 1999.

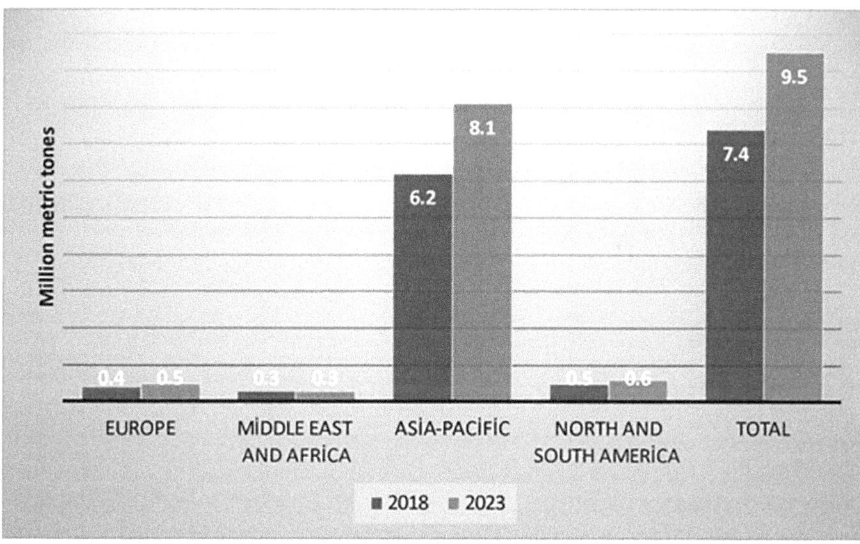

FIGURE 12.2 Current global distribution of coating material consumption.

Diverse material surfaces can be subjected to anti-corrosive coating processes in order to lessen interfacial friction, chemical corrosion, or corrosion caused by natural oxidation reactions, as well as the cost of such deleterious actions. The application of such processes uses millions of tons of coating materials, and per se results in large-scale economic contributions [15, 16]. Asia-Pacific countries appear to be the largest consumers, as given in Figure 12.2 [14].

12.3 CORROSION MECHANISM

Thermodynamic and electrochemical approaches have borne fruit in our understanding of corrosion. Metallurgical factors also play a significant role in corrosion resistance. Physical chemistry in general has been versatile in the study of corrosion reaction mechanisms [2]. Chemical corrosion takes place by conforming to the kinetics of heterogeneous reactions, not involving the generation of an electric current [17]. Etching can also be considered as acorrosive chemical attack [8, 18].

Solid-liquid metal interactions involve archetypal physical corrosion where liquid metal penetrates into the solid one [19]. Electrochemical corrosion involves either anodic, dissolution, orcathodic reactions. Cathodic reactions give rise to reduction and deposition. Oxidation involves both uniform surface reactions and specific regional reactions [8].

12.3.1 Protection mechanism of anti-corrosive coatings

Anti-corrosion coatings either work as barrier layers between the materials and their environment, or inhibit the corrosion processes, or act as a sacrificial layer. In a recent Schottky barrier approach, the coating layer forms a barrier and prevents the passive penetration of corrosive agents into the solid metal [20].

12.3.2 Active agent carriers

Active ingredient carriers are compatible additives loaded with functional species and used in the dispersal of the coating matrix. They hold functional species in certain locations until use. The carriers differ in shape and can be polymeric capsules, porous and hollow particles, or tubular. In the past two decades, active carriers also helped improve the self-healing functions of coating materials [21–23].

12.4 DIVERSITY IN ANTI-CORROSIVE COATING MATERIALS

12.4.1 Graphene-based anti-corrosive coating

The wettability and adhesion characteristics of metal surfaces can be adjusted using film-forming coating materials and fillers [24]. Graphene, known for its impermeability, offers remarkable resistance against abrasion and tearing forces. Employing graphene films serves to prevent oxidation reactions and enhances resistance to wetting on metal surfaces, effectively insulating the metal from its surrounding environment [25, 26]. Graphene layers can be applied to metal surfaces through various methods such as chemical vapor deposition (CVD), dip coating, spin coating, layer-by-layer self-assembly, sol-gel techniques, spray coating, in-situ polymerization, or electrophoretic deposition [27].

12.4.2 Urushiol-based polymers

Urushiol is a compound comprising trisubstituted catechol derivatives, which feature 0–3 olefins with $n = 15$ carbon chains, often contaminated by residual phenolics.

Utilizing the water-assisted assembly method, microporous films of Urushiol-formaldehyde polymer (UFP) can be created. These polymers, found in oriental varnish sap, exhibit natural resistance to damage from extremes of pH, heat, and various corrosive agents, rendering them significantly important [27].

12.4.3 Graphene oxide (GO)/urushiol-formaldehyde polymer (UFP) composite

Various modified graphene oxide (MGO) compounds are mixed with urushiol formaldehyde polymer (UFP). Graphene oxide, an anti-corrosive additive, thus participated in the recipes of polymer coatings [28, 29]. Modification of GO has been achieved by the addition of 3-aminopropyltrimethoxysilane (APS), vinyl tris(2-methoxyethoxy)silane(TMEVS), c-(2,3-epoxypropoxy)propyl trimethoxysilane (GPTMS), 3(aminopropyl), triethoxysilane (APTES), or 3-methacryloxy propyltrimethoxysilane (MPS) [30–32]. Raw lactan-modified graphene oxide/urushiol formaldehyde polymer (MGO/UFP) coatings have been reported to have improved physico-mechanical and corrosion resistance [27].

12.4.4 Graphene oxide/multi-walled carbon nanotube (MWCNTs)/urushiol formaldehyde polymer composite

Anti-corrosion and thermal stability features of UFP have been improved by adding to the in situ polymerization process the following substances: Fe_3O_4 [33], ZnO [34], TiO_2 [35], CuO [36], GO [37], and carbon nanotubes [38]. Nanofiller MWCNTs have also been exploited for the same ends [37–41]. CNT forms van der Waals interactions within the composite [42] and thus augments the overall coating features of UFP. Therefore, the ratio of the ingredients in the final composite product becomes very important [27, 43].

12.4.5 Multi-walled carbon nanotube (MWCNT)/epoxy nanocomposite

Axial flow impellers and ultrasonic waves have been utilized in the preparation of this coating composite as the formation of air bubbles in the process is detrimental to the coating structure. Chemical vapor deposition (CVD) is the method used in the preparation of MWCNTs.

12.5 THE LATEST ANTI-CORROSION POLYMER NANOCOMPOSITES

Through very small dimensions, nanocomposites effectively form very thin, smooth, and dense surface layers and yield highly adhesive coating materials. Polymer nanocomposites (PNCs) coatings thus effectively mitigate corrosion phenomenon and allow the insulation of both external and internal metal surfaces [44]. For the time being, PNCs are relatively expensive, but in the near future, this disadvantage can be overcome by enabling their large-scale use [45–47].

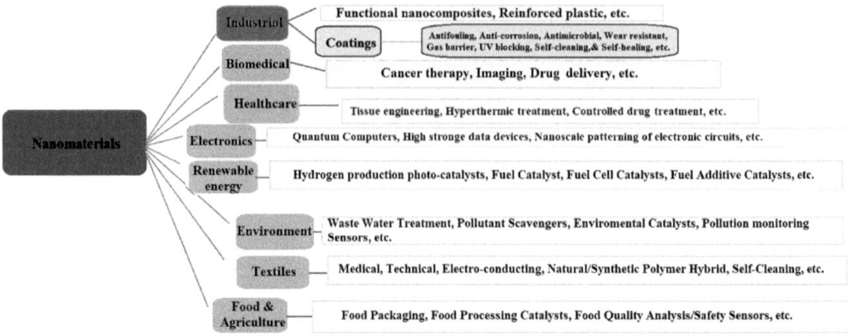

FIGURE 12.3 Classification of nanomaterials.

Modified PNCs, better with suppressing corrosion, fouling, and scratches, have been produced thanks to recent nanotechnological developments. These coatings have been versatile additives in shipyards [48]. They can be three-dimensional nanorods, nanotubes, and nanowires, or two-dimensional nanolayers, nanofilms, or nanoplatelets. These nanomaterials have much superior electronic, optical, mechanical, chemical, and physical properties, especially because of their very small size, enabling very high surface to volume ratios [49, 50].These materials form nanocrystalline coating layers and modify the metal surfaces, as shown in Figures 12.3 and 12.4 [47].

Though volatile organic compounds(VOC)or VOC-free coatings are still dominating the coating market, attempts have been focused on eliminating volatile organic compounds from anti-corrosive coating recipes, and on the production of waterborne, solid, UV curable coatings by the turn of the twentieth century. Besides, hyperbranched/hybrid polymer technologies, green corrosion inhibitors, bio-based materials, and superhydrophobic coatings have also been developed, as given in Figure 12.5 [14].

Using smart additives to improve the properties of an anti-corrosive coating is another trend under development. To illustrate, zeolites [51] have been shown to impart self-curing properties to epoxy resins. It also appears that highly versatile graphene-based technologies have been constituting one of the leading trends [52]. Graphene, for example, added to Zn-rich epoxy coatings has yielded highly superior anti-corrosion composites [53].

Green corrosion inhibitors, plant extracts, have recently been paid significant scientific attention. Polyphenols, for example, have been shown to increase anticorrosion activity of various coatings [54] by enlarging the impermeability potential of the coating layers. Another green coating material, shellac, has been recently reported [55]. Copper surfaces have been protected with modified shellac with a diamine crosslinker in ethanol. These superhydrophobic polymer coatings have been produced by incorporating nanomaterials into them and by creating the Lotus effect [56, 57]. These products are known to be environment-friendly, very efficient anti-fouling and anti-icing agents, and can also be used in the prevention of biocorrosion [14, 57].

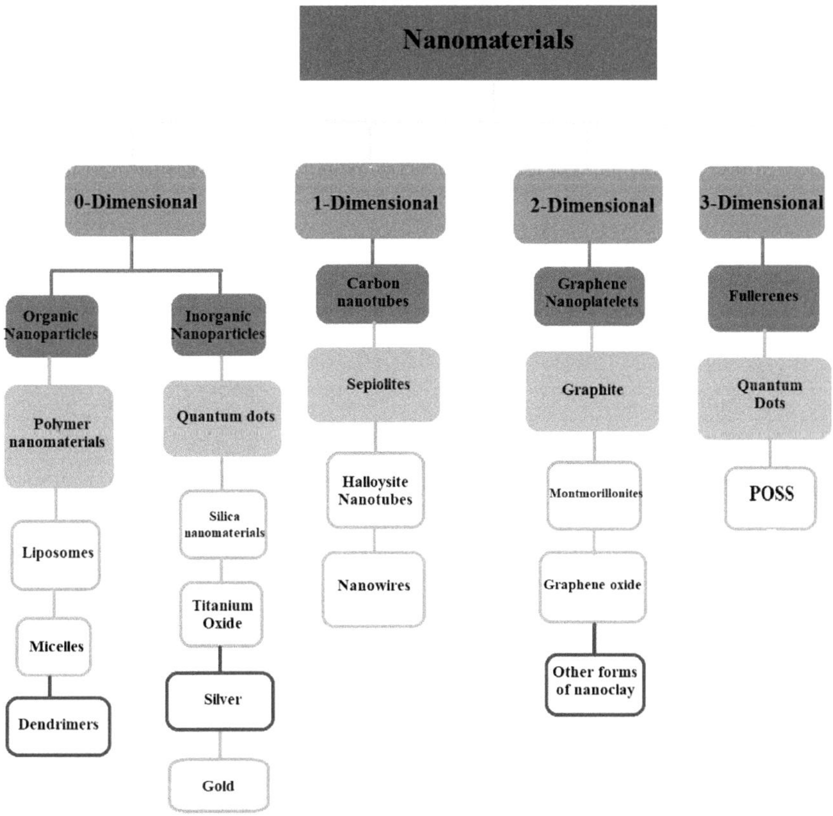

FIGURE 12.4 Applications of nanomaterials.

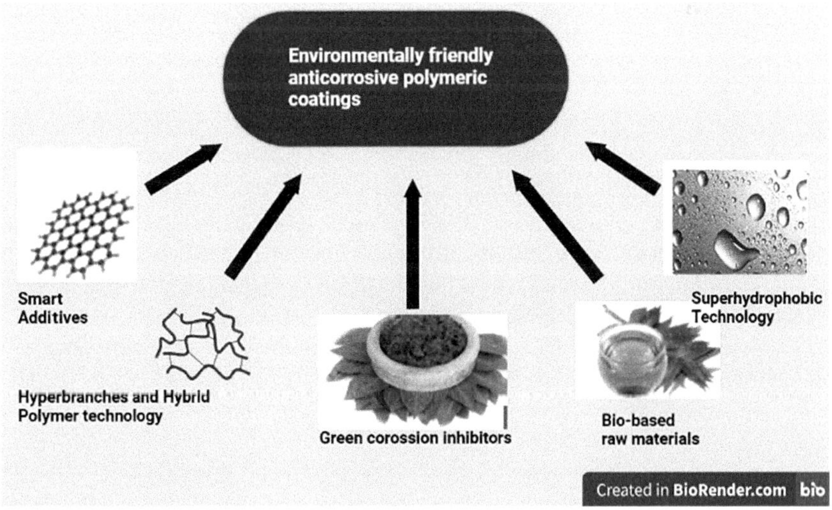

FIGURE 12.5 A schematic representation of trends in the development of new coatings.

12.6 APPLICATIONS

Immediate corrosive surroundings have been classified (ISO 12944–2:2017): atmospheric, immersing, and burying environments, and they are distinguished by their characteristic corrosion influences. And each of these corrosion characteristics has to be met with a range of compatible coating composites. The degree of corrosion in atmospheric environments resides between C1 (very low) and C5 (very high), and further CX "extreme corrosion.

12.7 SMART COATINGS

Smart coatings are an innovative concept activated by environmental cues upon physical contact. They significantly enhance the durability of coated materials, leading to extended operational life [58]. Their specialized composition includes corrosion inhibitors that enable responses to various environmental changes, such as alterations in pH, temperature, pressure, and more. These coatings incorporate specific functionalities like response to ionic strength, electric or magnetic fields, acoustics, light, and diverse chemical reactions [59]. The essential layers in coatings are typically categorized as pretreatment, primer, and topcoat, although in certain applications, the primer layer might serve as the pretreatment layer [60–63].

The multifunctionality of smart coatings in various industries offers a distinct advantage. In these coatings, internal stimuli refer to signals triggering responses within the coating itself, affecting its collective properties, particularly in corrosion-sensitive and self-healing coatings. External stimuli, illustrated in Figure 12.6, are environmental signals that prompt responses within the coatings, leading to changes on the coated surfaces [64, 65].

12.7.1 Corrosion-sensitive coatings

These coatings often integrate color-changing dyes or compounds within the film matrix. These elements can fluoresce or change color when exposed to high pH levels or interact with metal cations due to potential mechanical damage. Typically, transparent matrices are employed to showcase color changes or fluorescence when reacting with corroding substances. Colored dyes include anti-corrosive components and additives stored in microcapsules. Additionally, pigments can be combined with anti-corrosive agents. A variety of compounds like Schiff bases, hydroxyquinolines, fluorescein, phenolphthalein, auxins, bromothymol blue, 7-amino-4-methylcoumarin, and 7-diethylamino-4-methylcoumarin can be integrated into primer coatings, as shown in Figure 12.7 [66, 67].

The research conducted by Li and Calle [68] introduced a new microencapsulation technique aimed at corrosion detection and control. Their focus was on developing smart coatings sensitive to pH changes, which would detect corrosion and trigger the release of inhibitors. This pH-based mechanism, coupled with colored dyes, was used for sensing and releasing the encapsulated inhibitors. These microcapsules, housing both indicators and inhibitors, were integrated into conventional paint systems (Acrylic, Epoxy, Polyurethane, and Siloxane), showing significant

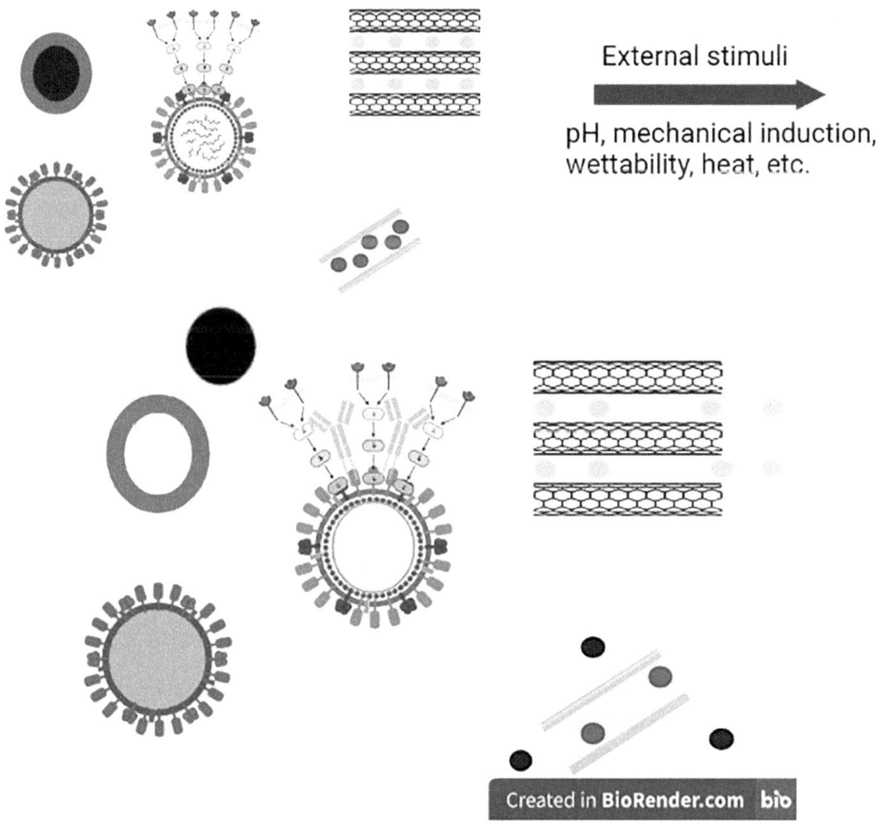

FIGURE 12.6 Environmental triggers of smart coatings.

Reproduced with permission.

FIGURE 12.7 Active corrosion sensing and inhibition.

improvements in paint adhesion properties. They found that increasing the cross-linking agent content and reducing microcapsule wall thickness enhanced the pH sensitivity of the microcapsules.

Exposing these microcapsules to basic pH solutions led to color changes, with thinner walls showing an approximate 80% change and thicker walls showing about 25%. Their corrosion-detecting ability was demonstrated by the color changes observed when dropping a pH indicator microcapsule into water on a carbon steel substrate [68].

A more comprehensive study [69] investigated active corrosion detection on two different substrates coated with AA2024 aluminum alloy and AZ31 Magnesium alloy. The study revealed a pink coloration due to pH sensitivity and an increase in cathodic sites, as depicted in Figure 12.7 [63].

The corrosion-sensitive coatings, applied to AA2024 and AZ31 alloys, were developed using polyurea microcapsules containing 12% phenolphthalein. These coatings showed active corrosion detection in an alkaline pH environment with the release of the indicator [31]. Additionally, Calle and colleagues at NASA worked on smart coatings designed to detect and prevent the initial stages of corrosion, as well as heal any damage to the coating. These coatings aimed to reduce maintenance costs and prolong the functional life of assets by incorporating eco-friendly inhibitors in microcapsules, which were released in response to pH changes indicating corrosion. The self-healing function targeted specific areas, preventing visible damage [70].

For evaluating corrosion in aerospace structures, fiber optic arrays were used for coated and uncoated surfaces, with compounds like 1,10-phenanthroline for steel and fluorescein for aluminum employed for detection. Modified bentonite clays and hydrotalcite were also utilized for corrosion protection and detection in organic coatings [71]. Recent advancements in nanotechnology have been applied to create corrosion-sensitive coatings, aiming to reduce the maintenance costs of metallic structures. These coatings contain nanocapsules with pH-sensitive additives, color dyes, and pigments, categorized as passive in terms of their mode of action (Figure 12.8) [67].

Figure 12.7 describes the active corrosion sensing of two different coated substrates due to the pH sensitivity and the increase in cathodic sites through the pink coloration. It was observed that the pH sensitivity of the microcapsules was increased, and a color change was observed with the effect of basic pH. This color change allowed corrosion detection. A corrosion-sensing coating was designed and used on (a-d) AA2024 aluminum alloy and (e-h) AZ31 magnesium alloy. Carbon steel is used in the base of both. It was observed that the corrosion was detected early after (a and e) immersion, 24 hours after immersion, and even 1 week after immersion. In the form (b and f), the growth parts 20 hours after and after the immersion process were examined. It was found that no corrosion prevention was observed, and the corrosion continued.

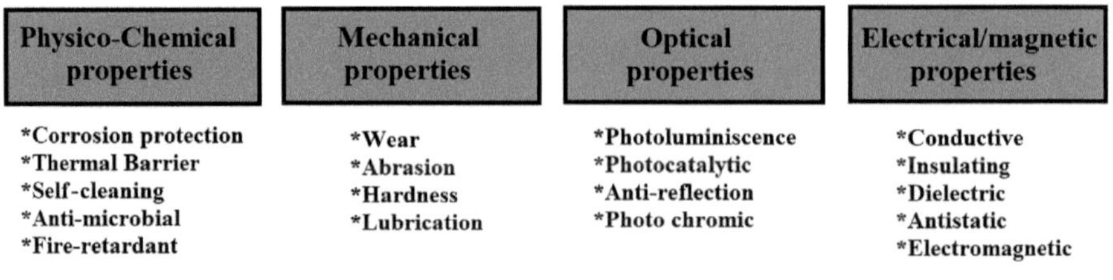

FIGURE 12.8 Overview of surface properties that can be enhanced or improved by functionalization with organic, inorganic, or hybrid coatings.

12.7.2 Applications, advantages, and limitations

Smart coatings have become integral across various sectors, offering protection in diverse environmental conditions. From mild to severe settings, they safeguard substrate materials from degradation and deterioration. They find application in safeguarding aircraft fuel tank interiors, outer and inner surfaces of oil and gas tanks, transport lines, and reservoirs, all of which present challenging environments. Additionally, these coatings are utilized in camouflaging military equipment, industrial gas turbines, and surfaces that require enhanced functionality [72].

In the automotive industry, self-cleaning coatings are employed for car bodies, mirrors, and windshields, providing a practical application of their self-cleaning properties. In the optical industry, smart coatings find use in sensors, lenses, cameras, and telescopes, contributing to their performance and longevity.

The application of nanomaterials in smart coatings offers various advantages, including prolonged equipment life, reduced maintenance hours, and the creation of functional surfaces with thermochromic and photochromic effects. Coatings comprising nanoparticles exhibit unique characteristics due to the increased surface area-to-volume ratio, affecting the coating properties by influencing the atomic behavior on the particle's surface [73].

Nanoparticles display size-dependent properties such as quantum constraints in semiconductors, surface plasmon resonance in metallic nanoparticles, and paramagnetism in magnetic nanoparticles. The use of nanomaterials and nanocontainers doped with inhibitors in smart coatings shows potential to outperform chromate-based inhibitors. The synergistic effects from these mixtures result in enhanced functionalities and reduced toxicity [74].

Furthermore, antibacterial coatings containing TiO2 nanoparticles have been applied to water vessels, equipment, and marine structures to minimize economic losses caused by microbial-induced corrosion. Barrier coatings play a significant role in inhibiting the generation of electromotive force (EMF) responsible for substrate corrosion [63, 75].

Certainly, certain limitations need consideration when dealing with smart coatings and nanomaterials. Organic coatings face temperature limitations, as temperatures above 350°C can cause these organic molecules to react with oxygen, resulting in their burning off and destroying the coating's effectiveness [76].

Excessive concentrations of nanomaterials in coatings can lead to agglomeration or coating degradation. The use of costly, unstable, and potentially toxic catalysts like the Grubbs catalyst should be minimized. Additionally, most reports have only demonstrated single curative effects, and their long-term or multiple recovery effects' economic feasibility remains uncertain [77, 78].

Stabilizers might be required in coatings to maintain the properties of nanoparticles, incurring additional costs. The absence of hydroxyl groups on surfaces affects the advantage of using organofunctional silanes as chromate substitutes in coatings. The incompatibility of nano-sized particles and pigments within coatings can reduce their barrier effect [79].

Furthermore, the lack of or insufficient encapsulation of active ingredients in coatings can lead to interactions among coating components, impacting the coating's functionality. Inadequate dispersion of charged nanocontainers within coatings can cause defects, although this issue can be managed by adjusting the disintegration time and agitation rate [80, 81].The self-healing feature was successful only in the case of initial damage to coatings with minor or moderate cracks. A sustainable multi-healing network through extensive fieldwork has not yet been established. Toohey et al. investigated the repetitive healing network interference of a cracked epoxy resin. The brittle epoxy resin was about 700 μm thick, and the Grubbs catalyst used to trigger healing in case of damage could not remain active throughout the sustained healing process. These reduced the curing efficiency despite being a constant source of monomers [82]. Therefore, the recovery of large damaged areas has not yet been successfully carried out. There is still a challenge in developing self-healing smart coatings with a very high glass transition temperature polymeric structure that offer self-healing properties when triggered by magnetic and/or electric fields, electromagnetic radiations, and atmospheric changes [83]. For the self-cleaning category, surfaces have yet to match their naturally occurring types, particularly in terms of their apparent multifunctionality [84].

12.8 CONCLUSIONS

Corrosion has been shown to be an important phenomenon leading to the deterioration of metal, ceramic, and polymeric surfaces. Coatings are clearly intended to protect abrasive materials, and there has been a lot of work done to improve their industrial quality and performance life. It is seen in the studies carried out that nanotechnological developments have increased the new applications of protective polymer nanocomposite (PNC) coatings. Nanotechnology has seen many developments involving the discovery, production, and use of nanoparticles of about 100 nm. It is seen that nanoparticles are used against corrosion with anti-corrosion, thermal stability, and mechanical strength to the adhered materials.

REFERENCES

[1] Ige, O. O. (2007). *Effects of tin and heat treatment on an aluminum-zinc-magnesium alloy as sacrificial anode in seawater.* Unpublished M.Sc Thesis, Obafemi Awolowo University, Nigeria.

[2] Fontana, M. G., & Greene, N. D. (2018). *Corrosion engineering.* McGraw-Hill.

[3] West, J. M. (1986). *Basic corrosion and oxidation.* https://www.osti.gov/biblio/6735035

[4] Trethewey, K. R., & Chamberlain, J. (1988). *Corrosion for students of science and engineering.* Longman Scientific & Technical, 382.

[5] Wang, S. (2009). *Effect of oxygen on CO2 corrosion of mild steel.* Doctoral dissertation, Ohio University.

[6] Tendayi, Z. (2010). *Hydrodynamics on carbon steel erosion-corrosion and inhibit or efficiency in simulated oil fields brines.* University of Glasgow.

[7] Popoola, A. P. I., Fayomi, O. S., & Popoola, M. O. (2012). Comparative studies of micro structural, tribological and corrosion properties of plated Zn and Zn-alloy coatings. *International Journal of Electrochemical Science*, 7, 4860–4870.

[8] Popoola, A. P. I., Olorunniwo, O. E., &Ige, O. O. (2014). Corrosion resistance through the application of anti-corrosion coatings. *Developments in Corrosion Protection*, 13(4), 241–270.

[9] Orman, S. (1976). Economic aspects of corrosion. In Shreir, L. L., ed. *Corrosion*, Vol. 2, Corrosion Control, Newnes—Butterworths, London, 10, 3–6.

[10] Gasem, Z.M. (2013). *ME 472: Corrosion engineering 1*, ocw.kfupm.edu.sa/ocw.courses/users062/ME4720102/LectureNotes/impact of Corrosion.

[11] Ige, O. O. (2013). *A study of the hydrodynamic effects of erosion corrosion in oil and gas industry*. Unpublished PhD Thesis, Obafemi Awolowo University, Nigeria.

[12] Jones, D.A. (1992). *Principles and prevention of corrosion*. Macmillan Publishing Company, 445–446.

[13] Koch, G.H., Varney, J., Thompson, N., Moghissi, O., Gould, M., & Payer, J. (2021). *Nace International Report*. International Measures of Prevention Application and Economics of Corrosion Technologies Study.

[14] Faccini, M., Bautista, L., Soldi, L., Escobar, A. M., Altavilla, M., Calvet, M., . . . & Domínguez, E. (2021). Environmentally friendly anticorrosive polymeric coatings. *Applied Sciences*, 11(8), 3446.

[15] Sud, A. (2015). Anticorrosive coatings. *European Coatings Journal*, 10, 8–9.

[16] Gagro, D. (2020). Protective coatings. *European Coatings Journal*, 8, 10–11.

[17] Köser, H. J. K. (1980). Corrosion of metals in carbon tetrachloride-containing solvents. *Analyst*, 105(1252), 696–704.

[18] Shreir, L. L., Jarman, R. A., & Burstein, G. T., eds. (1994). *Corrosion metal/environment reactions*. Elsevier, 1.

[19] Umoru, L. E. (2001). *A study of corrosion of materials in a tar sand digester*. A Ph.D. Thesis from Obafemi Awolowo University, Ile-Ife, Nigeria.

[20] Dennis, R. V., Lee, V., Viyannalage, L., Henderson, S. M., & Banerjee, S. (2013). *Anti-corrosive grapheme coatings: An active.passive alternative to hexavalent chromium coatings*. https://www.yumpu.com/en/document/view/5807673/anti-corrosive-graphene-coatings-an-active-passive-alternative-

[21] Balaskas, A. C., Kartsonakis, I. A., Tziveleka, L. A., & Kordas, G. C. (2012). Improvement of anti-corrosive properties of epoxy-coated AA 2024-T3 with TiO2 nanocontainers loaded with 8-hydroxyquinoline. *Progress in Organic Coatings*, 74(3), 418–426.

[22] Plawecka, M., Snihirova, D., Martins, B., Szczepanowicz, K., Warszynski, P., & Montemor, M. F. (2014). Self healing ability of inhibitor-containing nanocapsules loaded in epoxy coatings applied on aluminium 5083 and galvanneal substrates. *Electrochimica Acta*, 140, 282–293.

[23] Montemor, M. F. (2014). Functional and smart coatings for corrosion protection: A review of recent advances. *Surface and Coatings Technology*, 258, 17–37.

[24] Sastri, V. S. (2012). *Green corrosion inhibitors: Theory and practice*. John Wiley & Sons.

[25] Cui, G., Bi, Z., Zhang, R., Liu, J., Yu, X., & Li, Z. (2019). A comprehensive review on graphene-based anti-corrosive coatings. *Chemical Engineering Journal*, 373, 104–121.

[26] Yoo, B. M., Shin, H. J., Yoon, H. W., & Park, H. B. (2014). Graphene and graphene oxide and theiruses in barrier polymers. *Journal of Applied Polymer Science*, 131(1).

[27] Sharma, N., & Sharma, S. (2021). Anti corrosive coating of polymer composites: A review. *Materials Today: Proceedings*, 44, 4498–4502.

[28] Tang, J., Yao, W., Li, W., Xu, J., Jin, L., Zhang, J., & Xu, Z. (2018). Study on a novel composite coating based on PDMS doped with modified grapheneoxide. *Journal of Coatings Technology and Research*, 15, 375–383.

[29] Carrasco-Valenzuela, L., Zaragoza-Contreras, E. A., & Vega-Rios, A. (2017). Synthesis of grapheneoxide/poly (3, 4-ethylenedioxythiophene) composites by Fenton's reagent. *Polymer*, 130, 124–134.

[30] He, L., Zhao, Y., Xing, L., Liu, P., Wang, Z., Zhang, Y., & Liu, X. (2017). Preparation of phosphonic acid functionalized grapheneoxide-modified aluminum powder with enhanced anticorrosive properties. *Applied Surface Science*, 411, 235.
[31] Dun, Y., & Zuo, Y. (2017). Preparation and characterization of a GPTMS/graphene coating on AA-2024 alloy. *Applied Surface Science*, 416, 492–502.
[32] Kumari, S., Panigrahi, A., Singh, S. K., & Pradhan, S. K. (2018). Enhanced corrosion resistance and mechanical properties of nanostructured graphene-polymer composite coating on copper by electrophoretic deposition. *Journal of Coatings Technology and Research*, 15, 583–592.
[33] Wielant, J., Hauffman, T., Blajiev, O., Hausbrand, R., & Terryn, H. (2007). Influence of the ironoxide acid—base properties on the chemisorption of model epoxy compounds studied by XPS. *The Journal of Physical Chemistry C*, 111(35), 13177–13184.
[34] Wang, J., Gao, Z., Li, Z., Wang, B., Yan, Y., Liu, Q., . . . & Jiang, Z. (2011). Green synthesis of graphenenano sheets/ZnO composites and electrochemical properties. *Journal of Solid State Chemistry*, 184(6), 1421–1427.
[35] Chang, L., Zhang, Z., Breidt, C., & Friedrich, K. (2005). Tribological properties of epoxy nano composites: I. Enhancement of the wear resistance by nano-TiO2 particles. *Wear*, 258(1–4), 141–148.
[36] Cho, M. H., & Bahadur, S. (2005). Study of the tribological synergistic effects in nano CuO-filled and fiber-reinforced polyphenylenesulfide composites. *Wear*, 258(5–6), 835–845.
[37] Li, W., Shang, T., Yang, W., Yang, H., Lin, S., Jia, X., . . . & Yang, X. (2016). Effectively exerting there inforcement of dopamine reduced grapheneoxide on epoxy-based composites via strengthened interfacial bonding. *ACS Applied Materials & Interfaces*, 8(20), 13037–13050.
[38] Hsu, C. P., Chang, L. Y., Chiu, C. W., Lee, P. T., & Lin, J. J. (2013). Facile fabrication of robustsuperhydrophobicepoxy film with polyamine dispersed carbon nanotubes. *ACS Applied Materials & Interfaces*, 5(3), 538–545.
[39] Kaniyoor, A., Jafri, R. I., Arockiadoss, T., & Ramaprabhu, S. (2009). Nanostructured Pt-decorated graphene and multiwalled carbon nanotube based room temperature hydrogen gas sensor. *Nanoscale*, 1(3), 382–386.
[40] Kugler, S., Kowalczyk, K., & Spychaj, T. (2017). Transparent epoxy coatings with improved electrical, barrier and thermal features made of mechanically dispersed carbon nanotubes. *Progress in Organic Coatings*, 111, 196–201.
[41] Kumar, A., Ghosh, P. K., Yadav, K. L., & Kumar, K. (2017). Thermo-mechanical and anti-corrosive properties of MWCNT/epoxy nano composite fabricated by innovative dispersion technique. *Composites Part B: Engineering*, 113, 291–299.
[42] Gryshchuk, O., Karger-Kocsis, J., Thomann, R., Kónya, Z., & Kiricsi, I. (2006). Multi wall carbon nanotube modified vinylester and vinylester-based hybridresins. *Composites Part A: Applied Science and Manufacturing*, 37(9), 1252–1259.
[43] Zhang, L., Wu, H., Wei, M., Zheng, Z., Vu, D. D., Bui, T. T. X., & Huang, X. (2018). Preparation, characterization, and properties of grapheneoxide/urushiol-formal dehydepolymer composite coating. *Journal of Coatings Technology and Research*, 15, 1343–1356.
[44] Huang, W. F., Xiao, Y. L., Huang, Z. J., Tsui, G. C., Yeung, K. W., Tang, C. Y., & Liu, Q. (2020). Super-hydrophobic polyaniline-TiO2 hierarchical nanocomposite as anticorrosion coating. *Materials Letters*, 258, 126822.
[45] Amirou, S., &Pizzi, A. (2020). Biosourced heatresistant coatings by cross-linking of proteins with triethylphosphate. *Progress in Organic Coatings*, 138, 105403.

[46] Fan, H. Q., Shi, D. D., Ding, M. M., Li, M. C., Cheng, Y. F., & Li, Q. (2020). Preparation of (3-mercaptopropyl) trimethoxylsilane film on brass and its corrosion resistance in natural sea water. *Progress in Organic Coatings*, 138, 105392.
[47] Idumah, C. I., Obele, C. M., Emmanuel, E. O., & Hassan, A. (2020). Recently emerging nanotechnological advancements in polymer nanocomposite coatings for anti-corrosion, anti-fouling and self-healing. *Surfaces and Interfaces*, 21, 100734.
[48] Rajyalakshmi, T., Pasha, A., Khasim, S., Lakshmi, M., Murugendrappa, M. V., & Badi, N. (2020). Enhanced charge transport and corrosion protection properties of polyaniline-carbon nanotube composite coatings on mild steel. *Journal of Electronic Materials*, 49, 341–352.
[49] Shi, Q., & Zhu, A. (2020). Interface regulation of graphene/carbon nanotube on the thermal conductivity and anticorrosion performance of their nanocomposite. *Progress in Organic Coatings*, 140, 105480.
[50] Kumar, A., Sharma, K., & Dixit, A. R. (2020). Carbon nanotube-and graphene-reinforced multiphase polymeric composites: Review on their properties and applications. *Journal of Materials Science*, 55(7), 2682–2724.
[51] Calabrese, L., & Proverbio, E. (2019). A brief over view on the anticorrosion performances of sol-gel zeolite coatings. *Coatings*, 9(6), 409.
[52] Cui, G., Bi, Z., Zhang, R., Liu, J., Yu, X., & Li, Z. (2019). A comprehensive review on graphene-based anti-corrosive coatings. *Chemical Engineering Journal*, 373, 104–121.
[53] Teng, S., Gao, Y., Cao, F., Kong, D., Zheng, X., Ma, X., & Zhi, L. (2018). Zinc-reduced graphene oxide for enhanced corrosion protection of zinc-rich epoxy coatings. *Progress in Organic Coatings*, 123, 185–189.
[54] Veedu, K. K., Kalarikkal, T. P., Jayakumar, N., & Gopalan, N. K. (2019). Anticorrosive performance of *Mangiferaindica* L. Leafextract-based hybrid coating on steel. *ACS Omega*, 4(6), 10176.
[55] Wang, J., Chen, L., & He, Y. (2008). Preparation of environmentally friendly coatings based on natural shell acmodified by diamine and its applications for copper protection. *Progress in Organic Coatings*, 62(3), 307–312.
[56] Zhang, D., Wang, L., Qian, H., & Li, X. (2016). Super hydrophobic surfaces for corrosion protection: A review of recent progresses and future directions. *Journal of Coatings Technology and Research*, 13, 11–29.
[57] Montemor, M. F. (2014). Functional and smart coatings for corrosion protection: A review of recent advances. *Surface and Coatings Technology*, 258, 17–37.
[58] Tiwari, A., Hihara, L., & Rawlins, J. (Eds.). (2014). *Intelligent coatings for corrosion control*. Butterworth-Heinemann.
[59] He, X., & Shi, X. (2009). Self-repairing coating for corrosion protection of aluminum alloys. *Progress in Organic Coatings*, 65(1), 37–43.
[60] Wei, H., Wang, Y., Guo, J., Shen, N. Z., Jiang, D., Zhang, X., . . . & Guo, Z. (2015). Advanced micro/nanocapsulesfor self-healing smart anticorrosion coatings. *Journal of Materials Chemistry A*, 3(2), 469–480.
[61] White, S. R., Sottos, N. R., Geubelle, P. H., Moore, J. S., Kessler, M. R., Sriram, S. R., . . . & Viswanathan, S. (2001). Autonomic healing of polymer composites. *Nature*, 409(6822), 794–797.
[62] Parsa, M., Mohammad Ali Hosseini, S., Hassani, Z., & Jamalizadeh, E. (2014). Corrosion protective properties of coatings doped with inhibitors. *Anti-Corrosion Methods and Materials*, 61(6), 416–422.
[63] Ulaeto, S. B., Rajan, R., Pancrecious, J. K., Rajan, T. P. D., & Pai, B. C. (2017). Developments in smart anticorrosive coatings with multi functional characteristics. *Progress in Organic Coatings*, 111, 294–314.

[64] Makhlouf, A. S. H. (2014). Techniques for synthesizing and applying smart coatings for material protection. In *Handbook of smart coatings for materials protection*. Woodhead Publishing, 56–74.
[65] Dubbert, K., Völker, D., & Apel, P. (2014). *Factsheet-use of nanomaterials in coatings, federal environment agency*. Dessau-Roßlau, Germany.
[66] Yang, C., Wei, H., Guan, L., Guo, J., Wang, Y., Yan, X., . . . & Guo, Z. (2015). Polymer nanocomposites for energy storage, energy saving, and anticorrosion. *Journal of Materials Chemistry A*, 3(29), 14929–14941.
[67] Cole, I.S. (2014). Smart coatings for corrosion protection: An overview. In Makhlouf, A. S. H., ed. *Handbook of smart coatings for materials protection*. Wood head Publishing Limited, Cambridge, 29–55.
[68] Li, W., & Calle, L. M. (2007, April). Micro-encapsulation for corrosion detection and control. In *Proceedings of the first international conference on self healing materials*. Noordwijk aan Zee, The Netherlands, pp. 18–20.
[69] Maia, F., Tedim, J., Bastos, A. C., Ferreira, M. G., & Zheludkevich, M. L. (2014). Active sensing coating for early detection of corrosion processes. *RSC Advances*, 4(34), 17780–17786.
[70] Calle, L., Hintze, P., Li, W., & Buhrow, J. (2010). Smart coatings for autonomous corrosion detection and control. In *AIAA SPACE 2010 Conference & Exposition* (p. 8877). https://arc.aiaa.org/doi/abs/10.2514/6.2010-8877
[71] Feng, W., Patel, S. H., Young, M. Y., ZuninoIII, J. L., & Xanthos, M. (2007). Smart polymeric coatings—recent advances. *Advances in Polymer Technology: Journal of the Polymer Processing Institute*, 26(1), 1–13.
[72] Nicholls, J. R., Simms, N. J., Chan, W. Y., & Evans, H. E. (2002). Smart overlay coatings—concept and practice. *Surface and Coatings Technology*, 149(2–3), 236–244.
[73] Aliofkhazraei, M. (2014). Smart nano coatings for corrosion detection and control. In *Handbook of smart coatings for materials protection*. Elsevier, 198–223.
[74] Venila, R., Kamaraj, P., Arthanareeswari, M., & Devikala, S. (2014). Surface modification of mild steel using Agdoped SnO2 nanoparticles for corrosion inhibition. *International Journal of Advanced Chemical Science and Applications*, 1, 16–20.
[75] Musa, A. Y. (2012). Corrosion protection of Al alloys: Organic coatings and inhibitors. In *Recent researches in corrosion evaluation and protection*. Intech Open, 51–66.
[76] Rodič, P., Iskra, J., & Milošev, I. (2014). A hybridorganic-inorganic sol-gel coating for protecting aluminium alloy 7075-T6 against corrosion in Harrison's solution. *Journal of Sol-Gel Science and Technology*, 70, 90–103.
[77] Stankiewicz, A., Szczygieł, I., & Szczygieł, B. (2013). Self-healing coatings in anticorrosion applications. *Journal of Materials Science*, 48, 8041–8051.
[78] Mauldin, T. C., & Kessler, M. R. (2010). Self-healing polymers and composites. *International Materials Reviews*, 55(6), 317–346.
[79] Wei, H., Wang, Y., Guo, J., Shen, N. Z., Jiang, D., Zhang, X., . . . & Guo, Z. (2015). Advanced micro/nanocapsules for self-healing smart anticorrosion coatings. *Journal of Materials Chemistry A*, 3(2), 469–480.
[80] Zheludkevich, M. L., Ferreira, M.G.S., Raps, D., & Hack, T. (2011). Self-healing anticorrosion coatings. In Furbeth, L. F. W., & Montemor, F., eds. *Self-healing Properties of New Surface Treatments*. European Federation of Corrosion Publications No. 58, Maney Publishing on behalf of The Institute of Materials, Minerals Mining, 11–38.
[81] Haase, M. F., Grigoriev, D. O., Möhwald, H., & Shchukin, D. G. (2012). Development of nanoparticle-stabilized polymer nanocontainers with high content of the encapsulated active agent and their application in water-borne anticorrosive coatings. *Advanced Materials*, 24(18), 2429–2435.

[82] Toohey, K. S., Sottos, N. R., Lewis, J. A., Moore, J. S., & White, S. R. (2007). Self-healing materials with microvascular networks. *Nature Materials*, 6(8), 581–585.
[83] Yang, Y., & Urban, M. W. (2013). Self-healing polymeric materials. *Chemical Society Reviews*, 42(17), 7446–7467.
[84] Ganesh, V. A., Raut, H. K., Nair, A. S., & Ramakrishna, S. (2011). A review on self-cleaning coatings. *Journal of Materials Chemistry*, 21(41), 16304–16322.

Index

Note: Page numbers in *italics* indicate a figure, and page numbers in **bold** indicate a table on the corresponding page.

D

E

F

G

H

I

J

K

L

M